自己动手写分布式搜索引擎

罗 刚 崔智杰 编著

清华大学出版社
北京

内 容 简 介

本书介绍了分布式搜索引擎开发的原理与 Java 实现，主要包括全文检索的原理与实现、分布式算法与代码实现、SolrCloud 和 ElasticSearch 的使用与原理等内容，并着重介绍了一种实现分布式中文搜索引擎的方法。

本书适合有 Java 程序设计基础的开发人员或者对分布式搜索引擎技术感兴趣的从业人员使用。

本书封面贴有清华大学出版社防伪标签，无标签者不得销售。
版权所有，侵权必究。侵权举报电话：010-62782989　13701121933

图书在版编目(CIP)数据

自己动手写分布式搜索引擎/罗刚，崔智杰编著. —北京：清华大学出版社，2017
ISBN 978-7-302-47708-2

Ⅰ. ①自… Ⅱ. ①罗… ②崔… Ⅲ. ①搜索引擎—程序设计 Ⅳ. TP391.3

中国版本图书馆 CIP 数据核字(2017)第 162194 号

责任编辑：	杨作梅
装帧设计：	杨玉兰
责任校对：	张彦彬
责任印制：	李红英
出版发行：	清华大学出版社
网　　址：	http://www.tup.com.cn, http://www.wqbook.com
地　　址：	北京清华大学学研大厦 A 座　　邮　编：100084
社 总 机：	010-62770175　　邮　购：010-62786544
投稿与读者服务：	010-62776969, c-service@tup.tsinghua.edu.cn
质量反馈：	010-62772015, zhiliang@tup.tsinghua.edu.cn
印 装 者：	三河市金元印装有限公司
经　　销：	全国新华书店
开　　本：	185mm×260mm　印　张：26.75　字　数：650 千字
版　　次：	2017 年 9 月第 1 版　　印　次：2017 年 9 月第 1 次印刷
印　　数：	1～3000
定　　价：	59.00 元

产品编号：075226-01

前 言

搜索引擎成为人们获取信息不可或缺的工具。大数据技术的发展推动了多机集群的分布式搜索引擎技术走向成熟。普通的机器就可以搭建分布式搜索引擎。一些开源的分布式搜索引擎系统在数据存储、数据分析等方面的功能越来越强大。本书希望用通俗易懂的语言，让任何对分布式搜索引擎技术感兴趣的读者都能够有所收获。

本书的很多内容来源于搜索引擎、自然语言处理、金融等领域的项目开发和教学实践。在此感谢开源软件的开发者们，他们无私的工作丰富了本书的内容。

本书的第 1 章介绍开发分布式搜索引擎所需要的基本算法；第 2 章介绍如何从头开始自己动手写一个简单的全文检索软件包；第 3 章介绍 Lucene 的基本使用方法及其原理；第 4 章介绍使用 JSP 或者 Struts 2 开发搜索引擎用户界面，以及用户界面常用的 Taglib；第 5 章介绍 Solr 实现分布式搜索引擎的解决方案——SolrCloud，以及它对 SQL 查询的支持；第 6 章介绍如何使用基于 Lucene 的 ElasticSearch 实现分布式搜索引擎。

鉴于 ElasticSearch 处于快速发展中，一些新版本的具体使用情况可以加入 QQ 群 460405445，进行讨论。

本书配套的光盘中提供了相关的源代码，有的来源于猎兔搜索多年的开发经验积累，有的是经典算法实现。其中很多源代码都可以直接用于项目实践。

本书适合需要具体实现搜索引擎的程序员使用，对于信息检索等相关领域的研究人员也有一定的参考价值，同时猎兔搜索技术团队已经开发出以本书为基础的专门培训课程和商业软件。目前的一些分布式搜索引擎软件仍然有很多功能有待完善，作者真诚地希望通过本书把读者带入分布式搜索引擎开发的大门并认识更多的朋友。

感谢早期合著者、合作伙伴、员工、学员的支持，给我们提供了良好的工作基础。在将来，希望我们的分布式搜索引擎代码和技术能够像雨后春笋一样快速生长。

本书由罗刚、崔智杰编著，另外参与本书编写的还有张晓斐、石天盈、张继红、张进威、刘宇、何淑琴、任通通、高丹丹、徐友峰、孙宽，在此一并表示感谢。

编 者

目 录

第 1 章 搜索引擎 1
1.1 搜索引擎基本模块 2
1.2 开发环境 3
1.3 搜索引擎工作原理 4
1.3.1 网络爬虫 5
1.3.2 全文索引 5
1.3.3 搜索用户界面 8
1.3.4 分布式计算 9
1.3.5 文本挖掘 9
1.4 算法基础 9
1.4.1 折半查找 10
1.4.2 排序 10
1.4.3 最小生成树 12
1.5 软件工具 15
1.6 单元测试 15
1.7 本章小结 17
1.8 术语表 18

第 2 章 自己动手写全文检索 19
2.1 构建索引 22
2.2 生成索引文件 23
2.3 读入索引文件 25
2.4 查询 26
2.5 有限状态机 29
2.5.1 运算 29
2.5.2 编辑距离有限状态机 30
2.6 本章小结 32

第 3 章 Lucene 的原理与应用 33
3.1 Lucene 快速入门 34
3.1.1 创建索引 34
3.1.2 查询索引库 35
3.1.3 创建文档索引 36
3.1.4 查询文档索引 36
3.2 创建和维护索引库 37
3.2.1 设计索引库结构 37
3.2.2 创建索引库 38
3.2.3 向索引库中添加索引文档 40
3.2.4 删除索引库中的索引文档 43
3.2.5 更新索引库中的索引文档 44
3.2.6 关闭索引库 45
3.2.7 索引的优化与合并 45
3.2.8 灵活索引 46
3.2.9 索引文件格式 47
3.2.10 定制索引存储结构 49
3.2.11 写索引集成到爬虫 54
3.2.12 多线程写索引 56
3.2.13 分发索引 58
3.2.14 修复索引 61
3.3 查找索引库 61
3.3.1 查询过程 61
3.3.2 常用查询 64
3.3.3 基本词查询 65
3.3.4 模糊匹配 65
3.3.5 布尔查询 67
3.3.6 短语查询 69
3.3.7 跨度查询 71
3.3.8 FieldScoreQuery 74
3.3.9 排序 77
3.3.10 使用 Filter 筛选搜索结果 81
3.3.11 使用 Collector 筛选搜索结果 82
3.3.12 遍历索引库 85
3.3.13 关键词高亮显示 88
3.3.14 列合并 91
3.3.15 关联内容(BlockJoinQuery) 92
3.3.16 查询大容量索引 94
3.4 读写并发 95

3.5 Lucene 深入介绍 95
- 3.5.1 整体结构 96
- 3.5.2 索引原理 97
- 3.5.3 文档值 100
- 3.5.4 FST 102

3.6 查询语法与解析 102
- 3.6.1 JavaCC 104
- 3.6.2 生成一个查询解析器 114
- 3.6.3 简单的查询解析器 114
- 3.6.4 灵活的查询解析器 114

3.7 检索模型 119
- 3.7.1 向量空间模型 121
- 3.7.2 DFR 125
- 3.7.3 BM25 概率模型 130
- 3.7.4 BM25F 概率模型 136
- 3.7.5 统计语言模型 138
- 3.7.6 相关性反馈 140
- 3.7.7 隐含语义索引 140
- 3.7.8 学习评分 141
- 3.7.9 查询与相关度 142
- 3.7.10 使用 Payload 调整相关性 .. 142

3.8 查询原理 146
- 3.8.1 布尔匹配 147
- 3.8.2 短语查询 150
- 3.8.3 索引统计 150
- 3.8.4 相关性 152

3.9 分析文本 155
- 3.9.1 Analyzer 156
- 3.9.2 TokenStream 162
- 3.9.3 定制 Tokenizer 164
- 3.9.4 重用 Tokenizer 166
- 3.9.5 有限状态转换 167
- 3.9.6 索引数值列 168
- 3.9.7 检索结果排序 171
- 3.9.8 处理价格 171

3.10 Lucene 中的压缩算法 172
- 3.10.1 变长压缩 172
- 3.10.2 Gamma 174
- 3.10.3 PForDelta 176
- 3.10.4 VSEncoding 178
- 3.10.5 前缀压缩 179
- 3.10.6 差分编码 180
- 3.10.7 静态索引裁剪 182

3.11 搜索中文 182
- 3.11.1 Lucene 切分原理 185
- 3.11.2 Lucene 中的 Analyzer ... 186
- 3.11.3 自己写 Analyzer 188
- 3.11.4 Lietu 中文分词 191
- 3.11.5 字词混合索引 191

3.12 搜索英文 196
- 3.12.1 英文分词 196
- 3.12.2 词性标注 199
- 3.12.3 原型化 201

3.13 索引数据库中的文本 202

3.14 优化使用 Lucene 204
- 3.14.1 系统优化 204
- 3.14.2 查询优化 205
- 3.14.3 实现时间加权排序 207
- 3.14.4 词性标注 210
- 3.14.5 个性化搜索 213

3.15 实时搜索 213

3.16 语义搜索 215
- 3.16.1 发现同义词 215
- 3.16.2 垂直领域同义词 219
- 3.16.3 同义词扩展 219
- 3.16.4 语义标注 225

3.17 本章小结 225

3.18 术语表 226

第 4 章 搜索引擎用户界面 227

4.1 实现 Lucene 搜索 228
- 4.1.1 测试搜索功能 228
- 4.1.2 加载索引 229

4.2 搜索页面设计 231
- 4.2.1 Struts2 实现的搜索界面 .. 232
- 4.2.2 用于显示搜索结果的 Taglib 234

	4.2.3	实现翻页	235
4.3	实现搜索接口		238
	4.3.1	编码识别	238
	4.3.2	布尔搜索	241
	4.3.3	指定范围搜索	241
	4.3.4	搜索结果排序	242
	4.3.5	索引缓存与更新	243
4.4	实现分类统计视图		249
	4.4.1	单值列分类统计	255
	4.4.2	侧钻	256
4.5	实现相似文档搜索		257
4.6	实现 AJAX 搜索联想词		259
	4.6.1	估计查询词的文档频率 ...	259
	4.6.2	搜索联想词总体结构	259
	4.6.3	服务器端处理	260
	4.6.4	浏览器端处理	265
	4.6.5	拼音提示	267
	4.6.6	部署总结	267
4.7	推荐搜索词		268
	4.7.1	挖掘相关搜索词	268
	4.7.2	使用多线程计算相关搜索词	270
4.8	查询意图理解		271
	4.8.1	拼音搜索	271
	4.8.2	无结果处理	272
4.9	集成其他功能		272
	4.9.1	拼写检查	272
	4.9.2	分类统计	276
	4.9.3	相关搜索	281
	4.9.4	再次查找	284
	4.9.5	搜索日志	284
4.10	查询分析		286
	4.10.1	历史搜索词记录	286
	4.10.2	日志信息过滤	286
	4.10.3	信息统计	287
	4.10.4	挖掘日志信息	289
	4.10.5	查询词意图分析	290
4.11	部署网站		290

	4.11.1	部署到 Web 服务器	290
	4.11.2	防止攻击	292
4.12	手机搜索界面		295
4.13	本章小结		296

第 5 章 Solr 分布式搜索引擎 297

5.1	Solr 简介		298
5.2	Solr 基本用法		299
	5.2.1	Solr 服务器端的配置与中文支持	300
	5.2.2	数据类型	304
	5.2.3	解析器	306
	5.2.4	把数据放进 Solr	307
	5.2.5	删除数据	312
	5.2.6	查询语法	313
5.3	使用 SolrJ		313
	5.3.1	Solr 客户端与搜索界面 ...	313
	5.3.2	Solr 索引库的查找	315
	5.3.3	分类统计	317
	5.3.4	高亮	319
	5.3.5	同义词	322
	5.3.6	嵌入式 Solr	322
	5.3.7	Spring 实现的搜索界面 ...	323
	5.3.8	索引分发	331
	5.3.9	Solr 搜索优化	333
5.4	从 FAST Search 移植到 Solr ...		336
5.5	Solr 扩展与定制		337
	5.5.1	缺省查询	337
	5.5.2	插件	338
	5.5.3	Solr 中字词混合索引	338
	5.5.4	相关检索	340
	5.5.5	搜索结果去重	341
	5.5.6	定制输入输出	344
	5.5.7	聚类	348
	5.5.8	分布式搜索	348
	5.5.9	分布式索引	352
	5.5.10	SolrJ 查询分析器	353
	5.5.11	扩展 SolrJ	360

- 5.5.12 扩展 Solr 361
- 5.5.13 日文搜索 364
- 5.5.14 查询 Web 图 365
- 5.6 SolrNet 367
  - 5.6.1 使用 SolrNet 实现全文搜索 367
  - 5.6.2 实现原理 370
  - 5.6.3 扩展 SolrNet 371
- 5.7 Solr 的 PHP 客户端 373
- 5.8 Solr 的其他客户端 376
- 5.9 为网站增加搜索功能 376
- 5.10 SolrCloud 377
  - 5.10.1 Zab 协议 377
  - 5.10.2 ZooKeeper 377
  - 5.10.3 使用 SolrCloud 379
  - 5.10.4 SQL 查询 380
- 5.11 Solr 原理 381
  - 5.11.1 支持 Solr 的中文分词 381
  - 5.11.2 缓存技术 383
- 5.12 本章小结 384

第 6 章 ElasticSearch 分布式搜索引擎 387

- 6.1 安装 389
- 6.2 搜索集群 390
  - 6.2.1 Zen 发现机制 390
  - 6.2.2 JGroups 391
- 6.3 创建索引 393
- 6.4 Java 客户端接口 396
  - 6.4.1 创建索引 398
  - 6.4.2 插入数据 398
  - 6.4.3 索引库结构 400
- 6.5 查询 401
- 6.6 高亮显示 405
- 6.7 分页 406
- 6.8 中文搜索 407
  - 6.8.1 中文 AnalyzerProvider 407
  - 6.8.2 字词混合索引 409
- 6.9 分组统计 412
- 6.10 与爬虫集成 413
- 6.11 Percolate 413
- 6.12 权限 414
- 6.13 SQL 支持 415
- 6.14 本章小结 419

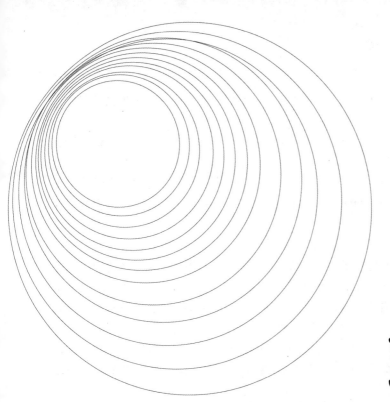

第1章

搜索引擎

每天都有很多人通过聊天软件进行交流。理想的聊天软件应该能够自动在服务器中永久记录聊天记录，而且有强大的搜索功能。

当用户只输入个别词时，搜索引擎系统可以猜测用户的查询意图，把用户查询意图扩展成整句。

搜索界面可以用 WebSocket 登录方式重新实现，这样可以给出个性化的查询结果。例如，如果已经知道某个用户在使用 Macbook，当他搜索"eclipse 下载"时，则给他提供 Mac 版本的 Eclipse 下载地址。

本章首先概要地介绍搜索引擎的总体结构和基本模块，然后介绍其中最核心的模块：全文检索的基本原理。为了尽快普及搜索引擎开发技术，本章介绍的搜索引擎结构可以采用开源软件实现。为了通过实践来深入了解相关技术，本章会介绍相关的开发环境。本书所介绍的搜索技术使用 Java 编程语言实现，之所以没有采用性能可能会更好的 C/C++，是因为 Java 代码的可维护性更好。C++开发团队的协作性更差，往往沦为手工作坊式的开发。另外，为了集中关注程序的基本逻辑，书中的 Java 代码去掉了一些错误和异常处理，实际可以运行的代码在本书附赠的资源中。在以后的各章中会深入探索搜索引擎的每个组成模块。

1.1 搜索引擎基本模块

一个最简单的搜索引擎由索引和搜索界面两部分组成，相对完整的搜索结构如图 1-1 所示。

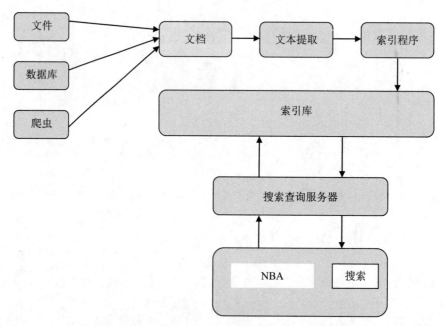

图 1-1 搜索引擎的简单结构

实现按关键字快速搜索的方法是建立全文索引库，所以最基础的程序是管理全文索引库的程序。搜索的数据来源可以是互联网或者数据库，也可以是本地路径等。搜索引擎的基本模块从底层至顶层的结构如图1-2所示。

图 1-2　搜索引擎中的主要模块

1.2　开　发　环　境

由于开源软件的迅速发展，可以借助开源软件简化搜索引擎开发工作。很多开源软件用 Java 语言编写，例如最流行的全文索引库 Lucene，所以本书采用 Java 来实现搜索。

为了实现一个简单的指定目录文件的搜索引擎，首先要准备好 JDK 和集成开发环境 Eclipse。当前可以使用 JDK 1.6。JDK 1.6 可以从 Java 官方网站 http://java.sun.com 下载得到，使用缺省方式安装即可。本书中的程序在附赠的资源中都能找到，可以直接导入 Eclipse 中。Eclipse 默认是英文界面，如果习惯用中文界面可以从 http://www.eclipse.org/babel/downloads.php 下载支持中文的语言包。

Lucene 是一个 Java 实现的 jar 包，用来管理搜索引擎索引库。可以从 http://lucene.apache.org/java/docs/index.html 下载到最新版本的 Lucene，当前最新的版本是 3.0。

如果需要用 Web 搜索界面，还要下载 Tomcat，可以从 http://tomcat.apache.org/下载，推荐使用 Tomcat 6 以上的版本。使用开源的全文检索包 Lucene 作索引后，要把实现搜索的界面发布到 Tomcat。

对于 Web 搜索界面，建议使用 MyEclipse 开发。对于其他普通的非 Web 开发工作则不建议使用 MyEclipse。例如，开发爬虫建议只使用 Eclipse，而不要用 MyEclipse。 MyEclipse 开发普通的 Java 项目时，速度慢。

Lucene 及一些相关项目的源代码由版本管理工具 SVN 管理，如果要构建源代码工程，可以使用 Ant 和 Maven 工具。

如果需要导出 Lucene 的最新开发版本，就需要用到 SVN 的客户端。"小乌龟"TortoiseSVN 是最流行的 SVN 客户端。TortoiseSVN 的下载地址是 http://tortoisesvn.tigris.org/。安装 TortoiseSVN 后，选择一个存放源代码的文件夹，单击右键，选择 TortoiseSVN 菜单中的 Export...选项导出源代码。

Ant 与 Maven 都是项目管理软件 Make 类似。虽然 Maven 正在逐步替代 Ant，但当前仍然有很多开源项目在继续使用 Ant。从 http://ant.apache.org/bindownload.cgi 可以下载 Ant 的最新版本。

在 Windows 下，ant.bat 和 3 个环境变量相关：ANT_HOME、CLASSPATH 和 JAVA_HOME。需要设置 ANT_HOME 和 JAVA_HOME 环境变量，并且路径不要以\或/结束，不要设置 CLASSPATH 环境变量。可以使用 echo 命令检查 ANT_HOME 环境变量：

```
>echo %ANT_HOME%
D:\apache-ant-1.7.1
```

如果把 Ant 解压到 c:\apache-ant-1.7.1，则需修改环境变量 PATH，增加当前路径 c:\apache-ant-1.7.1\bin。

如果项目的源代码根目录包括一个 build.xml 文件，则说明这个项目可能是用 Ant 构建的。大部分用 Ant 构建的项目只需要如下一个命令：

```
#ant
```

生成 jor 包。

可以从 http://maven.apache.org/download.html 下载最新版本的 Maven，当前最新版本是 Maven 2.2.1。解压下载的 Maven 压缩文件到 C 盘根目录下，将创建一个 c:\apache-maven-2.2.1 路径。修改 Windows 系统环境变量 PATH，增加当前路径 c:\apache-maven-2.2.1\bin。如果一个项目的源代码根目录包括一个 pom.xml 文件，则说明这个项目可能是用 Maven 构建的。大部分用 Maven 构建的项目只需要如下一个命令：

```
#mvn clean install
```

盖大楼的时候需要搭建脚手架，虽然最终不会交付使用。很多单元测试代码也不会在正式环境中运行，但是必须写。可以使用 JUnit 做单元测试。

如果参与开发的人很多，就需要一个项目管理软件，例如 Redmine 或者 JIRA。Lucene 和 Solr 的开发管理就是用的 JIRA。

搜索集群往往运行在 Linux 操作系统，需要熟悉一些 Linux 命令，例如使用 crash 命令分析 Linux 内核崩溃转储文件。

1.3　搜索引擎工作原理

一个基本的搜索包括采集数据的爬虫和索引库管理以及搜索页面展现等部分，如图 1-3 所示。

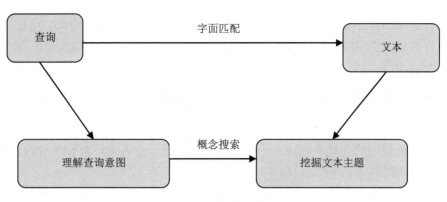

图 1-3　搜索的方向

1.3.1　网络爬虫

网络爬虫(Crawler)又被称作网络机器人(Robot)或者蜘蛛(Spider)，它的主要用途是获取互联网上的信息。只有掌握了"吸星大法"，才能源源不断地获取信息。网络爬虫利用网页中的超链接遍历互联网，通过 URL 引用从一个 HTML 文档爬行到另一个 HTML 文档。http://dmoz.org 可以作为整个互联网抓取的入口。网络爬虫收集的信息可有多种用途，如建立索引、HTML 文件的验证、URL 链接验证、获取更新信息、站点镜像等。为了检查网页内容是否更新过，网络爬虫建立的页面数据库往往包含根据页面内容生成的文摘。

在抓取网页时大部分网络爬虫会遵循 Robot.txt 协议。网站本身可以用两种方式声明不想被搜索引擎收入的内容：第一种方式是在站点的根目录增加一个纯文本文件 http://www.yourdomain.com/robots.txt；另外一种方式是直接在 HTML 页面中使用 Robots 的 <meta> 标签。

1.3.2　全文索引

假如你想看书，说出你的要求后，有经验的图书管理员可以从他们的书库中直接给你推荐几本。查询的基本过程与此类似，如图 1-4 所示。

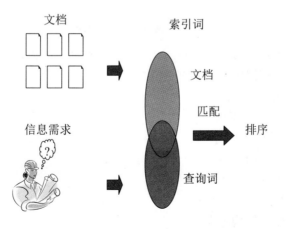

图 1-4　全文检索的基本过程

例如有10篇文档,编号为0~9。其中3篇文档中包含查询词,匹配出来文档集合{0,6,9}。对文档集合按相关性排序,得到文档数组{6,0,9}。返回结果中不仅存储文档,还存储分值。

```
public class ScoreDoc {
    Document doc;  //文档相关的信息,包括文档编号等
    public float score;  //表示这个文档和查询词有多相关
}
```

查找文档最原始的方式是通过文档编号查找。就像一个人对应有一个身份证号,一个文档从创建开始就有一个文档编号。

有的专业书籍末尾有名词术语索引,方便读者定位名词术语在书中出现的位置。为了按词快速查找文档,不是采用字符串匹配的方法在文档中查询词,而是按词建立文档的索引。以词为基础建立的全文索引,也称倒排索引,如图1-5所示。在这里,索引中的文档用编号表示。

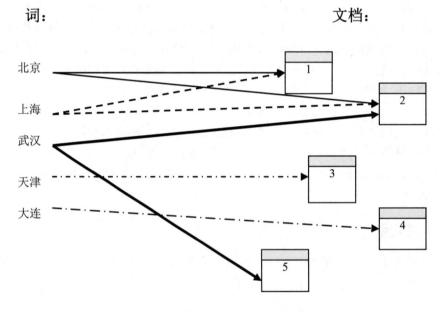

图1-5 以词为基础的全文索引

倒排索引是相对于正向索引来说的,首先用正向索引来存储每个文档对应的单词列表,然后再建立倒排索引,根据单词来索引文档编号。

例如要索引如下两个文档:

Doc Id 1: 自己动手写搜索引擎。

Doc Id 2: 自己动手写网络爬虫。

首先把这两个文档中的内容分成一个个的单词:

Doc Id 1: 自己/动手/写/搜索引擎;

Doc Id 2: 自己/动手/写/网络爬虫。

按词建立的倒排索引结构见表1-1。

表 1-1 倒排索引结构

词	（文档，频率）	在文档中出现的位置
动手	(1,1),(2,1)	(2),(2)
搜索引擎	(1,1)	(4)
网络爬虫	(2,1)	(4)
写	(1,1),(2,1)	(3),(3)
自己	(1,1),(2,1)	(1),(1)

每个词后面的文档编号(docId)列表称为投递列表(posting list)。除了在投递列表中记录文档编号外，还可以添加词频和位置信息。词频的添加有助于结果的排序。位置信息记录了一个索引词在文档中出现的位置，可以用于包含多个查询词的短语检索；此外，也可以用于快速高亮显示查询词。

索引子系统把要搜索的文档先建立好索引，如图 1-6 所示。

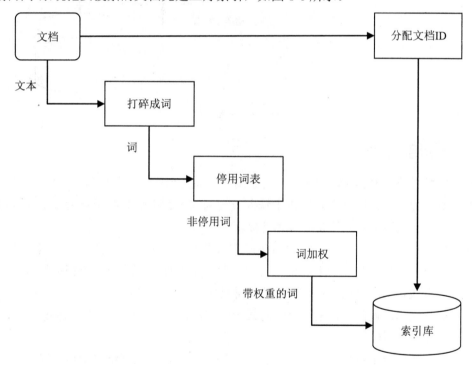

图 1-6 索引过程

待查询的很多文档放入同一个索引库，可以使用特征函数对文档中的词加权。
搜索子系统针对用户的即时查询找出相关文档，如图 1-7 所示。
每个词查询出来一个文档的集合。多个文档集合做 AND 或者 OR 这样的布尔操作。

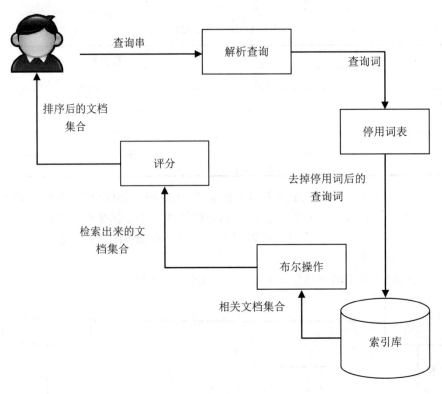

图 1-7 搜索过程

1.3.3 搜索用户界面

随着搜索引擎技术逐渐走向成熟,搜索用户界面也形成了一些比较固定的模式。

- 输入提示词:用户在搜索框中输入查询词的过程中随时给予查询提示词。对中文来说,当用户输入拼音时,也能提示。
- 相关搜索提示词:当用户对当前搜索结果不满意时,也许换一个搜索词就能够得到更有用的信息。一般会根据用户当前使用的搜索词给出多个相关的提示词。可以看成是协同过滤在搜索词上的一种具体应用。
- 相关文档:返回和搜索结果中的某一个文档相似的文档。例如:Google 搜索结果中的"类似结果"。
- 在结果中查询:如果返回结果很多,则用户可以在返回结果中再次输入查询词以缩小查询范围。
- 分类统计:返回搜索结果在类别中的分布图。用户可以按类别缩小搜索范围,或者在搜索结果中导航。有点类似于数据仓库中的向下钻取和向上钻取。
- 搜索热词统计界面:往往按用户类别统计搜索词,例如,按用户所属区域或者按用户所属部门等,当然也可以直接按用户统计搜索热词,如 Google 的 Trends。

搜索界面的改进都是以用户体验为导向。所以搜索用户界面往往还根据具体应用场景优化。所有这一切都是为了和用户的交互达到最大的效果。

1.3.4 分布式计算

算法设计是一件非常困难的工作，需要有很好的数据结构基础。本书中采用的算法设计技术主要有迭代法、分治法、动态规划法等。

互联网搜索经常面临海量数据。需要分布式的计算框架来执行对网页重要度打分等计算。有的计算数据很少，但是计算量很大；还有些计算数据量比较大，但是计算量相对比较小。例如，计算圆周率是计算密集型，互联网搜索中的计算往往是数据密集型。所以出现了数据密集型的云计算框架。MapReduce 是一种常用的云计算框架。但是 MapReduce 是批处理的操作方式。一般来说，直到完成上一阶段的操作后才能启动下一阶段的操作。

要有一种计算，可以尽快出结果，随着时间的延长，计算结果会越来越好。很多计算可以用迭代的方式做，迭代次数越多，结果往往越好，比如 PageRank 或者 KMeans、EM 算法。当然，这个应该不只需要迭代，还需要向最优解收敛。

1.3.5 文本挖掘

搜索文本信息需要理解人类的自然语言。文本挖掘是指从大量的文本数据中抽取隐含的、未知的、可能有用的信息。

常用的文本挖掘方法包括：全文检索、中文分词、句法分析、文本分类、文本聚类、关键词提取、文本摘要、信息提取、智能问答等。文本挖掘相关技术的结构如图 1-8 所示。

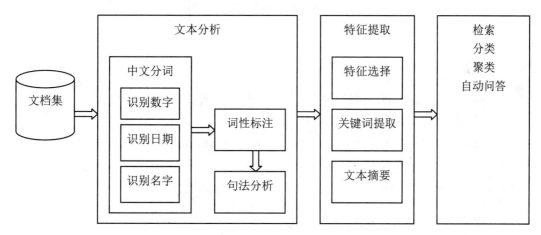

图 1-8　文本挖掘的结构

1.4　算法基础

搜索引擎开发已经发展成了一门复杂的技术，但是可以从简单的折半查找开始，理解搜索引擎的原理。

1.4.1 折半查找

为了快速地查找单词,可以先对单词列表排序,例如:《新华字典》和《现代汉语词典》按拼音排序。从排好序的词表中查找一个词,可以采用折半查找的方法快速查询。下面是实现折半查找的代码。

```java
int binarySearch(String[] a,String key){
    int low = 0; //开始位置
    int high = a.length - 1; //结束位置

    while (low <= high) {
        int mid = (low + high) >>> 1; //相当于mid = (low + high)/2
        String midVal = a[mid]; //取中间的值
        int cmp = midVal.compareTo(key); //中间值和要找的关键词比较

        if (cmp < 0)
            low = mid + 1;
        else if (cmp > 0)
            high = mid - 1;
        else
            return mid; // 查找成功,返回找到的位置
    }
    return -(low + 1);  // 没找到,返回负值
}
```

1.4.2 排序

如何得到一个排好序的词表?首先看一下如何对整数数组排序,然后再看一下如何对字符串数组排序。

排序可以看成是减少逆序的过程。通过交换值来消除逆序。检查任意两个位置的元素,如果是逆序,就交换这两个位置中的值。

```java
int[] scores = { 1, 6, 3, 8, 5 }; // 待排序的数组

// 比较任意两个数,将小数放在前面,大数放在后面
for (int i = 0; i < scores.length; i++) {
    for (int j = 0; j < scores.length; j++) {
        //如果是逆序,就交换这两个数
        if (i<j && scores[i] > scores[j]) { //同时满足两个条件
            int temp = scores[j];
            scores[j] = scores[i];
            scores[i] = temp;
        }
    }
}
```

这种随意消除逆序的方法并不能保证最后得到一个完全有序的数组。对于已经排好序的数组来说，最大的元素位于数组尾部。对于子数组来说，也是如此。也就是说，第二大的元素位于数组倒数第二的位置，第三大的元素位于数组倒数第三的位置，依次类推。冒泡排序正是基于这样的事实。

设想有一瓶汽水，溶解在水中的气体不断上浮，最后实现水气分离。每次让最重的一个元素沉到底。以后就不用再管它了。通过让轻的气泡不断上浮，从而达到有序。

将一个最重的元素沉底，顺便减少逆序。

```
int[] scores = { 1, 6, 3, 8, 5 }; //待排序的数组
for (int j = 0; j < scores.length - 1 ; j++) {
        //循环不变式是：scores[j]存储了数组从开始一直到j为止的最大的一个数
        //比较相邻的两个数，将小数放在前面，大数放在后面
        if (scores[j] > scores[j + 1]) {
            int temp = scores[j];
            scores[j] = scores[j + 1];
            scores[j + 1] = temp;
        }
}
```

完整的冒泡排序。

```
int[] scores = { 1, 6, 3, 8, 5 }; //待排序的数组

for (int i = 0; i < scores.length - 1; i++) { //每次搞定一个最大的元素
    // 最底下已经排好序，所以逐渐减少循环次数
    for (int j = 0; j < scores.length - 1 - i; j++) {
        //比较相邻的两个数，将小数放在前面，大数放在后面
        if (scores[j] > scores[j + 1]) {
            int temp = scores[j];
            scores[j] = scores[j + 1];
            scores[j + 1] = temp;
        }
    }
}
System.out.println("排序后的结果:");
for (int i = 0; i < scores.length; i++) {
    System.out.println(scores[i]);
}
```

两层循环，内层循环执行一遍后，让1个数沉底。如果数组中有 n 个元素，则外层循环执行 $n-1$ 遍。也就是说通过 $n-1$ 次循环让 $n-1$ 个元素沉底。

想象在和很多人喝酒，要把这些人都招待好。i 用来表示找每个人喝，j 表示这个人每次要喝很多杯才能喝好。

首先看一下如何比较两个字符串的大小，然后再看一下如何对字符串数组排序。比较两个对象用 compareTo()方法。字符串也是对象，所以也可以用 compareTo()方法比较两个字符串的大小。

```
String a = "北京";
String b = "广州";
System.out.print(a.compareTo(b));  //因为a小于b,所以返回负数
```

对字符串数组排序。

```
String[] words = {"北京", "广州", "上海"};

for (int i = 0; i < words.length - 1; i++) {
    // 最底下已经排好序,所以逐渐减少循环次数
    for (int j = 0; j < words.length - 1 - i; j++) {
        if (words[j].compareTo(words[j + 1])>0) {
            String temp = words[j];
            words[j] = words[j + 1];
            words[j + 1] = temp;
        }
    }
}
```

如果数组中的元素很多,则快速排序更快。所要搜索的词往往很多,所以应该使用快速排序。

1.4.3 最小生成树

在分布式系统中选举主节点时可能会用到最小生成树。一个网络可以用无向连通带权图表示。如图1-9所示为无向连通带权图G。

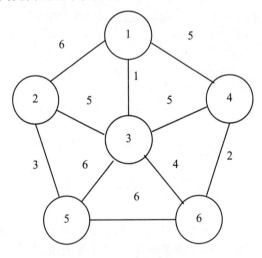

图1-9 无向连通带权图

遍历图G中所有的顶点的子图G'叫作G的生成树。生成树上各边权的总和称为生成树的耗费。考虑如何用最小的耗费遍历图G中所有的顶点。把得到的子图min(G')叫作G的最小生成树。

用贪心法生成最小生成树的方法是:首先设S = {1},然后,只要S是V的真子集,就

进行如下的贪心选择:选取满足条件 i∈S,j∈V - S 且 c[i][j]最小的边,将顶点 j 添加到 S 中。这个过程一直进行到 S = V 时为止。这称为 Prim 算法。

例如,如图 1-9 所示的图生成最小生成树的过程如下。

```
1 -> 3 : 1
3 -> 6 : 4
6 -> 4 : 2
3 -> 2 : 5
2 -> 5 : 3
```

得到的最小生成树如图 1-10 所示。

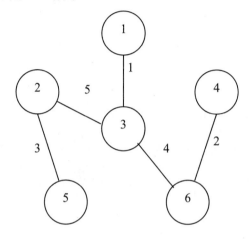

图 1-10 最小生成树

Prim 算法求解最小生成树的实现代码如下。

```java
public class PrimMST {
    public static void main(String[] args) {
        int N = 6; // 节点的数量
        int M = 10; // 边的数量
        HashMap<Integer, HashMap<Integer, Integer>> edgeMap =
                new HashMap<Integer, HashMap<Integer, Integer>>(); //图

        // 节点编号从 1 开始
        int[][] graph = { { 1, 6, 2 }, { 1, 1, 3 }, { 1, 5, 4 }, { 2, 5, 3 },
                { 2, 3, 5 }, { 3, 5, 4 }, { 3, 6, 5 }, { 3, 4, 6 },
                { 4, 2, 6 }, { 5, 6, 6 } };//边数组

        for (int i = 0; i < M; i++) {   //生成无向图
            int n1 = graph[i][0]; // 开始节点
            int n2 = graph[i][2]; // 结束节点
            int weight = graph[i][1]; // 权重

            if (edgeMap.containsKey(n1)) {
                edgeMap.get(n1).put(n2, weight);
```

```java
        } else {
            HashMap<Integer, Integer> edge = new HashMap<Integer, Integer>();
            edge.put(n2, weight);
            edgeMap.put(n1, edge);
        }

        if (edgeMap.containsKey(n2)) {
            edgeMap.get(n2).put(n1, weight);
        } else {
            HashMap<Integer, Integer> edge = new HashMap<Integer, Integer>();
            edge.put(n1, weight);
            edgeMap.put(n2, edge);
        }
    }

    int S = 1;// 开始节点

    Comparator<CostNode> cmp = new Comparator<CostNode>() {
        public int compare(CostNode node1, CostNode node2) {
            return (node1.cost - node2.cost);
        }
    };

    PriorityQueue<CostNode> costQueue = new PriorityQueue<CostNode>(cmp);
    boolean[] visited = new boolean[N + 1];
    int[] costArray = new int[N + 1];

    for (int i = 1; i <= N; i++) {
        costArray[i] = Integer.MAX_VALUE;
    }

    int numVisited = 0;
    costQueue.add(new CostNode(S, 0));
    costArray[S] = 0;

    while (numVisited < N) {
        CostNode costNode = costQueue.poll();
        int minNode = costNode.node;

        if (!visited[minNode]) {
            visited[minNode] = true;
            numVisited++;

            for (int neighbour : edgeMap.get(minNode).keySet()) {
                if (!visited[neighbour]
```

```
                && edgeMap.get(minNode).get(neighbour) <
costArray[neighbour]) {
                    costArray[neighbour] = edgeMap.get(minNode).get(
                            neighbour);
                    costQueue.add(new CostNode(neighbour,
                            costArray[neighbour]));
                }
            }
        }
    }

    int totalCost = 0;   //生成树的耗费
    for (int i = 1; i <= N; i++) {
        totalCost += costArray[i];
    }
}

class CostNode {
    int node;
    int cost;

    public CostNode(int node, int cost) {
        this.node = node;
        this.cost = cost;
    }
}
```

1.5 软件工具

要学会与开源软件共舞。如果发现开源代码有问题，可以打出一个新的代码分支(svn copy)，在这个分支里解决问题。然后可以通过 diff 命令获得一个 patch 文件，把它提交给官方。可以继续使用新分支编译出来的 jar 文件，作为你的第三方库。如果开源代码出了新版本，而又没有包含你的补丁，就需要再开出一个分支，这个分支是新版本和你修复的分支的合并。然后基于这个新分支编译出 jar 文件并使用。如果开源代码出了新版本并且包含了你的补丁，则可以丢弃自己的分支，采用官方版本即可。

1.6 单元测试

测试开发中的手机软件，往往先采用模拟器。搭建分布式系统往往比较麻烦，需要能够让其他的代码逻辑在没有真正的分布式系统的情况下也可以测试，也就是可以方便地进行单元测试。

通过依赖注入(Dependency Injection)解决单元测试问题。例如，有个类需要访问数据库：

```
public class Example {
  private DatabaseThingie myDatabase;  //依赖项

  public Example() {
    myDatabase = new DatabaseThingie();
  }

  public void doStuff() {
    ...
    myDatabase.getData();
    ...
  }
}
```

有的数据库安装起来比较麻烦，因为无法初始化 Example 类的实例，导致后续的测试无法进行。这时，可以使用从外部注入的实例。

```
public class Example {
  private DatabaseThingie myDatabase;

  public Example() {
    myDatabase = new DatabaseThingie();
  }

  public Example(DatabaseThingie useThisDatabaseInstead) {
    myDatabase = useThisDatabaseInstead;
  }

  public void doStuff() {
    ...
    myDatabase.getData();
    ...
  }
}
```

测试类中可以方便地注入数据库存根类。

```
public class ExampleTest {
  TestDoStuff() {
    MockDatabase mockDatabase = new MockDatabase();

    // MockDatabase 是 DatabaseThingie 的子类，因此可以在这里注入它
    Example example = new Example(mockDatabase);

    example.doStuff();
    mockDatabase.AssertGetDataWasCalled();
  }
}
```

1.7 本章小结

在 20 世纪 90 年代,计算机还不普及,图书馆为了方便借阅者检索图书,把每本书对应做了一个卡片,借阅者可以根据图书的分类编号手工查找分类放在盒子中的图书卡片。

20 世纪 50 年代中期,在 IBM 公司工作的 Luhn 提出利用词对文档构建索引并利用用户查询词与文档中词的匹配程度进行检索的方法,这种方法就是目前常用的倒排文档技术的雏形。倒排索引按词建立词到文档的映射。虽然词是最基本的语义单元,但是用户意图往往不仅仅只是用查询词就能完全代表。需要扩展查询词,猜测用户意图,根据用户意图展示信息。

本章介绍了互联网搜索及其创新原则。在 Google 出现之前,Yahoo 使用人工对网站分类,提供按目录导航和搜索目录数据库功能。在 Google 尚未占据互联网搜索绝对优势之前,也是在笔者第一次听人推荐 Google 之前,就出现了元搜索引擎(Meta Search Engine)。用户只需提交一次搜索请求,由元搜索引擎负责转换处理后提交给多个预先选定的独立搜索引擎,并将从各独立搜索引擎返回的所有查询结果,集中起来处理后再返回给用户。但 Google 开始独家垄断全球互联网搜索后,元搜索引擎逐渐被人遗忘。

Google 早期的时候使用 MapReduce 实现分布式索引。后来之所以放弃这种方式,是因为它并不能为 Google 提供它所想要的索引速度。工程师需要等待 8 个小时的计算时间才能够得到计算的全部结果,然后把它发布到索引系统中。随着实时检索时代的到来,Google 需要在几秒内刷新索引内容,而非 8 小时。

Hadoop 来源于开源的分布式搜索项目 Nutch。Powerset 公司在 Hadoop 的基础上开发了基于 BigTable 架构的数据库 Hbase(http://hbase.apache.org/)。2008 年,微软收购了 Powerset。

与文本挖掘技术对应的是包括语音识别、基于内容的图像检索等技术的流媒体挖掘技术。随着网络电视和视频网站的流行,流媒体挖掘技术正越来越引起人们的关注。

除了像 Google 的网页搜索这样的常规搜索引擎外,还有些特殊的搜索引擎。搜索的输入不一定是简单的关键词,例如,Wolfram|Alpha(http://www.wolframalpha.com/)是一个特殊的可计算的知识引擎。它可以根据用户的问句式的输入精确地返回一个答案。开放式信息提取搜索(http://openie.allenai.org/)是另外一个问答式的搜索。除了大的搜索引擎,还有些小的垂直搜索引擎,例如,专利搜索引擎 http://www.soopat.com/。

Prim 算法于 1930 年由捷克数学家沃伊捷赫·亚尔尼克(Vojtěch Jarník)发现,并在 1957 年由美国计算机科学家罗伯特·普里姆(Robert C. Prim)独立发现。1959 年,艾兹格·迪科斯彻再次发现了该算法。因此,在某些场合,Prim 算法又被称为 DJP 算法、亚尔尼克算法或普里姆-亚尔尼克算法。除了 Prim 算法,还可以用 Kruskal 算法求解最小生成树。

单元测试可以采用 JUnit 结合 Hamcrest 测试框架,或者 TestNG 结合 Hamcrest。

1.8 术 语 表

BubbleSort：冒泡排序。
BinarySearch：折半查找。
posting list：投递列表，包含某个索引词的所有文档列表。
Inverted index：倒排索引。
Meta search engine：元搜索引擎。
Minimum spanning tree：最小生成树。

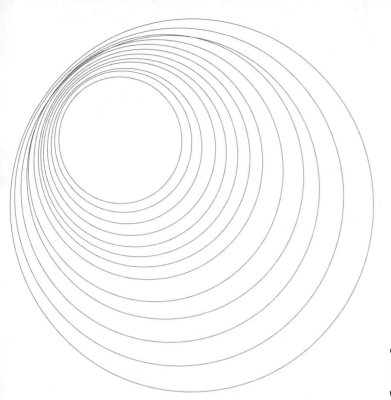

第 2 章

自己动手写全文检索

很多软件系统都需要对应的数据结构。信息检索中最常用的数据结构是倒排索引。全文索引如图2-1所示。

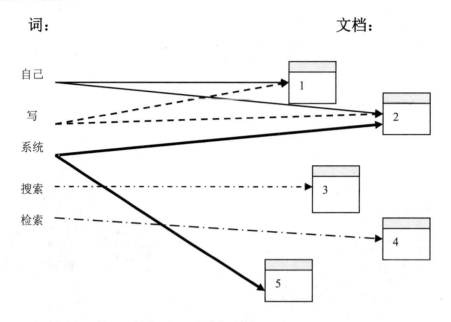

图2-1 以词为基础的全文索引

倒排索引就是一个词到文档列表的映射。用HashMap实现的一个简单的倒排索引代码如下。

```java
public class InvertedIndex {
    Map<String, List<Tuple>> index =
        new HashMap<String, List<Tuple>>(); //词和这个词在文档中出现的位置信息

    // 索引文档
    public void indexDoc(String docName, ArrayList<String> words) {
        int pos = 0;
        for (String word : words) {
            pos++; // 位置
            List<Tuple> idx = index.get(word);
            if (idx == null) {
                idx = new LinkedList<Tuple>();
                index.put(word, idx);
            }
            idx.add(new Tuple(docName, pos));
            System.out.println("indexed " + docName + " : " + pos + " words");
        }
    }

    // 搜索
    public void search(List<String> words) {
```

```
        for (String word : words) {
            Set<String> answer = new HashSet<String>();
            List<Tuple> idx = index.get(word);
            if (idx != null) {
                for (Tuple t : idx) { //找到了一些文档
                    answer.add(t.docName);
                }
            }
            System.out.print(word);
            for (String f : answer) {
                System.out.print(" " + f); //输出文件名
            }
            System.out.println("");
        }
    }

    private class Tuple { //<文档名，位置>元组
        private String docName; // 文档名
        private int position; // 位置

        public Tuple(String d, int position) {
            this.docName = d;
            this.position = position;
        }
    }
}
```

如果用户的查询中包含多个词，需要统计这些词在文档中出现的区间大小。区间越小的文档相关度越高。

```
public class Span {
    public int start; // 开始位置
    public int end;   // 结束位置

    public Span(int s, int e) {
        start = s;
        end = e;
    }

    public int length() {
        return end - start + 1;
    }

    public String toString() {
        return start + "-" + end;
    }
}
```

建立索引往往很耗时，所以应把建立好的倒排索引保存到文件。查询之前先读入建立

好的索引。

倒排索引由两个文件组成：一是文件存储倒排列表；另外一个是 B 树(Btree)存储词到倒排列表的映射。

索引的实现接口如下。

```java
/** 一个索引应该实现的功能模板 */
public interface Index {
    /** 使用数据库统计类构建索引 */
    public void build(DatabaseStatistics s);

    /** 从文件加载倒排索引 */
    public void read(String filename) throws IOException;

    /** 把内存中建好的倒排索引存入文件 */
    public void flush(String filename) throws IOException;
}
```

可以把创建索引和读入索引的方法分到不同的类实现，分别为 IndexWriter 和 IndexReader 类。

2.1 构建索引

为了按词建立索引，需要对文档分词。把这个分词类叫作 Analyzer。文档内容作为输入流 Reader 的实例传入。

```java
public class Analyzer {
    String[] split(Reader reader) {
        //对文档内容分词，并返回分词后的词数组
    }
}
```

首先建立文档索引，也就是正排索引，然后建立倒排索引。

正排索引用到的 DocConsumer 类如下。

```java
public class DocConsumer {
    public int docid; //文档编号
    //词到频率的映射，频率存在长度是 1 的整数数组中
    HashMap<String, int[]> frequencyList;
    int words;   //文档长度，也就是这个文档包含多少个词
}
```

然后建立倒排索引。倒排索引用到的 Posting 类如下。

```java
public class Posting implements Serializable {
    public int docid; //文档编号
    public int freq;  //这个词在文档中出现了多少次
```

```
    Posting(int doc, int freq) {
        this.docid = doc;
        this.freq = freq;
    }
}
```

这里把文档作为一个字符串,如果有多列,可以把文档作为一个自定义的对象。

```
public class Document {
    public String title; //标题列
    public String content; //内容列
}
```

为了更灵活地定义文档的结构,可以专门定义一个 Field 类。

```
public class Field {
    /** 列名 */
    public final String name;

    /** 列值 */
    public String fieldsData;
}
```

一个 Document 类的实例中可以包含多个 Field 类的实例。

在真正的信息检索中,都是有多个列的,而且很多列有同等重要的地位。对于多个列的检索,简单的方法是:给每个倒排列附加一个 bit-map,假设有 N 列,则 Bitmap 长度是 N。例如,title 出现,则第 1 位为 1,content 没出现,则第 2 位为 0。实际的做法是:把一个列中所有的词连续存放到一起,一个专门的列定义文件中包括某个列中这些词信息的开始位置,如图 2-2 所示。

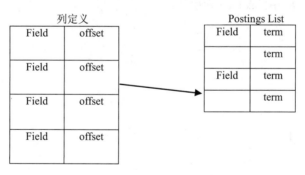

图 2-2 多列索引

2.2 生成索引文件

词表是类似这样的数据结构:SortedMap<Term,Postings>。如果词表很小,内存能够放下,则可以使用折半查找算法来查询一个词对应的文档序列。如果内存不能完全放下倒排索引中的词表，如何利用索引文件查找？理论上，可以采用 B 树存储词表。为了简化实现，Lucene 3 以前的版本使用了跳表。跳表把词组织成固定大小的块，例如每 32 个词放入一个新的块，然后在块上建立一个索引。记住每个块的开始词在文件中的位置。

但是，更好的分块方式是根据词共享了哪些前缀。按前缀组织词产生的词块大小是变化的。把相同前缀的词放到一起比按顺序放更好，这样在每个区块内，可最大化共享前缀的词，并且减少了产生的词索引。

这样做也可以加速词密集的查询。因为词索引成为了前缀 Trie 树。如果词典中不包含这个词，可以快速报告："没有找到。"不需要在查找某一个词块之后才能确定没有。

这种方法叫作 BlockTree。例如，词表中包含如下一些词。

```
able
above
apple
perfect
preface
prefecture
prefix
previous
profit
programmer
project
zoo
```

把 a 开头的 3 个词组织成一个词块：{able,above,apple}。p 开头的词有 8 个，超过 4 个。把 pre 开头的 4 个词组织成一个词块：{ preface, prefecture, prefix , previous }。把 pro 开头的 3 个词组织成一个词块：{ profit, programmer, project }。因为以 z 开头的词只有 1 个，所以 zoo 单独组成一个词块。BlockTree 索引如图 2-3 所示。

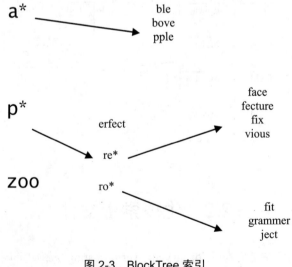

图 2-3　BlockTree 索引

根据有限状态转换找到词块在文件中存放的位置，如图 2-4 所示。

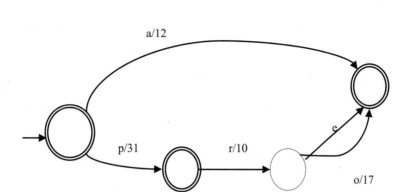

图 2-4　有限状态转换

BlockTreeTermsReader 类用于读索引。BlockTreeTermsWriter 类用于写索引。对于主键列来说，可以把所有的词和 postings 一起存到一个 FST 中。把 FST 保存到硬盘。可以利用 BlockTree 的结构加快排序的速度。

2.3　读入索引文件

如果词表太大，则只读入最上面几层的索引。为了批量读入数据，把数据分页存放。参考 B 树索引的分页读取。

```
public class Page{
    public Page() {
        data = new byte[MAX_SPACE];
    }

    public Page(byte[] apage) {
        data = apage;
    }

    /**
     *返回数据的字节数组
     *
     * @return 页面的字节数组
     */
    public byte[] getpage() {
        return data;
    }

    /**
     *用给定的字节数组设置页面
     *
     * @param array
     * 页面大小的字节数组
```

```
    */
    public void setpage(byte[] array) {
        data = array;
    }

    public byte[] getData() {
        return this.data;
    }

    /**
     *受保护的属性:字节数组
     *
     */
    protected byte[] data;
}
```

2.4 查询

查询一个词时，将直接把对应的文档编号列表找出来，然后按相关度返回文档列表。如果内存不能完全存下所有的文档列表，可以把用户最经常查询的词的文档列表放到内存中，不常查询的词的文档列表放到文件中。

查询两个词"NBA 视频"时，则对"NBA"这个词对应的文档编号列表和"视频"这个词对应的文档编号列表做交集(Intersection)运算后返回。例如，在倒排索引表中检索出包含"NBA"一词的文档列表为 docList("NBA")=(1, 5, 9, 12)，表示这 4 个文档编号的文档含有"NBA"这个词。包含"视频"的文档列表为 docList("视频")= (5, 7, 9, 11)。这样同时包含"NBA"和"视频"这两个词的文档为 docList("NBA")∩docList("视频")=(5, 9)。这里的"∩"表示文档列表集合求交集运算。

计算过程类似归并排序，如图 2-5 所示。

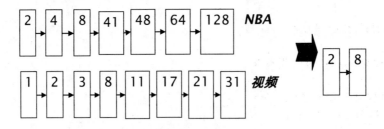

图 2-5 文档列表集合求交集

多文档列表求交集计算的示例代码如下。

```
public ArrayList<Integer> intersection(int[] docListA, int[] docListB) {
    ArrayList<Integer> ret = new ArrayList<Integer>();
    int aindex = 0;
    int bindex = 0;
```

```
            while (aindex < docListA.length && bindex < docListB.length) {
                if (docListA[aindex] == docListB[bindex]) {
                    // 找到在两个数组中都出现的值
                    ret.add(docListA[aindex]);
                    aindex++;
                    bindex++;
                } else if (docListA[aindex] < docListB[bindex]) {
                    aindex++;
                } else {
                    bindex++;
                }
            }
            return ret;
        }
```

使用构建好的倒排索引和文档索引执行搜索。

```
public class IndexSearcher {
    InvertedIndex index; //倒排索引
    DocumentIndex docIndex; //文档索引
}
```

根据一个或者多个查询词执行搜索,可以指定最多返回的搜索结果数量。如果不指定返回结果数量,则返回全部相关的文档。

只需要使用浮点数就能够区分相关文档和不太相关的文档,所以把 score 定义成 float 类型。

```
/** 用于搜索的辅助类。支持对文档按相关度排序 */
public class ScoreDoc implements Comparable<ScoreDoc> {
    DocumentData doc; //文档相关的信息,包括文档编号等
    public float score; //表示这个文档和查询词有多相关

    public ScoreDoc(DocumentData d, float b) { //构造方法
        doc = d;
        score = b;
    }

    public int compareTo(ScoreDoc other) { // 按相关度排序
        // 返回值并不重要,保证需要的顺序就行
        ScoreDoc o = other;
        return (o.score == score) ? (0) : ((score > o.score) ? (-1) : (1));
    }

    public String toString() { //格式化输出
        return doc.filename + ":\t" + doc.docid + "\n" + score;
    }
}
```

如果要对一个大的索引库排序,而且一个词有很多结果,则对所有这些结果排序然后

分页显示会很慢。

搜索时要排序，但实际上并不需要对所有的结果排序。如果只要第 1 页，就只对前 10 个结果排序。排序是和显示页相关的。前 10 个结果是怎么得到的呢？是用最小堆得到的，须先构建最小堆。

按相关度排序简介如下。

用优先队列记录前 n 个评分最高的文档。优先队列用最小堆实现。把前 n 个评分最高的文档放到一个最小堆中。堆虽然是一个二叉树，但是一般使用数组实现，不需要元素之间的指针，如图 2-6 所示。

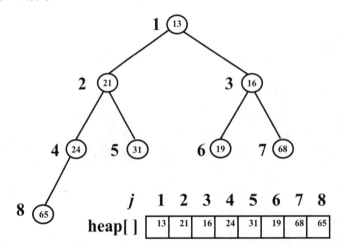

图 2-6　二叉堆

heap[0]左边的子元素是 heap[1]，右边的子元素是 heap[2]。heap[1]左边的子元素是 heap[3]，右边的子元素是 heap[4]。heap[2]左边的子元素是 heap[5]，右边的子元素是 heap[6]。则 heap[i]左边的子元素是 heap[2i+1]，右边的子元素是 heap[2i+2]。

如果要简化计算，第一个位置可以不用，则根节点位于 heap[1]，heap[1]左边的子元素是 heap[2]，右边的子元素是 heap[3]。则 heap[i]左边的子元素是 heap[2i]，右边的子元素是 heap[2i+1]。

因为没有使用第 0 个位置，所以数组要多分配一个位置。

```
private final T[] heap;

public PriorityQueue(int maxSize) {
  int heapSize = maxSize + 1;
  heap = (T[]) new Object[heapSize];   // 没有用 heap[0]
}
```

建立容量为 n 的最小堆，记录前 n 个评分最高的文档。对于每个和查询词相关的文档，如果它比堆顶文档的相关度更高，则删除堆顶元素，把当前文档插入堆中，然后自顶向下调节，维护最小堆。如果它比堆顶文档相关度低，则直接扔掉。遍历完整个相关文档集合，堆中的文档就是最相关的 n 个文档了。

```
public T insertWithOverflow(T element) {
  if (size < maxSize) {
    add(element);
    return null;
  } else if (size > 0 && !lessThan(element, heap[1])) {
    //堆中最小的元素用更大的元素替换了
    T ret = heap[1];
    heap[1] = element;
    updateTop();
    return ret;
  } else {
    return element;
  }
}
```

要在堆中增加一个元素,必须执行一个 upHeap。可以用 add(element) 方法调用 upHeap。要从最小堆中取得最小的元素就需要从堆中删除根节点,这叫作 downHeap。可以用 pop() 方法调用 downHeap。

最后把最小堆中的文档放入 ScoreDoc。

```
final ScoreDoc[] scoreDocs = new ScoreDoc[hq.size()];
for (int i = hq.size() - 1; i >= 0; i--) // 把最相关的文档放入数组
  scoreDocs[i] = hq.pop(); //先取出来最不相关的文档
```

2.5 有限状态机

可以用有限状态机实现模糊查询。例如,查找编辑距离相似的单词。

org.apache.lucene.util.automaton 包含有限状态机的实现。BasicAutomata.makeChar()方法生成接收单个字符的自动机。

```
Automaton a = BasicAutomata.makeChar('W'); //创建一个字符W组成的自动机
```

如果从同一个状态接收同样的输入后可以任意到达多个不同的状态,这样的有限状态机叫作非确定有限状态机。如果从一个状态接收一个输入后只能到达某一个状态,这样的有限状态机叫作确定有限状态机。上面的有限状态机 a 是一个确定的有限状态机。

可以用 BasicOperations.run 方法测试自动机是否能够接收输入字符串。

```
System.out.println(BasicOperations.run(a, "W")); //输出 true
```

2.5.1 运算

BasicOperations 中包含一些 Automaton 运算。Intersection()方法用于求交集。例如:

```
//创建一个可以接收A~W字符的自动机
Automaton a = BasicAutomata.makeCharRange('A','W');
```

```
Automaton b = BasicAutomata.makeCharRange('D','W');
Automaton c = BasicOperations.intersection(a,b);
System.out.println(BasicOperations.run(c, "A")); //输出 false
System.out.println(BasicOperations.run(c, "W")); //输出 true
```

union()方法用于求并集。例如:

```
Automaton a = BasicAutomata.makeCharRange('D', 'W');
Automaton b = BasicAutomata.makeCharRange('A', 'C');
Automaton c = BasicOperations.union(a, b);
System.out.println(BasicOperations.run(c, "A")); // 输出 true
System.out.println(BasicOperations.run(c, "W")); // 输出 true
```

concatenate()方法用于连接两个自动机。例如:

```
Automaton a = BasicAutomata.makeChar('W');
Automaton b = BasicAutomata.makeChar('W');
Automaton c = BasicOperations.concatenate(a, b);
System.out.println(BasicOperations.run(c, "WW")); // 输出 true
System.out.println(BasicOperations.run(c, "WWW")); // 输出 false
```

repeat()方法表示重复 0 次或者多次。例如重复字符"W"零次或者多次:

```
Automaton a = BasicAutomata.makeChar('W');
Automaton c = BasicOperations.repeat(a);
System.out.println(BasicOperations.run(c, "WW")); // 输出 true
System.out.println(BasicOperations.run(c, "WWW")); // 输出 true
```

determinize()方法表示把 NFA 转换成 DFA。

```
a.deberminize();
```

复杂的正则表达式难以维护。"[yY]es"这样的正则表达式可以写成如下等价的 Automaton:

```
Automaton a = BasicAutomata.makeChar('Y'); //创建一个字符 W 组成的自动机
Automaton b = BasicAutomata.makeChar('y');
Automaton c = a.union(b);
Automaton d = c.concatenate(BasicAutomata.makeString("es"));
//使用它
System.out.println(BasicOperations.run(d, "y yes")); //输出 true
```

2.5.2 编辑距离有限状态机

编辑距离自动机的基本想法是:构建一个有限状态自动机,准确地识别出和目标词在给定的编辑距离内的字符串集合。可以输入任何词,然后自动机可以基于是否和目标词的编辑距离最多不超过给定距离,从而接收或拒绝它。而且,由于 FSA 的内在特性,可以在 $O(n)$ 时间内判断是可以接收或应该拒绝。这里,n 是测试字符串的长度。而标准的动态规划编辑距离计算方法需要 $O(m*n)$ 时间,这里 m 和 n 是两个输入单词的长度。因此编辑距离自动机可以更快地检查许多单词和一个目标词是否在给定的最大距离内。

单词"food"的编辑距离自动机形成的非确定有限状态自动机,最大编辑距离是 2。开始状态在左下,状态使用 n^e 标记风格命名,这里 n 是目前为止正确匹配的字符数,e 是错误数量。垂直转换表示未修改的字符,水平转换表示插入,两类对角线转换表示替换(用*标记的转换)和删除(空转换),如图 2-7 所示。

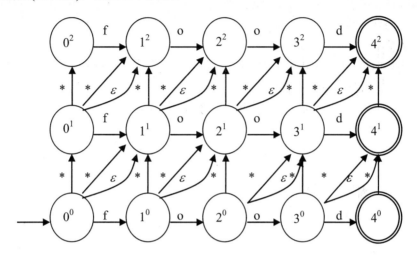

图 2-7 编辑距离自动机

单词"food"的长度是 4,所以图 2-7 有 5 列。允许 2 次错误,所以有 3 行。

LevenshteinAutomata 接收两个参数,输入字符串和一个布尔值,说明换位是否可以作为一个编辑操作。如果"GO"可以换位,则可以接收"OG"。使用 BasicOperations.run 方法测试。

```
//换位也可以看成是一个编辑操作
LevenshteinAutomata builder = new LevenshteinAutomata("GO", true);
Automaton a = builder.toAutomaton(1); //最大编辑距离是1
System.out.println(BasicOperations.run(a, "OG")); //输出 true
```

如果"GO"不能换位,则不能接收"OG"。

```
LevenshteinAutomata builder = new LevenshteinAutomata("GO", false);
Automaton a = builder.toAutomaton(1); //最大编辑距离是1
System.out.println(BasicOperations.run(a, "OG")); //输出 false
```

测试最大编辑距离是 2 的情况。

```
LevenshteinAutomata builder = new LevenshteinAutomata("EBER", true);
Automaton a = builder.toAutomaton(2); //最大编辑距离是2
```

测试可以接收哪些字符。

```
System.out.println(BasicOperations.run(a, "BR")); //输出 true
System.out.println(BasicOperations.run(a, "EB")); //输出 true
System.out.println(BasicOperations.run(a, "EBE")); //输出 true
System.out.println(BasicOperations.run(a, "EBER")); //输出 true
```

2.6 本章小结

　　计算机可以帮助人记忆，我们并不需要记住所有的细节，而只需要根据一些线索来查找相关细节。查找文档所需要的线索就是查询词。通过建立倒排索引，可以快速地找到查询词对应的文档。

　　org.apache.lucene.util.automaton 来源于丹麦奥尔胡斯大学的副教授 Anders Moller 开发的有限状态机实现 dk.brics.automaton。

　　除了基本的按关键词查找，还需要有高亮显示的功能。

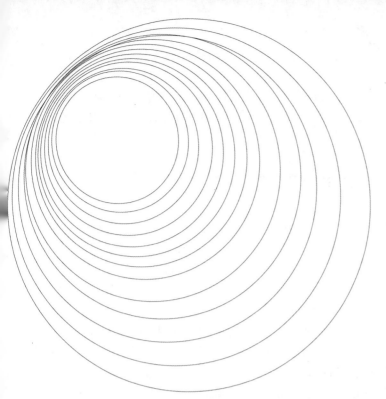

第 3 章

Lucene 的原理与应用

MangoDB 只是用来存储文档，实现全文检索还是要用到 Lucene。使用 Lucene 建立全文索引，然后就可以按关键词快速查找文档集合。下面先看一下使用 Lucene 最常用的方法，然后再看一下 Lucene 是如何实现的。

3.1 Lucene 快速入门

Lucene 是一个开放源代码的全文索引库，完成基本功能只有 1 个不依赖外部程序包的 jar 文件。因为这个文件是一个核心文件，所以称为 lucene-core-Version.jar。例如，Lucene 的 6.3.0 版本叫作 lucene-core-6.3.0.jar。可以从 http://lucene.apache.org/core/下载这个 jar 包。

待查询的文档集合按词组织成倒排索引。Lucene 中的索引库是位于一个目录下的一些二进制文件。Lucene 中的索引库叫作 Index。和一般的数据库不一样，Lucene 不支持定义主键。在 Lucene 中并不存在一个叫作 Index 的类。通过 IndexWriter 来写索引，通过 IndexReader 读索引。索引库在物理形式上一般是位于某个路径下的一系列文件。

先介绍如何创建索引库，然后介绍如何搜索索引库。总的来说，往 Lucene 中放的是文档，查询的是词，查询返回的也是文档。使用 Lucene 实现搜索的基本概念如图 3-1 所示。

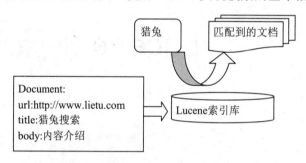

图 3-1 Lucene 搜索的基本概念

为了对 Lucene 有个大体了解，先用伪代码说明 Lucene 的核心概念。

3.1.1 创建索引

先准备好文档数据，然后往索引库中加入文档。写入索引的过程如图 3-2 所示。

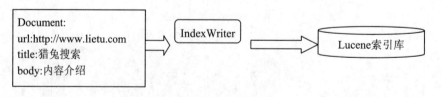

图 3-2 写入索引

创建索引部分骨干伪代码。

```
//打开索引库
IndexWriter writer=new IndexWriter(dir);
```

第 3 章 Lucene 的原理与应用

```
//写入一些文档
writer.addDocument(doc);
//关闭索引库
writer.close();
```

一个 Document 代表索引库中的一条记录,也叫作文档。要搜索的信息封装成 Document 后通过 IndexWriter 写入索引库。

一个文档有多个列,例如,标题或者内容列。Lucene 中的一个 Document 对象可以包含多个列对象,叫作 Field。例如,一个网页可以包含"网址""标题""正文""修改时间"等 Field。创建这些列对象以后,可以通过 Document 的 add()方法增加这些列。代码如下。

```
Document doc = new Document();
//创建网址列
Field f = new Field("url", "http://www.lietu.com" ,
      Field.Store.YES, Field.Index.UN_TOKENIZED,
      Field.TermVector.NO);
doc.add(f);
```

3.1.2 查询索引库

用户界面输入的查询词封装成 Query 对象后,由 IndexSearcher 执行查找过程。因为往往只需要最相关的一些文档,所以返回结果封装在一个叫作 TopDocs 的对象中,如图 3-3 所示。

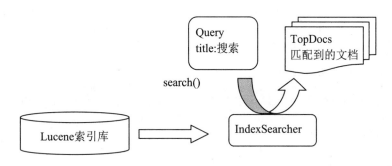

图 3-3 查询原理

查询部分骨干伪代码。

```
// DirectoryReader 读入一个目录下的索引文件
IndexReader ir = DirectoryReader.open(directory);

//打开索引库
IndexSearcher searcher = new IndexSearcher(ir);

//根据查询词搜索索引库
TopDocs docs = searcher.search(query, 10); //最多返回10个结果

//遍历查询结果
```

```
ScoreDoc[] hits = docs.scoreDocs;
for (ScoreDoc hit : hits){
    //…
}

//关闭索引库
searcher.close();
```

调用 Searcher 接口按关键词搜索后，返回和给定查询最相关的若干个文档以及匹配上的文档总数。TopDocs 封装了搜索结果以及匹配上的文档总数。

TopDocs.scoreDocs 就是一个封装后的 Document 的列表。TopDocs 主要包含如下两个属性。

```
public class TopDocs {
    public int totalHits; // 查询匹配上的文档总数
    public ScoreDoc[] scoreDocs; // 命中文档列表顶部的文档
}
```

3.1.3 创建文档索引

在 Eclipse 中创建一个 Java 控制台项目。创建 lib 目录，然后把 lucene-core-6.3.0.jar 文件复制到 lib 目录下。在项目属性中增加对 lucene-core-6.3.0.jar 文件的引用。

创建索引时需要指定切分文本用的 Analyzer，这里使用 StandardAnalyzer 切分文本。因为 StandardAnalyzer 位于 lucene-analyzers-common-6.3.0.jar 文件中，所以需要增加对这个文件的引用。把 lucene-analyzers-common-6.3.0.jar 文件复制到 lib 目录下，在项目属性中增加对 lucene-analyzers-common-6.3.0.jar 文件的引用。

新建一个测试类，实现在硬盘中创建索引。

```
//创建 StandardAnalyzer
Analyzer analyzer = new StandardAnalyzer();

// 把索引保存在硬盘上的一个目录中
Directory directory = FSDirectory.open(new File("d:/news")); //存放新闻的索引
IndexWriterConfig config = new IndexWriterConfig(analyzer);
IndexWriter iwriter = new IndexWriter(directory, config);
Document doc = new Document();
String text = "This is the text to be indexed."; //要索引的文本
doc.add(new Field("title", text, TextField.TYPE_STORED));
iwriter.addDocument(doc);
iwriter.close();
directory.close();
```

3.1.4 查询文档索引

QueryParser 这个类位于 org.apache.lucene.queryparser.classic 包，需要引用

lucene-queryparser-6.3.0.jar 文件。

新建一个测试类来查询索引，得到查询对象。

```
String defaultField = "title";
String queryString ="test";
Analyzer analyzer = new StandardAnalyzer();
QueryParser parser = new QueryParser(defaultField,
                                 analyzer);  //用于解析查询语法
//从字符串得到查询对象
Query query = parser.parse(queryString);
```

调用 IndexSearcher.search 方法执行搜索后，从 IndexSearcher 中根据文档编号取出文档对象。

```
TopDocs hits = searcher.search(query, 10);
System.out.println("hits.totalHits:"+ hits.totalHits);
for (int j = 0; j < hits.scoreDocs.length; j++) {
    //根据文档编号取出文档对象
    Document hitDoc = searcher.doc(hits.scoreDocs[j].doc);
    System.out.println(hitDoc.get("title"));  //输出文档
}
```

注意，这里的 hits.totalHits 和 hits.scoreDocs.length 的值往往不一样。对于网页搜索来说，hits.totalHits 表示索引中总的文档数，而 hits.scoreDocs.length 表示当前页中实际显示的文档数。

因为倒排索引中只存文档编号，不存全文，所以 IndexSearcher.search 方法搜出来的结果中只有文档编号序列。但 Lucene 的索引中不只是有倒排索引，还存储了文档内容，所以可以根据文档编号得到文档对象。

3.2　创建和维护索引库

先根据应用需求设计好索引库的结构，然后再建立索引并往索引库中加文档。最后为了提高查询速度，还可以优化索引，也就是把索引文件变得更紧凑。

3.2.1　设计索引库结构

假设需要按关键词搜索新闻，需要搜索的列包括标题和内容。每篇新闻文档本身还有发布日期和对应的 URL 地址，这两列不需要按关键词查询，只需要能够取出来对应的值就可以了。一个新闻索引库的结构见表 3-1。

表 3-1　索引库结构表

名　称	描　述	在 Lucene 中的定义

title	存储新闻标题，需要分词和关键词加亮	`new Field("title,`
		`        title ,`
		`        Field.Store.YES,`
		`        Field.Index.TOKENIZED,`
		`        Field.TermVector.WITH_POSITIONS_OFFSETS);`

body	存储新闻正文，需要分词和关键词加亮	`new Field("body",`
		`        body ,`
		`        Field.Store.YES, Field.Index.TOKENIZED,`
		`        Field.TermVector.WITH_POSITIONS_OFFSETS);`

url	存储新闻的来源 URL 地址	`new Field("url",`
		`        url ,`
		`        Field.Store.YES, Field.Index.UN_TOKENIZED,`
		`        Field.TermVector.NO);`

date	存储日期字段，在 Lucene 中需要转换成数值存储	`new LongField ("date").setLongValue(pubDate,` `        getTime(), Field.Store.NO);`

这个新闻索引库相当于 SQL 语句 create table news（"title"，"body"，"url"，"date"）。

3.2.2　创建索引库

索引一般存放在硬盘中的一个路径中。例如，在路径 d:/lucene/index 下创建索引。

```
//索引建立在硬盘上
Directory directory = FSDirectory.open(new File("d:/lucene/index"));
```

可以用 Lucene 提供的 API 创建和更新索引。在生成索引的过程中涉及几个类，它们之间的关系如图 3-4 所示。

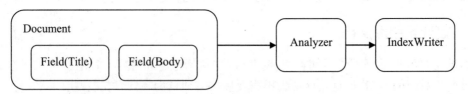

图 3-4　创建索引过程中用到的类

可以通过 IndexWriter 在指定的索引路径下创建一个新的索引库，IndexWriter 位于索引管理包 org.apache.lucene.index 中。虽然可以这样写：

```
org.apache.lucene.index.IndexWriter iw =
    new org.apache.lucene.index.IndexWriter(dir, iwc);
```

但是为了简化写法，往往使用 import 关键字导入这个类：

```
import org.apache.lucene.index.IndexWriter;
```

用类似的写法导入其他的类。

不同版本的索引格式不一样。可以按指定的版本号写入索引，例如，如果在项目中导入 lucene-core-6.3.0.jar 这个 jar 包，就指定版本号是 LUCENE_43：

```
static Version matchVersion=Version.LUCENE_43;
```

写入按词组织的索引时，要指定用哪个类把文档切分成词，Analyzer 类实现分词功能。各种文本分析器都是 Analyzer 的子类。例如，使用标准分析器：

```
Analyzer analyzer=new StandardAnalyzer();
```

有个专门配置 IndexWriter 参数的类 IndexWriterConfig。

```
Analyzer analyzer=new StandardAnalyzer();

IndexWriterConfig iwc=new IndexWriterConfig(analyzer);
iwc.setOpenMode(OpenMode.CREATE);   //总是新建索引
```

也可以根据指定的目录下是否已经有索引存在，决定是向已有的索引追加文档，还是重新建立索引。根据该文件夹下是否存在索引文件来决定是否追加索引的判断如下：

```
boolean createIndex = false;
String indexDir = "d:/index";
Directory indexDirectory = FSDirectory.open(new File(indexDir));
// 该文件夹下是否存在索引文件
if (!IndexReader.indexExists(indexDirectory)) {
    createIndex = true;
}
// 根据 createIndex 的值来决定是否追加索引
```

这些只需要一行 IndexWriterConfig 配置代码：

```
iwc.setOpenMode(OpenMode.CREATE_OR_APPEND);   //保存已有的索引中的数据
```

根据索引路径和配置信息创建一个 IndexWriter，代码如下：

```
//默认 create_or_append
iwc.setOpenMode(IndexWriterConfig.OpenMode.CREATE);//总是重新创建索引
Directory dir =FSDirectory.open(new File(indexPath));
IndexWriter iw=new IndexWriter(dir, iwc);   //指定索引路径和配置信息
iw.addDocument(doc);
iw.close();
```

如果有多个索引，一般不会放在同一个路径下。除了在可以长期保存的物理路径中，索引库还可以仅在内存中，也就是创建一个位于 RAMDirectory 的索引。

```
Directory dir = new RAMDirectory();   //内存路径
ndexWriter iw=new IndexWriter(dir, iwc);
```

在测试分词或索引的时候,可能会用到内存索引。

3.2.3 向索引库中添加索引文档

先增加文档,然后提交更新到索引库。和数据库表一样,索引库是结构化的,一个文档往往有很多列,例如标题和内容列等。往索引添加数据时涉及的几个类的关系如图 3-5 所示。

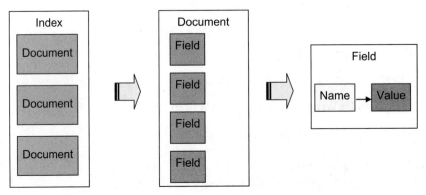

图 3-5 往索引中添加文档

TextField 是一个快捷类,它不存储词向量。如果你需要词向量,就只使用 Field 类。这需要更多的代码,因为首先要创建一个 FieldType 类的实例,然后设置 storeTermVectors 和 tokenizer 为 true,并在 Field 构造器中使用这个 FieldType 实例。

```
FieldType t = new FieldType();
t.setStoreTermVectorOffsets(true);
t.setTokenized(true);
fieldTitle = new Field("title", "标题", t);
```

使用 StringField:

```
field = new StringField("url", "bar", Store.YES);
```

使用 FieldType 实现的等价代码:

```
FieldType fieldType = new FieldType();
fieldType.setStored(true);
fieldType.setTokenized(false);
fieldType.setIndexed(true);
field = new Field("url", "lietu.com", fieldType);
```

下面这段程序是向索引库中添加网页地址、标题和内容列:

```
//如果初次使用Lucene,往索引中写入的每条记录最好都新创建一个Document与之对应,
//也就是说不要重复使用Document对象,否则可能会出现意想不到的错误
Document doc = new Document();
//创建网址列
Field f = new Field("url", news.URL ,
```

```
            Field.Store.YES, Field.Index.UN_TOKENIZED,
            Field.TermVector.NO);
doc.add(f);
//创建标题列
f = new Field("title", news.title ,
            Field.Store.YES, Field.Index.TOKENIZED,
            Field.TermVector.WITH_POSITIONS_OFFSETS);
doc.add(f);
//创建内容列
f = new Field("body", news.body.toString() ,
            Field.Store.YES, Field.Index.TOKENIZED,
            Field.TermVector.WITH_POSITIONS_OFFSETS);
doc.add(f);
index.addDocument(doc);
```

下面两种写法是等价的:

```
new Field("title", "标题", TextField.TYPE_STORED);
new TextField("title", "标题", Store.YES );
```

和一般的数据库不一样,一个文档的一个列可以有多个值。例如一篇文档既可以属于互联网类,又可以属于科技类。

Lucene 中的 API 相对数据库来说比较灵活,没有类似数据库先定义表结构后再使用的过程。如果前后两次写索引时定义的列名称不一样,Lucene 会自动创建新的列,所以 Field 的一致性需要我们自己掌握。

列选项组合见表 3-2。

表 3-2 Field 的类型

Index	Store	TermVector	用法举例
NOT_ANALYZED	YES	NO	文件名、主键等
ANALYZED	YES	WITH_POSITIONS_OFFSETS	标题、摘要
ANALYZED	NO	WITH_POSITIONS_OFFSETS	很长的全文
NO	YES	NO	文档类型
NOT_ANALYZED	NO	NO	隐藏的关键词

这里的 POSITIONS 表示以词为单位的位置,也就是语义位置,而 OFFSETS 表示词在文本中的实际物理位置。

FieldType 类设置这些组合。例如:

```
Analyzer ca = new CJKAnalyzer();
IndexWriterConfig indexWriterConfig = new IndexWriterConfig(ca);
FieldType nameType = new FieldType();
nameType.setIndexed(true);
nameType.setStored(true);
IndexWriter indexWriter = new IndexWriter(DIRECTORY, indexWriterConfig);
```

```
Document document = new Document();
document.add(new Field("name", "我购买了道具和服装", nameType));
indexWriter.addDocument(document);
```

在增加文档的阶段,给新的词分配 TokenID,新的文档分配 DocID。

增加索引后,记得提交索引,否则 reader 不一定能搜索到。

```
writer.commit();    //提交更新到索引库
```

索引创建完成后可以用索引查看工具 Luke(https://github.com/dmitrykey/luke)来查看索引内容并维护索引库。Luke 是一个可以执行的 jar 包,是用 Java 实现的 Windows 程序。在 Windows 下可以双击 lukeall-1.0.1.jar,启动 Luke。然后,可以选择菜单 File→Open Lucene index,打开 data/index 文件夹,然后可以在窗口看到索引创建的详细信息。

为了提高索引速度,可以重用 Field,而不是每次都创建新的。从 Lucene 2.3 开始,有新的 setValue()方法,可以改变 Field 的值。这样可以在增加许多 Document 的时候重用单个的 Field 实例,从而节省许多垃圾回收消耗的时间。

开始新建一个独立的 Document 实例,然后增加许多 Field 实例,并且增加每个文档到索引的时候都重用这些 Field。例如,有一个 idField、bodyField 和 nameField 等。当加入一个 Document 后,可以通过 idField.setValue()等直接改变 Field 的值,然后再增加文档实例。下面是一个重用 Field 的例子。

```
//创建索引
IndexWriter writer = new IndexWriter(directory, analyzer);
//创建文档结构
Document doc = new Document();
//定义 id 重用列
Field idField = new Field("id",
        Field.Store.YES, Field.Index.UN_TOKENIZED,
        Field.TermVector.NO);
doc.add(idField);

//定义标题重用列
Field titleField = new Field("title", null ,
        Field.Store.YES, Field.Index.ANALYZED,
        Field.TermVector.WITH_POSITIONS_OFFSETS);
doc.add(titleField);

//定义内容重用列
Field bodyField = new Field("body", null ,
        Field.Store.YES, Field.Index.ANALYZED,
        Field.TermVector.WITH_POSITIONS_OFFSETS);
doc.add(bodyField);

for(all document) {   //处理所有的文档,填充 Field 值,并索引文档
        Article a = parse(document);
        idField.setValue(a.id);
```

```
            titleField.setValue(a.title);
            bodyField.setValue(a.body);
            writer.addDocument(doc);  //doc 仅仅是一个容器
}
writer.close();
```

注意，不能在文档中重用单个 Field 实例，不应该改变列的值，直到包含这个 Field 的 Document 已经加入到索引库。

可以同时增加多个文档到索引：addDocuments()。

```
List<Document> docs = new ArrayList<Document>();
Document doc1 = new Document();
doc1.add(new Field("content", "猎兔搜索",  Field.Store.YES,
Field.Index.ANALYZED));
Document doc2 = new Document();
doc2.add(new Field("content", "中文分词",  Field.Store.YES,
Field.Index.ANALYZED));
Document doc3 = new Document();
doc3.add(new Field("content", "中国",  Field.Store.YES,
Field.Index.ANALYZED));
Document doc4 = new Document();
doc4.add(new Field("content", "NBA",  Field.Store.YES,
Field.Index.ANALYZED));

docs.add(doc1); docs.add(doc2); docs.add(doc3); docs.add(doc4);

writer.addDocuments(docs);
```

当在 Windows 系统下使用的时候，最好取消勾选杀毒软件的"自动删除已感染病毒文件"选项，否则当索引带病毒特征的文档时，杀毒软件可能会破坏 Lucene 的索引文件。

每一个添加的文档都被传递给 DocConsumer 类，它处理该文档并且与索引链表(indexing chain)中其他的 consumers 相互发生作用。确定的 consumers，就像 StoredFieldWriter 和 TermVectorsTermsWriter，提取一个文档中的词，并且马上把字节写入文件。

IndexWriter.setRAMBufferSizeMB 方法可以设置当更新文档使用的内存达到指定大小之后才写入硬盘。这样可以提高写索引的速度，尤其是在批量创建索引的时候。

在 Lucene 2.3 版本之前，存入索引的每个 Token 都是新创建的。重复利用 Token 可以加快索引速度。新的 Tokenizer 类可以回收利用已用过的 Token。

3.2.4 删除索引库中的索引文档

索引过一段时间后可能会过期。例如，索引的一些网页链接会失效。可从 IndexWriter 中删除文档。

```
indexWriter.delete(new Term("id", "1"));
```

Term 的值在索引库中并不一定是唯一的，比如要把某个类别的商品下架：

```
indexWriter.delete(new Term("cat", "book"));  //把图书类商品从索引库中删除
```

IndexReader 和 IndexWriter 都能删除文档，区别是：当 IndexWriter 打开索引的时候，IndexReader 的删除操作会抛出 LockObtainFailedException 异常。

Lucene 在删除索引时，经常会出现代码成功执行，但索引并未真正删除的现象。在创建 Term 时，注意 Term 的值必须是不可再拆分的词，否则删除不成功。

若需要批量删除某个网站的网页，则可以根据查询条件删除：

```
PrefixQuery query = new PrefixQuery(new Term("url", "http://www.lietu.com"));
indexWriter.deleteDocuments(query);
```

3.2.5 更新索引库中的索引文档

对于分词列来说，直接更新索引中的值很困难。可以采用先删除旧文档，然后再向索引增加传入的新文档的方法实现更新索引。

通过 IndexWriter.updateDocument 方法更新文档。

```
indexWriter.updateDocument(new Term("url", "http://www.lietu.com"),document);
```

如果只希望更新个别列而保持其他的列不动就会有问题。为了解决这个问题，可以搜索索引中的当前文档，改变要改变的列，然后把修改后的文档作为参数传给 updateDocument()。

```
public void searchAndUpdateDocument(IndexWriter writer,
                    IndexSearcher searcher,
                    Document updateDoc,
                    Term term) throws IOException {
    TermQuery query = new TermQuery(term);

    TopDocs hits = searcher.search(query, 10);

    if (hits.scoreDocs.length == 0) {
        throw new IllegalArgumentException("索引中没有匹配的结果");
    } else if (hits.scoreDocs.length > 1) {
        throw new IllegalArgumentException("Given Term matches
          more than 1 document in the index.");
    } else {
        int docId = hits.scoreDocs[0].doc;

        //找出旧的文档
        Document doc = searcher.doc(docId);

        List<Field> replacementFields = updateDoc.getFields();
        for (Field field : replacementFields) {
            String name = field.name();
            String currentValue = doc.get(name);
            if (currentValue != null) {
```

```
                        //替换列值
                        Field = new Field(name, term.text());
                        //删除所有出现的值旧列
                        doc.removeFields(name);

                        //插入替换列
                        doc.add(field);
                    } else {
                        //新加到列
                        doc.add(field);
                    }
                }

                //把旧文档更改后重新写入索引
                writer.updateDocument(term, doc);
            }
}
```

3.2.6 关闭索引库

关闭索引库，保证 IndexWriter 实际更新索引。在增加、删除和更新文档后，需要调用 IndexWriter.close()方法关闭索引库。

重新打开索引库：

```
writer.close(); //关闭索引库
writer = new IndexWriter(); //打开索引库
```

IndexWriter.ensureOpen()方法用来检查 IndexWriter 的状态是否为开启。如果它已经关闭了，则 IndexWriter 抛出一个 AlreadyClosedException 异常。

3.2.7 索引的优化与合并

在完成了创建最终倒排文件和词典后，全部倒排索引文件的创建工作完毕。从某种角度上看，这些都是一种预先计算(PreComputation)。这种预先计算都是为了节省查询时间，对海量数据完成一次最终倒排索引文件的制作是非常耗时的，这些尽可能预先完成的计算为查询争取了宝贵时间。优化索引很耗时，但索引优化后，查询速度快。

在优化索引阶段，大约需要索引大小 2 倍的临时空间。因此需要大约 400GB 的硬盘来容纳 200GB 的索引。

索引很多文档的过程通常比较慢。为了加快索引速度，可以多台机器同时索引不同内容，然后合并。要合并的几个不同的索引结构要一致。下面的程序可以把多个目录下的索引合并到一个目录下。

```
IndexWriter writer = new IndexWriter(args[0], null, true);
writer.setMergeFactor(50); //参数越大，用到的内存越多。影响内存的使用
//影响索引文件的数量
```

```
writer.setUseCompoundFile(false);
Directory[] dirs = new Directory[args.length -1];
System.out.println("begin :"+args[0]);
for (int i=1 ;i<args.length;i++)    {
            dirs[i-1]= FSDirectory.getDirectory( args[i], false);
            if (dirs[i-1]==null)
                System.out.println("Directory is null:"+i);
}
writer.addIndexes(dirs);
writer.close();
```

3.2.8 灵活索引

以前，可以通过声明列属性来说明是否存储词在文档中的位置信息，如 Field.TermVector.WITH_POSITIONS_OFFSETS。但这样的方式不够灵活，需要能够用任意有效的组合来说明倒排索引中存什么内容。例如，有时候也许并不关心词在文档中出现了多少次。有下面4种有效的索引格式：

(1) <doc>+
(2) <doc, boost>+
(3) <doc, freq, <position>+ >+
(4) <doc, freq, <position, boost>+ >+

例如可以只存文档编号，如(1)所示；或者还附加文档的加权信息，如(2)所示；或者还存储位置信息，如(3)所示；或者给不同位置的词设置不同的加权，如(4)所示。

这要求每个列设置如下的布尔值：

(1) freq。
(2) document boost。
(3) position (要求频率)。
(4) position boost (要求位置)。

索引选项枚举类型定义如下：

```
public static enum IndexOptions {
    DOCS_ONLY,   //仅索引文档
    DOCS_AND_FREQS,   //索引文档和频率
    DOCS_AND_FREQS_AND_POSITIONS,   //索引文档、频率和词的位置
    DOCS_AND_FREQS_AND_POSITIONS_AND_OFFSETS,   //索引文档、频率、词的位置
//以及字符偏移量
};
```

可以在设置列类型的时候设置索引选项。

```
FieldType fieldType = new FieldType();
fieldType.setStoreTermVectors(true);
fieldType.setStoreTermVectorPositions(true);
fieldType.setIndexed(true);
fieldType.setIndexOptions(IndexOptions.DOCS_AND_FREQS);   //索引文档和频率
```

```
fieldType.setStored(true);
Document doc = new Document();
doc.add(new Field("content", content, fieldType));
```

对于写入索引，如何写入很重要。应用程序应该可以很容易地存储新的东西到索引。或者更改现有东西的编码，例如更改文档编号、位置、payloads 等的编码方式。

3.2.9 索引文件格式

在 Lucene 中，倒排索引结构存储在二进制格式的多个索引文件中，其中以 tis 为后缀的文件中包含单词信息；frq 后缀的文件记录单词的文档编号和这个单词在文档中出现了多少次，也就是频率信息；prx 后缀的文件包含单词出现的位置信息。

Lucene 把和一个索引相关的文件全部放在一个目录下。其中既存储描述索引结构的元数据，又包含索引数据。每类信息放在不同后缀的二进制文件中。元数据包含以 fnm 为后缀的列信息文件。列信息文件的格式如下：

字段数量，<字段名称，字段的二进制位描述>

其中，"字段数量"表示索引中字段的数量；"字段名称"是用字符串表示的字段的名称；"字段的二进制位描述"是个 byte 值和 int 值。byte 值用 1 个最低位表示是否索引这个字段。

例如：

```
doc.add(new Field("text", "content", Field.Store.NO, Field.Index.TOKENIZED));
```

在列信息文件中存储成：

1, <content, 0x01>

词典文件以 tis 为后缀。这个文件中的词是按顺序存放的。词首先按词对应的字段名称排序，然后按词的正文排序。词典文件的格式如下：

词的数量，<词，词的文档频率>

词用前缀压缩的方式存储，格式是：前缀的长度，后缀，字段编号。"词的文档频率"指词在多少个文档中出现过。排好序的词表中，前后两个词往往包括共同的前缀。"前缀的长度"变量是表示与前一项相同的前缀的字数。例如，如果前一个词是"bone"，后一个是"boy"的话，前缀的长度值为 2，后缀值为"y"。

例如有两个文档，内容如下。

文档 1：Penn State Football …football。

文档 2：Football players … State。

在索引中的存储形式如下：

4,<<0,football,1>,2> <<0,penn,1>, 1> <<1,layers,1>,1> <<0,state,1>,2>

词典文件太大，为了能把词信息完整地读入内存，设计出了词信息索引文件(.tii)。词信息索引文件不存储词本身，但是保存随机读取的文件的位置信息。

词在文档中出现的频率文件以 frq 为文件后缀。词频率首先是按照 tis 中的词序来排列，在每个词存储信息，存储这个词在文档中出现的频率，按照 docID 排序。频率文件中并没有存储 docID，而是存储与前后两个 docID 的增量相关的一个值 DocDelta。频率文件的格式如下：

```
<DocDelta[, Freq?]>
```

DocDelta 同时决定了文档号和频数。详细地说，DocDelta/2 表示当前 docID 相对于前一个 docID 的偏移量(或者是 0，表示这是 TermFreqs 里面的第一项)。当 DocDelta 是奇数时表示在该文档中频数为 1；当 DocDelta 是偶数时，则下一个整数表示在该文档中出现的频数。

例如，假设某一项在文档 7 中出现 1 次，在文档 11 中出现 3 次，在 TermFreqs 中就存在如下的整数序列：15, 8, 3。

在这里：

15 = 2 * 7 + 1，在文档 7 中出现频率是 1。

8 = 2 * (11 - 7)，在文档 11 中出现频率 > 1。

3，在文档 11 中出现频率 = 3。

Posting List 见表 3-3，其对应的频率文件存储内容如下：

```
<<2, 2, 3> <3> <5> <3, 3>>
```

存储词在文档中出现过的位置的位置文件以 prx 为后缀。TermPositions 按照词来排序(依据 tis 文件中词的位置)。Positions 数值按照 docID 升序排列。实际存储的是 PositionDelta 值，PositionDelta 是当前位置相对于前一个出现位置(或者为 0，表示这是第一次在这个文档中出现)的增量值。例如，假设某词在某文档第 4 项出现，在接下来的一个文档中第 5 项和第 9 项出现，将表示为如下的整数序列：4, 5, 4。表 3-3 对应的位置文件的存储内容如下：

```
<<3, 64> <1>>  <<1> <0>>  <<0> <2>>  <<2> <13>>
```

从索引数据可以看出，索引数据是以词为中心组织的。TermPositions 类提供了用来遍历一个词的<文档, 频率, <位置>* >元组的方法。

表 3-3 Posting List

Posting id	词	docID	offset
1	football	1	3
		1	67
		2	1
2	penn	1	1
3	players	2	2
4	state	1	2
		2	13

Lucene 的查询过程访问的文件如图 3-6 所示。

复合文件格式的索引优化后只有 3 个文件，其中 segments_N 和 segments.gen 是固定不变的，因为这两个文件是在索引级别存在的文件，还有一个是复合索引文件格式(.cfs)。可以通过 setUseCompoundFile()方法设定是否使用复合文件格式。

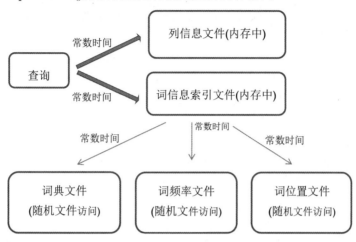

图 3-6　查询访问文件

3.2.10　定制索引存储结构

开发一个高效的搜索引擎是一项有挑战性的工作。Lucene 底层代码读起来往往很费劲。可以自己通过 codec 来定制编码和索引的结构。

Lucene 4 相对更早的版本，一个很大的变化就是提供了可插拔的编码器架构，可以自行定义索引结构，包括词元、倒排列表、存储字段、词向量、已删除的文档、段信息、字段信息。codec 在 Lucene 中的结构如图 3-7 所示。

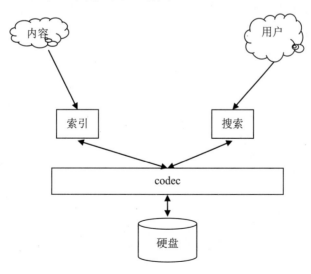

图 3-7　codec 在 Lucene 中的结构

codec 直接传递给 SegmentReader 来编码索引格式。提供枚举类的实现给 SegmentReader。提供索引文件的写入器给 IndexWriter。

Lucene 4 中已经提供了多个 codec 的实现，其中 Lucene40 是默认编码器 Lucene40Codec。为了兼容更早的版本，提供了只读的 Lucene3xCodec，可以用来读取 Lucene 3.x 创建的索引，但不能使用该编码器创建 Lucene3.x 的索引。

PerFieldCodec 用来支持不同的列使用不同的读写格式。

lucene-codecs-6.3.0.jar 中包含一些额外的 codec，和其他的 codec 写入到压缩的二进制文件不一样，SimpleTextCodec 把所有的投递列表写到可读的文本文件。SimpleTextCodec 适合用来学习，不建议在生产环境中使用。

```
Analyzer analyzer = new StandardAnalyzer();
IndexWriterConfig iwc = new IndexWriterConfig(analyzer);

iwc.setCodec(new SimpleTextCodec());
iwc.setUseCompoundFile(false);

Directory directory = FSDirectory.open(new File("F:/lucene/index"));

IndexWriter writer = new IndexWriter(directory, iwc);

// index a few documents
writer.addDocument(createDocument("1", "青菜鸡肉"));
writer.addDocument(createDocument("2", "老鸭粉丝汤"));
writer.addDocument(createDocument("3", "辣子鸡丁"));
writer.close();
```

主要产生 5 个文件：_0.len、_0.fld、_0.inf、_0.pst 和_0.si。其中，pst 文件保存倒排索引；fld 文件保存存储到索引的原值；inf 文件保存文件是如何索引的。

倒排索引在_0.pst 文件中。先存某一列的倒排索引，然后再存另外一列的倒排索引。例如，像如下方式写入 contents 列和 id 列的倒排索引：

```
field contents
  term 丁
    doc 2
      freq 1
      pos 3
  term 丝
    doc 1
      freq 1
      pos 3
  term 子
    doc 2
      freq 1
      pos 1
  term 汤
    doc 1
```

```
      freq 1
      pos 4
    term 粉
      doc 1
      freq 1
      pos 2
    term 老
      doc 1
      freq 1
      pos 0
    term 肉
      doc 0
      freq 1
      pos 3
    term 菜
      doc 0
      freq 1
      pos 1
    term 辣
      doc 2
      freq 1
      pos 0
    term 青
      doc 0
      freq 1
      pos 0
    term 鸡
      doc 0
      freq 1
      pos 2
      doc 2
      freq 1
      pos 2
    term 鸭
      doc 1
      freq 1
      pos 1
  field id
    term 1
      doc 0
    term 2
      doc 1
    term 3
      doc 2
END
```

_0.fld 文件的内容如下：

```
doc 0
  numfields 2
  field 0
    name id
    type string
    value 1
  field 1
    name contents
    type string
    value 青菜鸡肉
doc 1
  numfields 2
  field 0
    name id
    type string
    value 2
  field 1
    name contents
    type string
    value 老鸭粉丝汤
doc 2
  numfields 2
  field 0
    name id
    type string
    value 3
  field 1
    name contents
    type string
    value 辣子鸡丁
END
```

_0.inf 文件的内容如下：

```
number of fields 2
  name id
  number 0
  indexed true
   index options DOCS_ONLY
   term vectors false
   payloads false
   norms false
   norms type false
   doc values false
   attributes 0
  name contents
  number 1
  indexed true
```

```
index options DOCS_AND_FREQS_AND_POSITIONS
term vectors false
payloads false
norms true
norms type NUMERIC
doc values false
attributes 0
```

codec 事实上就是由多组 format 构成的，一个 codec 共包含 8 个 format，即 PostingsFormat、DocValuesFormat、StoredFieldsFormat、TermVectorsFormat、FieldInfo Format、SegmentInfoFormat、NormsFormat 和 LiveDocsFormat。例如，StoredFieldsFormat 用来处理存储数据的列；TermVectorsFormat 用来处理词向量。在 Lucene 4 中可以自行定制各个 format 的实现。

其他的 codec 可以转换成这样的标准输出。SimpleTextCodec 这样的文件格式没有索引，所以无法快速查找某个词，但是可以用于调试和学习。

在 IndexWriterConfig 中有 setCodec() 方法可以设置编解码器，可以用这个 IndexWriterConfig 创建一个 IndexWriter。但在 IndexReader 类中没有这样的方法。写索引的时候需指定要使用的 codec，并且把所使用的 codec 的名字写入索引的每个段中。

在读索引的时候(当打开一个 IndexReader 的时候)，不能再改变编解码器，只能保证索引使用的所有 codec 都在 CLASSPATH 中。IndexReader 将检查每个段，以确定它是用哪种 codec 写的，在 CLASSPATH 中找到这个 codec，并用它来打开该段。

```
IndexWriterConfig iwc = new IndexWriterConfig(analyzer);

System.out.println(iwc.getCodec().availableCodecs());
String name = "Lucene42";
iwc.setCodec(iwc.getCodec().forName(name));
IndexWriter writer = new IndexWriter(directory, iwc);
```

可以在索引中保存和每个词相关的字节数组信息，叫作 Payload。首先在分析文本期间生成 Payload 信息。可以使用 PayloadAttribute 达到这一点，只需要在分析过程中将该属性添加到 Token 属性中。使用 PayloadHelper 将数字编码为 Payload，然后就可以设置到 PayloadAttribute。例如，编码浮点数：

```
Payload p = new Payload(PayloadHelper.encodeFloat(42));
```

注意，这里的 PayloadHelper 类不在核心包中，而在 contrib/common/lucene-analyzers-3.x 中。PayloadHelper 中的 decodeInt()方法从字节数组中得到一个整数。

```
public static final int decodeInt(byte [] bytes, int offset){
    return ((bytes[offset] & 0xFF) << 24) | ((bytes[offset + 1] & 0xFF) << 16)
        | ((bytes[offset + 2] & 0xFF) <<  8) |  (bytes[offset + 3] & 0xFF);
}
```

这里的 bytes[offset] & 0xFF 是为了得到整数结果，然后参与后续的位移运算。

3.2.11 写索引集成到爬虫

爬虫把抓取的信息写入索引:

```java
public class IndexDao {
    private IndexWriter indexWriter;

    public IndexDao(){
        try {
            Directory directory = FSDirectory.open(new File("d:/lietu/index"));
            Analyzer analyzer = new StandardAnalyzer();
            indexWriter = new IndexWriter(directory, analyzer,
                    MaxFieldLength.LIMITED);
        } catch (IOException e) {
            e.printStackTrace();
        }
    }

    public void save(GoodsInfo goodsInfo){
        Document doc = goodsInfo2Document(goodsInfo);

        try{
            indexWriter.addDocument(doc);
        }catch(Exception e){
            e.printStackTrace();
        }
    }

    public void close(){
        try {
            indexWriter.close();
        } catch (Exception e) {
            e.printStackTrace();
        }
    }

    public Document goodsInfo2Document(GoodsInfo ti) {
        Document doc = new Document();
        Field f = new Field("url", ti.getGoodsNameURL(), Field.Store.YES,
                Field.Index.NOT_ANALYZED, Field.TermVector.NO);
        doc.add(f);

        f = new Field("title", ti.getGoodsName(), Field.Store.YES,
                Field.Index.ANALYZED,
            Field.TermVector.WITH_POSITIONS_OFFSETS);
        doc.add(f);
```

```java
if (ti.getGoodsDescription() != null) {
    f = new Field("body", ti.getGoodsDescription(), Field.Store.YES,
            Field.Index.NOT_ANALYZED, Field.TermVector.NO);
    doc.add(f);
}

f = new Field("date", DateTools.dateToString(new Date(),
        DateTools.Resolution.DAY), Field.Store.YES,
        Field.Index.NOT_ANALYZED, Field.TermVector.NO);
doc.add(f);

f = new Field("priceInt", String.valueOf(ti.getPriceInteger()),
        Field.Store.YES, Field.Index.ANALYZED,
        Field.TermVector.WITH_POSITIONS_OFFSETS);
doc.add(f);

if (ti.getMoneyUnit() != null) {
    f = new Field("moneyUnit", ti.getMoneyUnit(), Field.Store.YES,
            Field.Index.NOT_ANALYZED, Field.TermVector.NO);
    doc.add(f);
}

try {
    URL website = new URL(ti.getGoodsNameURL().toString());
    f = new Field("fromwebsite", website.getHost(), Field.Store.YES,
            Field.Index.NOT_ANALYZED, Field.TermVector.NO);
    doc.add(f);
} catch (MalformedURLException e1) {
    System.out.println("error url =" + ti.getGoodsNameURL().toString());
    e1.printStackTrace();
}

// 分类
f = new Field("category", ti.getGoodsType(), Field.Store.YES,
        Field.Index.NOT_ANALYZED, Field.TermVector.NO);
doc.add(f);

// img
if (ti.getImage() != null) {
    f = new Field("img", ti.getImage(), Field.Store.YES,
            Field.Index.NOT_ANALYZED, Field.TermVector.NO);
    doc.add(f);
}
// 制造厂名称
if (ti.getMfrName() != null) {
    f = new Field("brand", ti.getMfrName(), Field.Store.YES,
```

```
                    Field.Index.NOT_ANALYZED, Field.TermVector.NO);
            doc.add(f);
        }
        // 商品型号 序列号
        if (ti.getMfrNumber() != null) {
            f = new Field("type", ti.getMfrNumber(), Field.Store.YES,
                    Field.Index.NOT_ANALYZED, Field.TermVector.NO);
            doc.add(f);
        }

        // 价格
        if (ti.getGoodsPrice() != null) {
            f = new Field("price", ti.getGoodsPrice(), Field.Store.YES,
                    Field.Index.NOT_ANALYZED, Field.TermVector.NO);
            doc.add(f);
        }
        return doc;
    }
}
```

3.2.12 多线程写索引

Lucene 默认只使用一个线程写索引,而且一个索引只能由一个进程打开。

```
public class ThreadedIndexWriter extends IndexWriter {

 private ExecutorService threadPool;
 private Analyzer defaultAnalyzer;

 private class Job implements Runnable {              //保留要加入索引的一个文档
   Document doc;
   Analyzer analyzer;
   Term delTerm;
   public Job(Document doc, Term delTerm, Analyzer analyzer) {
     this.doc = doc;
     this.analyzer = analyzer;
     this.delTerm = delTerm;
   }
   public void run() {                                //实际增加和更新文档
     try {
       if (delTerm != null) {
         ThreadedIndexWriter.super.updateDocument(delTerm, doc, analyzer);
       } else {
         ThreadedIndexWriter.super.addDocument(doc, analyzer);
       }
     } catch (IOException ioe) {
       throw new RuntimeException(ioe);
```

```java
      }
    }
  }

  public ThreadedIndexWriter(Directory dir, Analyzer a,
                      boolean create, int numThreads,
                      int maxQueueSize, IndexWriter.MaxFieldLength mfl)
    throws CorruptIndexException, IOException {
   super(dir, a, create, mfl);
   defaultAnalyzer = a;
   threadPool = new ThreadPoolExecutor(            //创建线程池
        numThreads, numThreads,
        0, TimeUnit.SECONDS,
        new ArrayBlockingQueue<Runnable>(maxQueueSize, false),
        new ThreadPoolExecutor.CallerRunsPolicy());
  }

  public void addDocument(Document doc) {           //让线程池增加文档
   threadPool.execute(new Job(doc, null, defaultAnalyzer));
  }

  public void addDocument(Document doc, Analyzer a) {   //让线程池增加文档
   threadPool.execute(new Job(doc, null, a));
  }

  public void updateDocument(Term term, Document doc) {  //让线程池更新文档
   threadPool.execute(new Job(doc, term, defaultAnalyzer));
  }

  //让线程池更新文档
  public void updateDocument(Term term, Document doc, Analyzer a) {
   threadPool.execute(new Job(doc, term, a));
  }

  public void close()
     throws CorruptIndexException, IOException {
   finish();
   super.close();
  }

  public void close(boolean doWait)
     throws CorruptIndexException, IOException {
   finish();
   super.close(doWait);
  }

  public void rollback()
```

```java
        throws CorruptIndexException, IOException {
    finish();
    super.rollback();
}

private void finish() {            //关闭线程池
    threadPool.shutdown();
    while(true) {
        try {
            if (threadPool.awaitTermination(Long.MAX_VALUE, TimeUnit.SECONDS)) {
                break;
            }
        } catch (InterruptedException ie) {
            Thread.currentThread().interrupt();
            throw new RuntimeException(ie);
        }
    }
}
```

如果是多线程写索引，则让每个线程使用一个不同的 Document，不要在多个线程之间共享同一个 Document。

3.2.13 分发索引

为了实现分布式的搜索，需要把索引从一台服务器备份到另外一台服务器。最简单的方法是：在生成索引的服务器上用 tar 命令压缩索引目录，然后在另外一台搜索服务器上执行 wget 获得压缩成一个文件的索引。在 Windows 下，可以使用 7-zip 软件中的 7z 压缩格式备份正在写入的索引。

当增加、修改索引时，Lucene 索引文件一般不会发生太大的变化。优化索引时，索引文件才会有大的变化。所以，实际中可以通过只传送修改的文件(也就是同步索引文件目录)的方式实现索引分发。rsync(http://samba.anu.edu.au/rsync/)是一个开源的工具，提供了快速的增量文件传输功能。使用 rsync 可以保持主服务器中的索引目录能够定期同步到从服务器的索引目录中。rsync 支持通过 SSH 进行网络加密传输，也可以利用 SSH 客户端密钥建立服务器之间的信任关系。当在两台服务器之间保持大型、复杂目录结构同步的时候，使用 rsync 比 tar 或 wget 等方式都要快，而且可以做到精确同步。使用 rsync 的分布式垂直搜索架构如图 3-8 所示。

rsync 命令的基本形式如下：

rsync [OPTION]... 来源地址路径 目的地址路径

rsync 可以通过两种不同的方式连接一个远程系统：使用一个远程 shell 程序(例如 ssh 或 rsh)作为中转，或者通过 TCP 直接连接一个 rsync 守护进程。当地址路径信息包含 "::" 分隔符时启动服务器传输模式；当地址路径包含单个冒号 ":" 分隔符时启动远程 shell 传输模式。

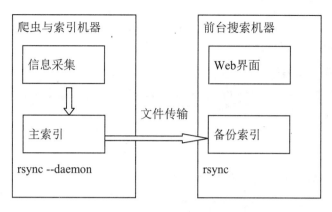

图 3-8　分布式垂直搜索架构

首先可以在主服务器和从服务器上都安装最新的 rsync 软件。rsync 的编译安装非常简单，只需要以下简单的几步：

```
# ./configure
# make
# make install
```

然后用一个命令就可以一次性同步索引目录：

```
#rsync -ave ssh master:/home/index/ /home/index/
```

这个命令把远端 master 机器上的/home/index/目录中的内容同步到本地的/home/index/目录下。

在创建索引的机器上，通过在 rsync 命令中声明--daemon 参数可以运行 rsync 守护进程。在索引机器上建立配置文件 rsyncd.conf 和密码文件 rsyncd.secrets。前台搜索机器通过 crontab 定时运行 rsync 获取最新的索引文件。

配置 rsync 守护进程的方法是修改/etc/rsyncd.conf 并把 rsync 守护进程设置成自启动方式。Linux 中的大部分网络服务都是由 inetd 启动的，修改/etc/xinetd.d/rsync 文件把其中的 disable = yes 改成 disable = no。在/etc/目录下建立配置文件 rsyncd.conf。

```
#cat /etc/rsyncd.conf

use chroot = yes  # 使用 chroot
max connections = 4  # 最大连接数为 4
pid file = /var/run/rsyncd.pid
lock file = /var/run/rsync.lock
log file = /var/log/rsyncd.log  # 日志记录文件

[index]  # 这里是认证的模块名，在 client 端需要指定
path = /home/index  # 需要做镜像的目录
auth users = index  # 认证的用户名
uid = index
gid = index
secrets file = /etc/rsyncd.secrets  # 认证文件名
read only = yes  # 只读
```

在/etc/目录下建立配置文件 rsyncd.secrets。rsyncd.secrets 是 rsyncd 的密码文件，保存有登录 Linux 系统的用户名和密码。

```
#cat /etc/rsyncd.secrets

index:abcde # 用户名 index 密码 abcde
```

然后配置客户端。为了避免以交互的方式输入密码，大多使用密码文件。也可以将两台服务器生成密钥互相设为信任认证，这样做的麻烦是程序不通用，而且每两台服务器都需要生成证书。假设密码文件是 rsyncd.secrets。

```
#cat /home/index/rsyncd.secrets
abcde
```

需要把密码文件改为只有所属人有权限。

```
#chmod 600
```

把文件同步命令写到一个可执行的脚本：

```
#cat index_rsyncd.sh
#!/bin/bash
rsync -tvzrp --progress --password-file=/home/index/rsyncd.secrets --delete
--exclude /home/index/logs index@master::index  /home/index/
```

通过 crontab 设定，让这个脚本每 10 分钟运行一次。

```
#echo "0,10,20,30,40,50 index_rsyncd.sh">>/etc/crontab
```

为了实现搜索从服务器和索引主服务器的索引同步，在索引主服务器上，为索引做阶段性的检查点。每分钟的时候，关闭 IndexWriter，并且从 Java 中执行 cp -lr index index.DATE 命令，这里 DATE 是指当前时间。这样通过构建硬连接树，而不是制作完全的备份，就有效地制作了一个索引的备份。如果 Lucene 重写了任何文件(例如 segments 文件)，将会创建新的 inode，而原来的备份不变。

在每个搜索从服务器上，定期检查新的检查点。当发现一个新的 index.DATE 时，使用 cp -lr index index.DATE 准备一份备份，然后使用 rsync -W -delete master:index.DATE index.DATE 得到增量的索引改变，最后使用原子性的符号连接操作安装更新的索引 (ln -fsn index.DATE index)。

在从服务器上，当索引版本改变的时候将重新打开"index"。最好是在一个独立的线程中定时检查索引版本。当索引版本改变时，打开新的索引版本，执行一些热门查询操作，预加载 Lucene 缓存。然后在一个同步块中，替换在线服务的 Searcher 变量。

在主索引服务器的一个 crontab 中，定时移走最旧的检查点索引。

这样可以实现每分钟的同步，将主索引服务器上的 mergeFactor 设为 2，以最小化产品中的 segments 数量。主服务器有一个热备份。

当增加文档到已经存在的索引库时，会生成新的.cfs 文件。这个文件需要全部传输(而不是差量传输)，因为文件名改了。所以，为了使增量更新更有效率，尽量不要让索引使用复合文件格式。

3.2.14 修复索引

当索引崩溃时，可以重建索引，但是重建大的索引往往比较耗时，所以还可以考虑修复索引。CheckIndex 是 Lucene 包中的一个工具。它检查文件并创建新的不包含有问题的入口的段。这意味着这个工具以很小的数据丢失为代价来修复坏索引。这个工具要一个字节一个字节地分析索引，因此对于大的索引，分析和修复的时间可能比较长。

可以先使用 CheckIndex 来检查索引的完整性。例如：

```
CheckIndex D:\index
```

如果有问题再修复索引。修复索引的命令如下：

```
java -cp lucene-core-2.9.3.jar org.apache.lucene.index.CheckIndex d:\index\ -fix
```

3.3 查找索引库

查询条件分结构化查询条件和非结构化查询条件。例如，按价格区间查找是结构化查询条件，按关键词查询为非结构化查询条件。

可以按关键词查询指定的列，根据相关度返回结果，也可以自定义搜索结果排序方式。检索管理包(org.apache.lucene.search) 根据查询条件，检索得到结果。

3.3.1 查询过程

Lucene 中使用 IndexSearcher 类对索引进行检索。IndexSearcher 类继承自抽象类 Searcher。查询依赖一个或者多个索引库，IndexSearcher 依赖一个或者多个 IndexReader。查询执行过程如图 3-9 所示。

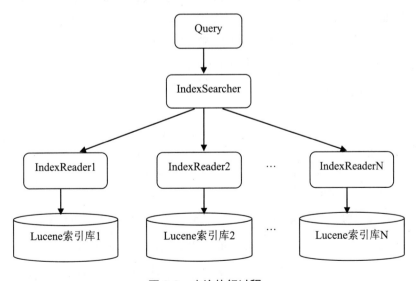

图 3-9　查询执行过程

在初始化一个 IndexSearcher 类时，重要的就是要告诉它使用哪个索引完成查找任务。可以用 IndexReader 初始化一个 IndexSearcher 对象。

```
Directory directory = FSDirectory.open(new
File("E:/luceneTest/fileindex"));
IndexReader indexReader = DirectoryReader.open(directory);
IndexSearcher indexSearcher = new IndexSearcher(indexReader);
```

DirectoryReader 是 IndexReader 的子类。使用 DirectoryReader.open(directory)方法得到 IndexReader 的实例。

Directory 类型的对象包含索引存放的路径信息，从而可以定位索引。这里使用 Directory 类型的对象来创建 IndexReader，然后再用 IndexReader 构建 IndexSearcher。要查询的索引往往只有一个，所以也可以直接使用 Directory 类型的对象来构建 IndexSearcher。

```
String indexDir = "D:/indexdir"; //索引库路径
Directory directory = FSDirectory.open(new File(indexDir)); //打开一个文件路径
//只读的方式使用 directory，所以 read-only=true
IndexSearcher searcher = new IndexSearcher(directory, true); //隐式创建了
//IndexReader
```

上面只是一个隐式创建 IndexReader 的简单写法。使用 Directory 构造的 IndexSearcher 实例仍然各自持有一个 IndexReader 实例，若系统中存在同一个索引的多个 IndexSearcher 实例时，会占用过多的内存空间。这时，应该是一份索引用一个 IndexReader 实例打开，用 IndexReader 构造 IndexSearcher 实例。

可以用 DirectoryReader.getVersion()取得版本号：

```
System.out.println("版本号 :"+reader.getVersion());
```

DirectoryReader 类是 IndexReader 的子类。

把 111.doc 分成 111，1111.doc 分成 1111，11111.doc 分成 11111。这样输入 111，就只能搜索到 111.doc。一段有意义的文字需要通过 Analyzer 分割成一个个词语后才能按关键词搜索。Analyzer 就是分析器，StandardAnalyzer 是 Lucene 中最常用的分析器。

```
Analyzer analyzer = new StandardAnalyzer();
```

为了达到更好的搜索效果，不同的语言可以使用不同的分析器，例如 CnAnalyzer 就是一个主要处理中文的分析器。

Analyzer 返回的结果就是一串 Token。Token 包含一个代表词本身含义的字符串和该词在文章中相应的起止偏移位置，Token 还包含一个用来存储词类型的字符串。

Term 是搜索语法的最小单位，复杂的搜索语法会分解成一个个的 Term 查询，它表示文档中的一个词语。Term 由两部分组成：它表示的词语和这个词语所出现的 Field。例如：

```
new Term("url","http://www.lietu.com");
```

最简单的 Query 对象是 TermQuery。根据 TermQuery 查询索引库。

```
Query query = new TermQuery(new Term("url","http://www.lietu.com"));
```

```
//搜索索引库并且返回最相关的10个文档
TopDocs tds = searcher.search(query, 10);
```

要是用户输入一个逗号，也会搜出文档来，逗号是停用词。一般停用词进索引，但是搜索的时候会过滤掉。指用户输入逗号，系统识别为停用词就给提示说它不能进行搜索。

查询串中可能包括一些高级查询语法，例如，要查找包含java的PDF文件，可以使用查询串"java filetype:pdf"。所以用查询分析器QueryParser来解析查询串，也就是根据查询串生成Query对象。例如：

```
Analyzer analyzer = new StandardAnalyzer();
QueryParser qp = new QueryParser(fields, analyzer);
query = qp.parse(queryString);
```

使用IndexSearcher的search()方法执行搜索，返回一个TopDocs对象。TopDocs对象中的totalHits属性记录搜索返回结果的总条数。基本的关键词查询代码如下：

```
String indexDir = "D:/indexdir"; //索引库路径
Directory directory = FSDirectory.open(new File(indexDir)); //打开一个文件路径
// read-only=true
IndexSearcher searcher = new IndexSearcher(directory, true); //搜索

String defaultField = "title"; //默认查询列
String queryString ="NBA"; //查询词
Analyzer analyzer = new StandardAnalyzer();
QueryParser parser = new QueryParser(defaultField,
                                     analyzer);

Query query = parser.parse(queryString);
TopDocs docs = searcher.search(query, 10);   //查询最多只返回前10条结果
```

这里的FSDirectory表示硬盘中的索引，RAMDirectory表示内存中的索引，可以用来测试索引和查找过程。

遍历查询结果：

```
ScoreDoc[] hits = docs.scoreDocs; //从TopDocs取得查询结果

// 遍历结果
for (ScoreDoc hit : hits){
    Document hitDoc = searcher.doc(hit.doc);
    System.out.println(hitDoc.get("title") +hit.score); //输出标题和文档相关度分值
}
```

ScoreDoc中的score属性就是相关度。相关度得分是1～0的值。1表示相关度最高，而0则表示不相关。文档的相关度跟很多因素有关。比如字段的长短、里面词条的权重等。决定score的因素简单概括如下：

(1) 项频率，即查询项在某个文档中出现的次数；
(2) 文档频率，即查询项在很多文档中出现的次数。

完整的查询代码：

```
Directory directory = FSDirectory.open(new File("d:/testindex"));
// DirectoryReader 读入一个目录下的索引文件
IndexReader ir = DirectoryReader.open(directory);
//打开索引库
IndexSearcher searcher = new IndexSearcher(ir);

//根据查询词搜索索引库
TopDocs docs = searcher.search(new TermQuery(new Term("title","text")), 10);
//遍历查询结果
ScoreDoc[] hits = docs.scoreDocs;
for (ScoreDoc hit : hits){
    System.out.print("doc:"+hit.doc+" score:"+hit.score+"\n");
}
//关闭索引库
ir.close();
```

3.3.2 常用查询

Query 是一个用于查询的抽象基类，有各种具体查询实现类，比如实现基本词查询的 TermQuery、实现布尔逻辑查询的 BooleanQuery、实现短语查询的 PhraseQuery、实现前缀匹配查询的 PrefixQuery、实现区间查询的 RangeQuery、实现多词查询的 MultiTermQuery、实现过滤条件查询的 FilteredQuery、约束多个查询词密集度的 SpanQuery 等。

对于复杂查询，需要调用 rewrite()方法使其变成简单的查询。例如，一个 PrefixQuery 将被重写成由 TermQuery 组成的 BooleanQuery。

可以使用 QueryParser 查询分词列。一般不需要查询没有使用分词的列，如果要查询，可以使用 TermQuery 查询，例如 url 列。

最基本的词条查询使用 TermQuery，用来查询不切分的字段。例如查询一个杂志：

```
Term term = new Term("journal_id","1672-6251");
TermQuery query = new TermQuery(term);
```

为了方便测试查询，使用 createDocument()方法索引测试内容。

```
private static Document createDocument(String id, String content) { //创建文档
    Document doc = new Document();
    doc.add(new Field("id", id, StringField.TYPE_STORED)); //索引不分词的字符串列
    //索引分词的文本列
    doc.add(new Field("contents", content, TextField.TYPE_STORED));
    return doc;
}
```

组合条件查询使用布尔逻辑查询 BooleanQuery。下面举一个布尔逻辑查询的例子：同时查询标题列和内容列。使用 BooleanClause.Occur.SHOULD 的查询效果如图 3-10 所示。

第 3 章　Lucene 的原理与应用

图 3-10　BooleanClause.Occur.SHOULD 查询效果

```
QueryParser parser = new QueryParser(Version.LUCENE_CURRENT, "body",
analyzer);
Query bodyQuery =  parser.parse("NBA");//查询内容列
parser = new QueryParser(Version.LUCENE_CURRENT, "title", analyzer);
Query titleQuery = parser.parse("NBA");//查询标题列

BooleanQuery bodyOrTitleQuery = new BooleanQuery();
//用 OR 条件合并两个查询
bodyOrTitleQuery.add(bodyQuery, BooleanClause.Occur.SHOULD);
bodyOrTitleQuery.add(titleQuery, BooleanClause.Occur.SHOULD);

//返回前 10 条结果
ScoreDoc[] hits = isearcher.search(bodyOrTitleQuery, 10).scoreDocs;
```

3.3.3　基本词查询

TermQuery 是最简单、也是最基本的 Query。TermQuery 可以理解成为"词条搜索"，在搜索引擎中最基本的搜索就是在索引中搜索某一词条，而 TermQuery 就是用来完成这项工作的。

查询某个字段中所包含的某个关键字，例如，查询类别列中为"服装"的商品。

```
Query query = new TermQuery(new Term("cat","服装"));

//搜索索引库并且返回最相关的 10 个商品
TopDocs tds = searcher.search(query, 10);
```

3.3.4　模糊匹配

关于字符串的前缀匹配问题：只查询产品名字为字母 A 打头的搜索，可以使用前缀匹配(PrefixQuery)实现。查询语法是 A*。注意，查询的不一定是整个字段以 A 开头的记录，而只是其中的单词以 A 开头。

```
Query query = new WildcardQuery(new Term(FIELD, "cut*")); // 通配符
```

等价于：

```
Query query = new PrefixQuery(new Term(FIELD, "cut")); // 自动在结尾添加 *
```

测试 WildcardQuery()：

```
// 索引一些文档
writer.addDocument(createDocument("1", "foo bar baz"));
writer.addDocument(createDocument("2", "red green blue"));
writer.addDocument(createDocument("3",
        "The Lucene was made by Doug Cutting"));
writer.close();

IndexReader reader = DirectoryReader.open(directory);

IndexSearcher searcher = new IndexSearcher(reader);

Query query = new WildcardQuery(new Term(FIELD, "cut*"));

TopDocs topDocs = searcher.search(query, 10);
```

这里的 cut 只能是小写，否则就匹配不上。如果用户输入的是大写，如 Cut，可以用 QueryParser()转换成小写。

使用反转 Token 的方法允许通配符出现在开头。ReversedWildcardsTokenFilter 可以实现 Token 反转。ReversedWildcardsTokenFilter 返回原来的 Token 和反转的 Token，其中反转的 Token 的 positionIncrement 值是 0。

可以用一个标识字符来避免正常 Token 和反转 Token 之间的冲突。例如："DNA"反转后就变成"and"了，和正常的词"and"有冲突。但是使用标识字符后，"DNA"就变成了"\u0001and"。

可以把 ReversedWildcardsTokenFilter 加入分析器链，这样在做索引的时候就可以使用了。

有些英文单词需要查询扩展，如搜索"dog"的同时查找"dogs"。FuzzyQuery()有限状态查询(Finite-State Query)用编辑距离衡量相似度，例如"dog"和"dogs"的编辑距离是 1。FuzzyQuery()内部使用编辑距离有限状态机实现，所以性能很好。

```
int maxEdits = 2;  //编辑距离最多不能超过 2
new FuzzyQuery(new Term("title","dog"), maxEdits);
```

可以设置相同前缀的长度。例如，相同前缀的长度是 1：

```
FuzzyQuery query = new FuzzyQuery(new Term("field", "WEBER"), 2, 1);
```

query 要求匹配的词必须以 W 开头，而且匹配的词与查询词之间的编辑距离不超过 2。这个 FuzzyQuery 等价于如下的有限状态自动机：

```
LevenshteinAutomata builder = new LevenshteinAutomata("EBER", true);
Automaton a = builder.toAutomaton(2); //最大编辑距离是 2
Automaton b = BasicAutomata.makeChar('W'); //创建一个字符 W 组成的自动机
Automaton c = BasicOperations.concatenate(b, a); //连接两个自动机 b 和 a
```

这里的 BasicOperations.concatenate()方法把 b 的结束状态和 a 的开始状态用空转换连接起来，也就是按顺序走过 a 中的状态，然后走 b 中的状态。例如，自动机 b 可以接收字符

W，然后自动机 a 继续接收 EB，就可以结束了。这说明自动机 c 可以接收 WEB。

测试自动机 c 可以接收哪些字符：

```
System.out.println(BasicOperations.run(c, "WBR"));  //输出 true
System.out.println(BasicOperations.run(c, "WEB"));  //输出 true
System.out.println(BasicOperations.run(c, "WEBE")); //输出 true
System.out.println(BasicOperations.run(c, "WEBER"));//输出 true
```

在中文人名中使用自动机：

```
LevenshteinAutomata builder = new LevenshteinAutomata("杰伦", true);
Automaton a = builder.toAutomaton(1);  //最大编辑距离是 1
Automaton b = BasicAutomata.makeChar('周');
Automaton c = BasicOperations.concatenate(b, a);  //连接两个自动机 b 和 c
System.out.println(BasicOperations.run(c, "周杰伦")); //匹配写错的人名
```

在中文人名中使用模糊查询：

```
FuzzyQuery query = new FuzzyQuery(new Term("field", "周杰伦"), 1, 1);
```

"dogs~" 这样的模糊查询语法使用 FuzzyQuery()。FuzzyQuery() 还可以用于拼写检查。可以根据 Automaton 对象构建 AutomatonQuery。

```
Term term= new Term("yourfield", "周~*");
LevenshteinAutomata builder = new LevenshteinAutomata("杰伦", true);
Automaton a = builder.toAutomaton(1);  //最大编辑距离是 1
Automaton b = BasicAutomata.makeChar('周');
Automaton c = BasicOperations.concatenate(b, a);  //连接两个自动机 b 和 c
AutomatonQuery query = new AutomatonQuery(term, c);
```

再给出一个星闭包的例子：

```
//查询的 term 表示，包含列
Term term= new Term("yourfield", "bla~*");
//对所有的和"bla"编辑距离在 2 以内的字符串构建一个确定性有限状态自动机
Automaton fuzzy = new LevenshteinAutomata("bla", false).toAutomaton(2);
//串联 fuzzy 和另外一个 DFA 等于"*"操作符，也就是星闭包
Automaton fuzzyPrefix=
            BasicOperations.concatenate(fuzzy,BasicAutomata.makeAnyString());
//构建一个查询，用它搜索以得到结果
AutomatonQuery query = new AutomatonQuery(term, fuzzyPrefix);
```

Lucene 的 NumericRangeQuery 采用了 Trie 树结构的索引，可以模仿 NumericRangeQuery 编写字符串的前缀匹配实现。

3.3.5 布尔查询

可以使用 BooleanQuery 组合多个查询条件。合取查询使用 BooleanClause.Occur.MUST 连接多个查询词。

```
Term t1 = new Term(FIELD, "lucene");
TermQuery q1 = new TermQuery(t1);

Term t2 = new Term(FIELD, "doug");
TermQuery q2 = new TermQuery(t2);

//合取查询
BooleanQuery query = new BooleanQuery();
query.add(q1,BooleanClause.Occur.MUST); //必须包含这个条件
query.add(q2,BooleanClause.Occur.MUST);
```

选择出包含任何一个查询词的文档叫作析取(Disjunction)。用真值表描述析取如下。

```
p    q    (p v q)
T    T       T
T    F       T
F    T       T
F    F       F
```

析取查询使用 BooleanClause.Occur.SHOULD 连接多个查询词。

```
//析取查询
BooleanQuery q = new BooleanQuery();
q.Add(q1,BooleanClause.Occur.SHOULD); //可以只包含这个条件
q.Add(q2,BooleanClause.Occur.SHOULD);
```

例如去餐馆点菜，有人不喜欢吃辣的东西，所以他找出所有不辣的菜，这叫作否定(Negation)。用真值表描述否定如下。

```
p      ~p
T      F
F      T
```

BooleanClause.Occur.MUST_NOT 表示不能包括符合这个条件的文档。例如找出不包含"辣"的所有文档。

```
Term t1 = new Term("body", "辣");
TermQuery q1 = new TermQuery(t1);

BooleanQuery mbq = new BooleanQuery();
MatchAllDocsQuery alldocs = new MatchAllDocsQuery();   //匹配所有文档
mbq.Add(q1, BooleanClause.Occur.MUST_NOT);   //不包括满足q1条件的文档
mbq.Add(alldocs, BooleanClause.Occur.MUST);
```

打麻将三缺一不行，但是如果同时查询多个词，最好只差一个查询词的文档也能匹配。

对于长句搜索，则提取其中的主要查询词，只要大部分词在文档中出现就可以了。例如搜索联系方式列：

- "Stanford University School of Medicine, Palo Alto, CA USA",
- "Institute of Neurobiology, School of Medicine, Stanford University, Palo Alto, CA",

- "School of Medicine, Harvard University, Boston MA",
- "Brigham & Women's, Harvard University School of Medicine, Boston, MA"
- "Harvard University, Cambridge MA"

搜索联系方式使用如下的长查询词：

```
"School of Medicine, Stanford University, Palo Alto, CA"
```

需要找到所有和 Stanford 相关的文档。

PhraseQuery 要求短语中所有的词都存在才能匹配。这里需要一个更加宽松的 PhraseQuery 版本，要求它对词出现的顺序敏感，但是允许缺少个别词，缺少词的文档分值低，但是仍然能匹配。BooleanQuery.setMinimumNumberShouldMatch(int)方法可以定义需要匹配的条件的最小数量。

例如，BooleanQuery 中有 2 项，调用 BooleanQuery.setMinimumNumberShouldMatch(1)方法可以匹配其中任意的 1 项或者更多。

```
Term t1 = new Term(FIELD, "鸡");
TermQuery q1 = new TermQuery(t1);

Term t2 = new Term(FIELD, "鸭");
TermQuery q2 = new TermQuery(t2);

BooleanQuery mbq = new BooleanQuery();
mbq.add(q1 , BooleanClause.Occur.SHOULD);
mbq.add(q2 , BooleanClause.Occur.SHOULD);
mbq.setMinimumNumberShouldMatch(1);
```

3.3.6 短语查询

如果按字索引中文文档，则查询的时候往往要求文档中的这些字是连续出现的。使用 PhraseQuery 可以查询连续出现的几个关键词。PhraseQuery 称为短语匹配查询，用于要求精确匹配的查询。PhraseQuery 使用了词保存在索引中的位置信息，因此需要索引中的相关列已经保存了位置信息。例如按字索引，按字查询"开封"：

```
PhraseQuery query = new PhraseQuery();
query.add(new Term("subject","开"));
query.add(new Term("subject","封"));
```

搜索"软件工程师"时，标题为"安卓软件开发工程师"排第一位，而标题为"软件工程师"反而排在后面。词库为："软件""工程师"。为了解决这个问题，可以增加短语查询：

```
PhraseQuery query = new PhraseQuery();
query.add(new Term("subject","软件"));
query.add(new Term("subject","工程师"));
```

有时候要匹配上的词之间有间隔，匹配上的词之间的距离称为 slop。默认情况下，slop

的值是 0，可以通过调用 setSlop()方法设置这个值。例如，用户搜索"西红柿牛腩"时，会匹配上"西红柿炖牛腩"。

```
PhraseQuery query = new PhraseQuery();
query.setSlop(1);
query.add(new Term("subject","西红柿"));
query.add(new Term("subject","牛腩"));
```

使用 PhraseQuery 完整的例子：

```
public static void main(String[] args) throws Exception {
    //在内存中建立索引
    Directory directory = new RAMDirectory();
    Analyzer analyzer = new StandardAnalyzer();
    IndexWriterConfig iwc = new IndexWriterConfig(analyzer);
    IndexWriter writer = new IndexWriter(directory, iwc);

    // 索引一些文档
    writer.addDocument(createDocument("1", "foo bar baz"));
    writer.addDocument(createDocument("2", "red green blue"));
    writer.addDocument(createDocument("3", "test foo bar test"));
    writer.close();

    // 查找包含 foo bar 这个短语的文档
    String sentence = "foo bar";
    IndexReader reader = IndexReader.open(directory);
    // 根据 IndexReader 创建 IndexSearcher
    IndexSearcher searcher = new IndexSearcher(reader);
    PhraseQuery query = new PhraseQuery();
    String[] words = sentence.split(" ");
    for (String word : words) {
        query.add(new Term("contents", word));
    }

    // 显示搜索结果
    TopDocs topDocs = searcher.search(query, 10);
    for (ScoreDoc scoreDoc : topDocs.scoreDocs) {
        Document doc = searcher.doc(scoreDoc.doc);
        System.out.println(doc);
    }
}

private static Document createDocument(String id, String content) {
    Document doc = new Document();
    doc.add(new Field("id", id, Store.YES, Index.NOT_ANALYZED));
    doc.add(new Field("contents", content, Store.YES, Index.ANALYZED,
            Field.TermVector.WITH_POSITIONS_OFFSETS));
    return doc;
}
```

第 3 章 Lucene 的原理与应用

}

3.3.7 跨度查询

若查询词是"吃饭",而文档中出现了"吃完饭"这样的词。把查询词和文档都按最小粒度分词,分成"吃"和"饭"两个词。文档中的这两个词虽然不是连续出现,但只间隔了一个词。

近似查询假设同时查询的几个词之间的匹配点很近,则这样的文档可能也是要查找的。匹配词在文档中的位置信息用跨度(Spans)描述。抽象类 Spans 封装了一次匹配的文档和位置信息。span 是一个<文档编号,开始位置,结束位置>的三元组。

Spans 的接口定义如下:

```
boolean next() //移动到下一个匹配的文档
boolean skipTo(int target)//跳到指定的文档
int doc()//当前的文档编号
int start()//匹配区域的开始位置
int end()  //匹配区域的结束位置
```

TermSpans 是 Spans 的具体子类。

SpanQuery 称为跨度查询,用于查询多个词时考虑几个词在文档中的匹配位置。SpanQuery 和 PhraseQuerys 或者 MultiPhraseQuerys 很相似,因为都是通过位置限制匹配,但是 SpanQuery 更灵活。

SpanNearQuery 用来查询在比较近的区域内出现的多个查询词。例如,下面这个查询既可以匹配"吃饭",又可以匹配"吃完饭"。

```
new SpanNearQuery(new SpanQuery[] {
  new SpanTermQuery(new Term(FIELD, "吃")), //第1个词
  new SpanTermQuery(new Term(FIELD, "饭"))}, //第2个词
  1, //在1个位置以内
  true); //有序
```

SpanTermQuery 是一个最基础的跨度搜索实现类,因为可以通过它得到一个词的位置信息。在 SpanNearQuery 的构造方法中可以指定一些查询词需要在多近的区域出现。

同样的几个词在一起,但是如果顺序不一样,意思可能就不一样。例如,"花的中间"与"中间的花"意义就不一样,类似的例子还有"妈妈爱自己的儿女"和"儿女爱自己的妈妈"。所以,SpanNearQuery 有一个参数表示是否要求按数组中的顺序匹配。

例如,需要查找 lucene 和 doug 在 5 个位置以内,doug 在 lucene 之后,也就是说,词出现的顺序是需要考虑的。可以使用如图 3-11 所示的 SpanNearQuery。

```
SpanQuery[] baseQueries = new SpanQuery[] {
            new SpanTermQuery(new Term(FIELD, "lucene")),
            new SpanTermQuery(new Term(FIELD, "doug"))};
//这里的 true 表示词必须按数组中的顺序出现在文档中
new SpanNearQuery(baseQueries, 5, true);
```

在这个例子中，lucene 和 doug 的间隔在 3 以内，也就是间隔了 3 个词。

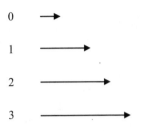

图 3-11 SpanNearQuery

SpanNearQuery 会查找互相在给定距离内的一些数量的 SpanQuery。可以指定 span 是按顺序或者不按顺序查询。SpanNearQuery 根据 SpanQuery 对象组成的数组来构建。因为 SpanTermQuery 是 SpanQuery 的子类，所以可以根据 SpanTermQuery 构造出 SpanNearQuery。

SpanNearQuery 构造方法接收一个 SpanQuery 数组。span 之间的距离，一个布尔值表明是否要求按 SpanQuery 数组中的顺序出现。下面是另一个嵌套的例子。这次，查找 doug 在 lucene 之后的 5 个间隔内，而 hadoop 在 lucene → doug 之后的 4 个间隔内，如图 3-12 所示。

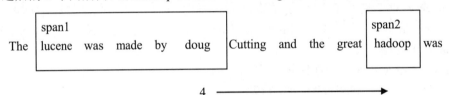

图 3-12 嵌套的 SpanNearQuery

```
SpanNearQuery spanNear = new SpanNearQuery(new SpanQuery[] {
    new SpanTermQuery(new Term(FIELD, "lucene")),
    new SpanTermQuery(new Term(FIELD, "doug"))},
    5,
    true);

new SpanNearQuery(new SpanQuery[] {
    spanNear,
    new SpanTermQuery(new Term(FIELD, "hadoop"))},
    4,
    true);
```

有些列，例如标题，头几个词可能更重要，可以使用 SpanFirstQuery 限定只查询前几个词。例如，查找名称以"玩具"开头的产品。

```
SpanTermQuery toy = new SpanTermQuery(new Term("title","玩具"));
SpanFirstQuery first = new SpanFirstQuery(toy, 1); //限定在头一个位置出现
```

可以把 SpanFirstQuery 看成是 SpanNearQuery 的特例，第一个词是一个虚拟的开始词，

从第二个词开始才是真正的匹配词。

例如，有两个文档"I love Lucene"和"Lucene is nice"，想要能够查询 Lucene 出现在最开始的文档，也就是匹配正则表达式"^Lucene .*"的文档。

```
SpanTermQuery lucene = new SpanTermQuery(new Term("title","Lucene"));
SpanFirstQuery first = new SpanFirstQuery(lucene, 1); //限定在头一个位置出现
```

SpanRegexQuery 使用标准的正则表达式语法。例如，查询年：2000, 2001, 2002, ..., 2009。

```
SpanRegexQuery srq = new SpanRegexQuery(new Term("year","200.?"));
```

SpanNotQuery 包含必须满足的 SpanQuery 和必须排除的 SpanQuery。例如，要查找 Microsoft Windows，但是不匹配 Microsoft Windows 之后的文档是大写的，伪代码如下：

```
SpanNot:
   include:
      SpanNear(in-order=true, slop=0):
         SpanTerm: "Microsoft"
         SpanTerm: "Windows"
   exclude:
      SpanNear(in-order=true, slop=0):
         SpanTerm: "Microsoft"
         SpanTerm: "Windows"
         SpanRegex: "^\\p{Lu}.*"
```

实际的代码：

```
SpanNearQuery includeQuery = new SpanNearQuery(new SpanQuery[] {
 new SpanTermQuery(new Term(FIELD, "Microsoft")),
 new SpanTermQuery(new Term(FIELD, "Windows"))},
 0,
 true);

SpanRegexQuery upperCaseQuery = new SpanRegexQuery( new Term(FIELD,
"^\\p{Lu}.*"));
SpanNearQuery excludeQuery = new SpanNearQuery(new SpanQuery[] {
 new SpanTermQuery(new Term(FIELD, "Microsoft")),
 new SpanTermQuery(new Term(FIELD, "Windows")),
 upperCaseQuery },
 0,
 true);
```

若想查找包含"喜欢"一词，并且在其前面没有"不"字的文档。可以这样：

```
SpanNotQuery(like, SpanNearQuery(not,like))
```

再扩展出同类查询词，也就是：

```
SpanNotQuery(
      SpanOrQuery(喜欢,爱...)
```

```
        SpanNearQuery(
            SpanOrQuery(不)
            SpanOrQuery(喜欢,爱...)
            )
        )
```

SpanOrQuery 可以匹配符合任何一个条件的文档。

```
SpanTermQuery term1 = new SpanTermQuery(new Term("field", "thirty"));
SpanTermQuery term2 = new SpanTermQuery(new Term("field", "three"));
SpanNearQuery near1 = new SpanNearQuery(new SpanQuery[] {term1, term2},
                          0, true);
SpanTermQuery term3 = new SpanTermQuery(new Term("field", "forty"));
SpanTermQuery term4 = new SpanTermQuery(new Term("field", "seven"));
SpanNearQuery near2 = new SpanNearQuery(new SpanQuery[] {term3, term4},
                          0, true);

SpanOrQuery query = new SpanOrQuery(new SpanQuery[] {near1, near2});
```

例如，说一个人坏透了，用"头上长疮，脚下流脓"来形容。这个条件可以这样描述：

```
SpanAndQuery(
       SpanFirstQuery(疮)
       SpanLastQuery(脓)
)
```

SpanAndQuery 和 SpanLastQuery 都还没有现成的实现例子。

3.3.8 FieldScoreQuery

FieldScoreQuery 叫作函数查询(通过数字型的字段影响排序结果)，时间加权排序时会用到。

除了文本列上使用 TF-IDF 相似性的标准词查询之外，打分还参考数值列的相似性。相似性依赖于查询对象中的值和文档中的数值列中的值之间的距离。(例如，高斯函数，使用参数：m= [user input], s= 0.5)

例如，"猎头"使用搜索引擎找人。表示人的文档，有两列：

- description (文本列)；
- age (数值列)。

想要找这样的文档：

```
description:(x y z) age:30
```

但是 age 不是过滤条件，而是 score 的一部分(对于 30 岁的人，乘积因子是 1.0；对于 25 岁的人，乘积因子是 0.8；等等)。

把 ValueSourceQuery 和 TermQuery 包装在 CustomScoreQuery 中实现。

```
public class AgeAndContentScoreQueryTest extends TestCase{
  public class AgeAndContentScoreQuery extends CustomScoreQuery {
```

第 3 章　Lucene 的原理与应用

```
    protected float peakX;
    protected float sigma;

    //接收 4 个参数，其中 subQuery 表示文本列查询，valSrcQuery 表示值查询
    public AgeAndContentScoreQuery(Query subQuery, ValueSourceQuery
valSrcQuery, float peakX, float sigma) {
        super(subQuery, valSrcQuery);
        this.setStrict(true); // 不要归一化从 ValueSourceQuery 得到的分值！
        this.peakX = peakX;    // 哪个年纪的相关性最好
        this.sigma = sigma;
    }

    @Override
    public float customScore(int doc, float subQueryScore, float valSrcScore){
        // subQueryScore 来源于内容查询的 td-idf 分值
        float contentScore = subQueryScore;

        // valSrcScore 是生日字段的值，表示成一个浮点数
        // 把年纪值转换成高斯型年纪相关分值
        float x = (2011 - valSrcScore); // age
        float ageScore = (float) Math.exp(-Math.pow(x - peakX, 2) / 2*sigma*sigma);

        float finalScore = ageScore * contentScore;

        System.out.println("#contentScore: " + contentScore);
        System.out.println("#ageValue:     " + (int)valSrcScore);
        System.out.println("#ageScore:     " + ageScore);
        System.out.println("#finalScore:   " + finalScore);
        System.out.println("+++++++++++++++++");

        return finalScore;
    }
}

protected Directory directory;
protected Analyzer analyzer = new WhitespaceAnalyzer();
protected String fieldNameContent = "content";
protected String fieldNameDOB = "dob";

protected void setUp() throws Exception  {
    directory = new RAMDirectory();
    analyzer = new WhitespaceAnalyzer();

    //索引文档
    String[] contents = {"foo baz1", "foo baz2 baz3", "baz4"};
    int[] dobs = {1991, 1981, 1987}; // 生日
```

```java
        IndexWriter writer = new IndexWriter(directory, analyzer,
IndexWriter.MaxFieldLength.UNLIMITED);
        for (int i = 0; i < contents.length; i++) {
            Document doc = new Document();
            doc.add(new Field(fieldNameContent, contents[i], Field.Store.YES,
Field.Index.ANALYZED));  // store 并且 index
            doc.add(new NumericField(fieldNameDOB, Field.Store.YES,
true).setIntValue(dobs[i]));         // store & index
            writer.addDocument(doc);
        }
        writer.close();
    }

    public void testSearch() throws Exception {
        String inputTextQuery = "foo bar";
        float peak = 27.0f;
        float sigma = 0.1f;

        QueryParser parser = new QueryParser(fieldNameContent, analyzer);
        Query contentQuery = parser.parse(inputTextQuery);

        ValueSourceQuery dobQuery = new ValueSourceQuery( new
IntFieldSource(fieldNameDOB) );
        // 或者 FieldScoreQuery dobQuery = new
// FieldScoreQuery(fieldNameDOB,Type.INT);

        CustomScoreQuery finalQuery = new AgeAndContentScoreQuery(contentQuery,
dobQuery, peak, sigma);

        IndexSearcher searcher = new IndexSearcher(directory);
        TopDocs docs = searcher.search(finalQuery, 10);

        System.out.println("\n发现的文档:\n");
        for(ScoreDoc match : docs.scoreDocs)   {
            Document d = searcher.doc(match.doc);
            System.out.println("CONTENT: " + d.get(fieldNameContent) );
            System.out.println("D.O.B.: " + d.get(fieldNameDOB) );
            System.out.println("SCORE:   " + match.score );
            System.out.println("-----------------");
        }
    }
}
```

日期加权：

```
if (x >= now) {
    score = now / x;
} else {
    score = (float) (now / (now+sigma) - sigma / x);
}
```

3.3.9 排序

为了实现情景搜索，根据用户 IP 判断用户所在地区，然后把与用户所在区域相同的文档排在前面，即通过排序把地名相同的搜索结果放在前面。

若采用自定义排序对象的方法实现把和用户所在地域相同的结果放在前面，需要实现 FieldComparatorSource 和 FieldComparator 接口。FieldComparatorSource 功能是返回一个用来排序的比较器。这个接口只定义一个方法 newComparator()，用来生成 FieldComparator 对象。生成的 FieldComparator 对象可作为指定文档列的排序比较器。

当使用 TopFieldCollector 收集最前面的结果时，要用 FieldComparator 比较命中的文档，从而确定它们的排列顺序。

具体的 FieldComparator 类对应不同的 SortField 类型。

FieldComparator 对象需要实现 compare()、compareBottom()、copy()、setBottom()、setNextReader()、value()等方法，如果需要用到文档的评分值，还需要实现 setScorer()方法。

- compare(int, int)：比较槽 a 和槽 b 的命中文档。
- setBottom(int)：通过 FieldValueHitQueue 调用此方法，以通知 FieldComparator 当前最底部的槽。
- compareBottom(int)：新的命中文档(docID)和队列中的最弱的底部进行比较。
- copy(int, int) 安排一个新的命中文档到优先队列中去。当一个新的命中文档具有竞争力时，FieldValueHitQueue 调用这个方法。
- setNextReader(org.apache.lucene.index.IndexReader, int)：当搜索切换到下一个段时调用这个方法。可能需要更新比较器的内部状态，例如从 FieldCache 检索新值。
- value(int)：返回存储在指定位置的值。当返回顶部结果时，为了填充 FieldDoc.fields，才在搜索结束时调用它。

首先要定义一个继承抽象类 FieldComparatorSource 的实现类 CityFieldComparator，重写其中的抽象方法：

```
public FieldComparator newComparator(String field, int numHits,
        int sortPos, boolean reversed) throws IOException;
```

该方法返回一个 FieldComparator 实例，实现自定义排序。

然后定义一个类 CitySortComparator 继承自 FieldComparator 抽象类。重写其中的抽象方法，在 public int compare(int slot1, int slot2)方法中实现排序业务逻辑。

TopFieldCollector 是一个用 SortField 排序的 Collector。SortField 使用 FieldComparator 构造。

最后再调用该自定义排序类。

```
//这里设置为true是正序,false为倒序。调用CityFieldComparator的构造方法初始化
//城市名,城市名称可以根据用户IP获得。
SortField citySort = new SortField("city", new CityFieldComparator(city),
true);
//按时间排序,true为时间减序,false为时间增序。
SortField dateSort = new SortField("date", SortField.INT, true);
//首先按城市排序,然后按日期排序
Sort sort = new Sort(new SortField[] {citySort,dateSort});

TopFieldCollector collector = TopFieldCollector.create(sort, offset+rows,
            false, true, false, false);
//isearcher是IndexSearcher的实例
isearcher.search(bq, collector);
ScoreDoc[] hits = collector.topDocs().scoreDocs;
```

排序速度很慢,因为排序列的值都缓存在FieldCache中。对于要用来排序的字段,先从索引中将每篇文档中该字段的值都读出来,放到一个大小为maxDoc的数组中。maxDoc是文档编号的最大值。倒排索引的词列是字符串或者字节数组类型,从倒排索引的词列解析出值的过程叫作反倒排。

有两点需要注意:

(1) FieldCache中的字段值是从倒排表中读出来的,它不是索引文件中存储的字段值,所以排序的字段必须是索引字段。

(2) 用来排序的字段在索引的时候不能拆分,因为FieldCache数组中,每个文档只对应一个字段值,拆分的话,缓存中只会保存词典中靠后的值。

FieldCache是Lucene最占用内存的部分,大部分内存溢出的错误都是由它引起的,需要特别注意。

每个商品只有一个价格,也就是说,商品的价格是唯一的。用按列存储值的方法依价格排序。

每个文档有几个数值类型的字段。可根据这些列的加权和来排序。例如:

```
field1=100
field2=002
field3=014
```

加权函数类似:

```
f(d) = field1 * 0.5 + field2 * 1.4 + field3 * 1.8
```

结果通过f(d)的值排序,这里的变量d表示文档。排序方法应该是非静态的,并且在不同的搜索之间是不同的,因为对执行搜索的不同用户来说,常量是不一样的。也就是说,排序方法是个性化的。

```
public class ScaledComparator extends FieldComparator {
    private String[] fields;
```

```java
    private float[] scalars;
    private int[][] slotValues;
    private int[][] currentReaderValues;
    private int bottomSlot;

    public ScaledComparator(int numHits, String[] fields, float[] scalars)
{
        this.fields = fields;
        this.scalars = scalars;

        this.slotValues = new int[this.fields.length][];
        for (int fieldIndex = 0; fieldIndex < this.fields.length; fieldIndex++)
{
            this.slotValues[fieldIndex] = new int[numHits];
        }

        this.currentReaderValues = new int[this.fields.length][];
    }

    protected float score(int[][] values, int secondaryIndex) {
        float score = 0;

        for (int fieldIndex = 0; fieldIndex < fields.length; fieldIndex++) {
            int value = values[fieldIndex][secondaryIndex];
            float scalar = scalars[fieldIndex];
            score += (value * scalar);
        }

        return score;
    }

    protected float scoreSlot(int slot) {
        return score(slotValues, slot);
    }

    protected float scoreDoc(int doc) {
        return score(currentReaderValues, doc);
    }

    @Override
    public int compare(int slot1, int slot2) {
        float score1 = scoreSlot(slot1);
        float score2 = scoreSlot(slot2);
        return Float.compare(score1, score2);
    }

    @Override
```

```java
    public int compareBottom(int doc) throws IOException {
        float bottomScore = scoreSlot(bottomSlot);
        float docScore = scoreDoc(doc);
        return Float.compare(bottomScore, docScore);
    }

    @Override
    public void copy(int slot, int doc) throws IOException {
        for (int fieldIndex = 0; fieldIndex < fields.length; fieldIndex++) {
            slotValues[fieldIndex][slot] =
currentReaderValues[fieldIndex][doc];
        }
    }

    @Override
    public void setBottom(int slot) {
        bottomSlot = slot;
    }

    @Override
    public void setNextReader(IndexReader reader, int docBase, int numSlotsFull) throws IOException {
        for (int fieldIndex = 0; fieldIndex < fields.length; fieldIndex++) {
            String field = fields[fieldIndex];
            currentReaderValues[fieldIndex] =
                    FieldCache.DEFAULT.getInts(reader, field);
        }
    }

    @Override
    public int sortType() {
        return SortField.CUSTOM;
    }

    @Override
    public Comparable<?> value(int slot) {
        float score = scoreSlot(slot);
        return Float.valueOf(score);
    }
}
```

使用 ScaledComparator 实现排序的功能。FieldComparatorSource 的匿名类重写 newComparator()方法。newComparator()方法返回 ScaledComparator 类的一个实例。

```
//列名数组
final String[] fields = new String[]{ "field1", "field2", "field3" };
final float[] scalars = new float[]{ 0.5f, 1.4f, 1.8f }; //列的权重
```

```
Sort sort = new Sort(
    new SortField(
        "",
        new FieldComparatorSource() {
            public FieldComparator newComparator(String fieldname, int
numHits, int sortPos, boolean reversed) throws IOException {
                return new ScaledComparator(numHits, fields, scalars);
            }
        }
    )
);
```

3.3.10 使用 Filter 筛选搜索结果

可以定义 Filter 类过滤查询结果，也可以缓存和重用 Filter。如下条件可用 Filter 来实现：
- 根据不同的安全权限显示搜索结果。
- 仅查看上个月的数据。
- 在某个类别中查找，例如在分类统计查询中，根据分类列的有效取值缩小查询范围。

Filter 返回一个 DocIdSet，其中包含符合条件的文档编号。DocIdSet 是一个存储文档编号的集合。例如：

```
IndexReader reader= …
OpenBitSet bitSet = new OpenBitSet(reader.maxDoc()); // 假设所有的文档都不在集合内
bitSet.set( 0 ); //让第一个文档在集合内
```

Filter 过滤整个索引，得到 DocIdSet。

```
public abstract class Filter implements java.io.Serializable {
  public abstract DocIdSet getDocIdSet(IndexReader reader) throws
IOException;
}
```

下面定义一个 BestDriversFilter，把搜索结果限定到 score 是 5 的司机。

```
public class BestDriversFilter extends Filter{
    @Override
    public DocIdSet getDocIdSet(IndexReader reader) throws IOException {
        OpenBitSet bitSet = new OpenBitSet( reader.maxDoc() );
        //查询出 score 是 5 的文档
        TermDocs termDocs = reader.termDocs( new Term( "score", "5" ) );
        while ( termDocs.next() ) {
            bitSet.set( termDocs.doc() );// 把符合条件的文档对应的位置设为1
        }
        return bitSet;
    }
}
```

在查询中使用这个 Filter：

```
Hits h = searcher.search(tq,filter);
```

完整的代码：

```
Filter bestDriversFilter = new BestDriversFilter();
//query 不变，增加 bestDriversFilter
ScoreDoc[] hits = isearcher.search(query, bestDriversFilter, 10).scoreDocs;
//因为不是每个司机都能得 5 分，所以返回的结果可能比以前少了
```

3.3.11 使用 Collector 筛选搜索结果

Collector 主要用来从一次查询中收集原始结果。可以通过它遍历查询结果中的每个文档。执行搜索的 Searcher.search()方法可以接收 Collector 对象作为参数。查询类每次匹配到一个文档都会调用 Collector 对象上的 collect(int doc)方法。下面是一个收集文档，但不对文档评分的例子：

```
private static final String FIELD = "contents";

public static void main(String[] args) throws Exception {
    // 建立使用内存索引的 Lucene
    Directory directory = new RAMDirectory();

    Analyzer analyzer = new StandardAnalyzer();
    IndexWriterConfig iwc = new IndexWriterConfig(analyzer);
    IndexWriter writer = new IndexWriter(directory, iwc);

    // 索引一些文档
    writer.addDocument(createDocument("1", "foo bar baz"));
    writer.addDocument(createDocument("2", "red green blue"));
    writer.addDocument(createDocument("3",
            "The Lucene was made by Doug Cutting"));

    writer.close();

    IndexReader reader = DirectoryReader.open(directory);

    IndexSearcher searcher = new IndexSearcher(reader);

    Term t1 = new Term(FIELD, "lucene");
    TermQuery query = new TermQuery(t1);

    final BitSet bits = new BitSet(reader.maxDoc());
    searcher.search(query, new Collector() {
        private int docBase;

        //忽略评分器
```

```java
        @Override
        public void setScorer(Scorer scorer) {
        }

        //允许文档乱序
        public boolean acceptsDocsOutOfOrder() {
            return true;
        }

        public void collect(int doc) {
            bits.set(doc + docBase);
        }

        public void setNextReader(AtomicReaderContext context) {
            this.docBase = context.docBase;
        }
    });

    System.out.println("结果数: " + bits.cardinality());
    for (int i = bits.nextSetBit(0); i >= 0; i = bits.nextSetBit(i + 1)) {
        System.out.println("文档编号: " + i);
    }
}

private static Document createDocument(String id, String content) {
    Document doc = new Document();
    doc.add(new Field("id", id, StringField.TYPE_STORED));
    doc.add(new Field(FIELD, content, TextField.TYPE_STORED));
    return doc;
}
```

例如,有个 Collector 的子类 TopDocsCollector 处理原始的查询结果,并且返回最相关的 N 个文档。

它是一个基类,用于做所有返回 TopDocs 输出的类的父类。TopDocsCollector.topDocs() 方法返回最相关的 N 个文档。

例如:

```
Query q = ...;
IndexSearcher searcher = ...;
TopDocsCollector<ScoreDoc> tdc = new MyTopsDocCollector(numResults);
searcher.search(q, tdc); //传递一个查询对象和一个 Collector 对象
//给 Searcher.search 方法
```

TopDocsCollector 接收一个优先队列和一个返回结果的总数作为参数。TopScoreDocCollector 是 TopDocsCollector 的子类,也是可以直接使用的类。使用 TopScoreDocCollector 返回最相关的 10 个文档的例子如下:

```
int hitsPerPage = 10;
IndexSearcher searcher = new IndexSearcher(index, true);
TopScoreDocCollector collector = TopScoreDocCollector.create(hitsPerPage,
true);
searcher.search(q, collector);
ScoreDoc[] hits = collector.topDocs().scoreDocs;
```

一个会员可以发布多条信息,搜索的时候如何将会员发布的信息按每个会员一条的方式显示。也就是说每个会员编号最多只显示一条。每个会员最多只显示和查询词最相关的一条信息。这个功能是在结果集而不是在索引库上遍历有效文档,所以应该采用 Collector 而不是 Filter。

可以按指定列排序,然后过滤信息。但是 Searcher.search()方法不能同时接收 Collector 对象和 Sort 对象。只存在:

```
searcher.search(query, collector);
```

对搜索结果首先按会员列排序,然后按相关度排序。找出每个会员最多只显示和查询词最相关的一条信息。让 Collector 把这样的结果放入一个 TopDocs 对象。

可以扩展 TopDocsCollector 类。在 collect()方法中构造一个新的 ScoreDoc 对象。

```
private static final class MyTopsDocCollector extends
TopDocsCollector<ScoreDoc> {
  private int idx = 0;
  private int base = 0;

  public MyTopsDocCollector(int size) {
    super(new HitQueue(size, false));
  }

  @Override
  protected TopDocs newTopDocs(ScoreDoc[] results, int start) {
    if (results == null) {
      return EMPTY_TOPDOCS;
    }

    float maxScore = Float.NaN;
    if (start == 0) {
      maxScore = results[0].score;
    } else {
      for (int i = pq.size(); i > 1; i--) { pq.pop(); }
      maxScore = pq.pop().score;
    }

    return new TopDocs(totalHits, results, maxScore);
  }

  @Override
  public void collect(int doc) throws IOException {
```

```
    ++totalHits;
    pq.insertWithOverflow(new ScoreDoc(doc + base, scores[idx++]));
}

@Override
public void setNextReader(IndexReader reader, int docBase)
    throws IOException {
  base = docBase;
}

@Override
public void setScorer(Scorer scorer) throws IOException {
    //不做任何事。随机分配评分值
}

@Override
public boolean acceptsDocsOutOfOrder() {
    return true;
}
}
```

Collector 类的 setScorer()方法是一个方法，当 IndexSearcher 实际执行搜索时，通过 IndexSearcher 传入 score。

对每个匹配的文档，调用 collect()方法，传递一个索引段内部的文档编号给 collect() 方法。

会员发布的信息要一轮一轮地显示，不是说折叠起来就显示一条。比如会员 1 有两条数据：信息 1，信息 2；会员 2 也有两条数据：信息 1，信息 2，要的结果是会员 1 的信息 1，会员 2 的信息 1，会员 1 的信息 2，会员 2 的信息 2。

把每个会员的相关信息都存到一个优先队列。所有的搜索结果就是每个会员组成的优先队列。也就是说，优先队列有很多。

3.3.12 遍历索引库

IndexReader 提供了遍历索引库的接口。在遍历索引库之前可以先看一下索引库中文档的数量。

```
Directory directory = getDir();   //得到路径对象
IndexReader indexReader = DirectoryReader.open(directory);  //取得 IndexReader 对象
System.out.println(indexReader.maxDoc());  // 打印文档数量
```

可以根据文档编号遍历索引库中的每个文档。

```
IndexReader reader = DirectoryReader.open(directory);
int totalDocs = reader.numDocs();//取得所有文档的数量
for(int m=0;m<totalDocs;m++){
    Document thisDoc = reader.document(m);  //取得索引库中的每个文档
}
```

检查文档是否已经被删除了，可以调用 MultiFields.getLiveDocs(reader)方法。

```
Bits liveDocs = MultiFields.getLiveDocs(reader); //返回值可能为空
for (int i=0; i<reader.maxDoc(); i++) {
   if (liveDocs != null && !liveDocs.get(i))
       continue;
   Document doc = reader.document(i);
}
```

文档编号是一个非负整数。在重新做索引后，文档的文档编号可能会改变。所以在重新打开 IndexReader 以后，原来的文档编号就失效了。

IndexSearcher 也是通过 IndexReader 取得文档对象。

```
public class IndexSearcher {
  final IndexReader reader;

  public Document doc(int docID) throws IOException {
    return reader.document(docID); //通过 IndexReader 取得文档对象
  }
}
```

IndexSearcher.doc(int)方法只是为了方便从搜索结果中得到文档对象。

下面介绍与遍历索引库相关的几个类。TermEnum 用来枚举一个给定的域中的所有项，而不管这个项在哪个文档中。例如：

```
String indexDir = "D:/test/chatindex";
FSDirectory directory = FSDirectory.open(new File(indexDir));
DirectoryReader reader = DirectoryReader.open(directory);

// 读取索引文件里所有的 Term
Terms terms = SlowCompositeReaderWrapper.wrap(reader).terms("field");

TermsEnum termsEnum = terms.iterator(null);

BytesRef term;
while ((term = termsEnum.next()) != null) {
    String s = new String(term.bytes, term.offset, term.length);
    System.out.println("词: "+s);
}
```

TermDocs 和 TermEnum 不同，TermDocs 用来识别哪个文档包含指定的项，并且它也会给出该项在文档中的词频。

TermFreqVector(即 Term Frequency Vector 或者简称 Term Vector)是一个数据结构,包含一个指定文档的项和词频信息,并且当在索引期间存储项向量的时候,才能通过 IndexReader 检索出 TermFreqVector。

所谓 Term Vector，就是对于文档的某一列，如 title、body 这种文本类型的列，建立词频的多维向量空间。一个词就是一维，该维的值就是这个词在这个列中的频率。

getTerms()和 getTermFrequencies()是并列的数组。也就是说,getTerms()[i]有一个文档频率 getTermFrequencies()[i]。

例如,源文本见表 3-4。

表 3-4 源文本

词位置	0	1	2	3	4	5	6	7	8
词	the	quick	brown	fox	jumps	over	the	lazy	dog

词排序后的位置见表 3-5。

表 3-5 排序后位置

词索引	0	1	2	3	4	5	6	7
词	brown	dog	fox	jump	lazy	over	quick	the

用下面的代码发现 the 出现的位置:

```
int index = termPositionVector.indexOf("the"); // 7
int positions = termPositionVector.getTermPositions(index); // {0, 6}
```

这里使用 TermEnum 按词遍历索引库,代码如下:

```
public static void getTerms(IndexReader reader) throws IOException {
    TermEnum terms = reader.terms(); // 读取索引文件里所有的 Term

    while (terms.next()) {// 取出一个 Term 对象
        String field = terms.term().field(); //列名
        String text = terms.term().text(); //词
        System.out.println(text+":"+field);
    }
}
```

经常需要统计索引库中哪些词出现的频率最高。例如,需要统计旅游活动索引库中的热门目的地。可以先对目的地列做索引,索引列不分词,然后取得该列中最常出现的几个词,也就是热门目的地。实现方法是:可以用 TermEnum 遍历索引库中所有的词,取出每个词的文档频率,然后使用优先队列找出频率最高的几个词。

```
public class TermInfo {
  public Term term;    //索引库中的词
  public int docFreq; //文档频率,也就是这个词在多少个文档中出现过

  public TermInfo(Term t, int df) {
    this.term = t;
    this.docFreq = df;
  }
}

//找出频率最高的 numTerms 个词
```

```java
public static TermInfo[] getHighFreqTerms(IndexReader reader,
                        int numTerms, String field){
    //实例化一个 TermInfo 的队列
    TermInfoQueue tiq = new TermInfoQueue(numTerms);
    TermEnum terms = reader.terms(); //读取索引文件里所有的 term
    int minFreq = 0; //队列最后一个 term 的频率即当前最小频率值
    while (terms.next()) {//取出一个 term 对象
        String field = terms.term().field();
        if (fields != null && fields.length > 0) {
            boolean skip = true; //跳过标识
            for (int i = 0; i < fields.length; i++) {
                //当前 Field 属于 fields 数组中的某一个则处理对应的 term
                if (field.equals(fields[i])) {
                    skip = false;
                    break;
                }
            }
            if (skip) continue;
        }
        //当前 term 的内容是过滤词, 则直接跳过
        if (junkWords != null && junkWords.get(terms.term().text()) != null) continue;

        /*获取最高频率的 term。基本方法是:
        队列底层是最大频率 term, 顶层是最小频率 term,
        当插入一个元素后超出初始化队列大小则取出最上面的那个元素,
        重新设置最小频率值 minFreq*/
        if (terms.docFreq() > minFreq) {//当前 Term 的频率大于最小频率则插入队列
            tiq.insertWithOverflow(new TermInfo(terms.term(), terms.docFreq()));
            if (tiq.size() >= numTerms) { //当队列中的元素个数大于 numTerms
                tiq.pop(); // 取出最小频率的元素, 即最上面的一个元素
                minFreq = ((TermInfo)tiq.top()).docFreq; //重新设置最小频率
            }
        }
    }
    //取出队列元素, 最终存放在数组中元素的词频率按从大到小排列
    TermInfo[] res = new TermInfo[tiq.size()];
    for (int i = 0; i < res.length; i++) {
        res[res.length - i - 1] = (TermInfo)tiq.pop();
    }
    return res;
}
```

四维枚举 API 指通过列、文档、词、位置四个维度来遍历索引库。

3.3.13 关键词高亮显示

因为搜索出来的文档内容可能比较长,所以不仅要检索出命中的文本,还要提供查询

词在文本中出现的位置,方便用户直接看到想要找的信息。最终高亮显示的是一个片段(包含高亮词),而不是一个完整的列值。将来的浏览器应该可以支持显示全文的同时,先定位到查询词所在的位置。

在搜索结果中一般都有和用户搜索关键词相关的摘要。关键词一般都会高亮显示。从实现上说,就是把要突出显示的关键词前加上<B>标签,关键词后加上</B>标签。Lucene 的 highlighter 包可以做到这一点。

Lucene 有两个高亮显示的实现:一个是 org.apache.lucene.search.highlight;还有一个是 org.apache.lucene.search.vectorhighlight。下面是使用 org.apache.lucene.search.highlight 的例子:

```
doSearching("汽车");
//使用一个查询初始化 Highlighter 对象
Highlighter highlighter = new Highlighter(new QueryScorer(query));
//设置分段显示的文本长度
highlighter.setTextFragmenter(new SimpleFragmenter(40));
//设置最多显示的段落数量
int maxNumFragmentsRequired = 2;
for (int i = 0; i < hits.length(); i++) {
    //取得索引库中存储的原始文本
    String text = hits.doc(i).get(FIELD_NAME);
    TokenStream tokenStream=analyzer.tokenStream(FIELD_NAME,
                                    new StringReader(text));

    //取得关键词加亮后的结果
    String result = highlighter.getBestFragments(tokenStream,
                                    text,
                                    maxNumFragmentsRequired,
                                    "...");
    System.out.println("\t" + result);
}
```

QueryScorer()设置查询的 query,这里还可以加上对字段列的限制,比如只对 body 条件的 term 高亮显示,可以使用 new QueryScorer(query, "body")。对于模糊匹配,需要先找出要高亮显示的词。可以使用 SpanScorer 和 SimpleSpanFragmenter,或者使用 QueryScorer 和 SimpleFragmenter。

使用 SpanScorer 和 SimpleSpanFragmenter 生成高亮段落的代码如下:

```
TokenStream stream = TokenSources.getTokenStream(fieldName, fieldContents,
                analyzer);
SpanScorer scorer = new SpanScorer(query, fieldName,
                new CachingTokenFilter(stream));
Fragmenter fragmenter = new SimpleSpanFragmenter(scorer, 100);

Highlighter highlighter = new Highlighter(scorer);
highlighter.setTextFragmenter(fragmenter);
String[] fragments = highlighter.getBestFragments(stream, fieldContents, 5);
```

为了实现关键词高亮显示，必须知道关键词在文本中的位置。对英文来说，可以在搜索的时候实时切分出位置。但是中文分词的速度一般来说相对慢很多。在 Lucene 1.4.3 以后的版本中，Term Vector 支持保存 Token.getPositionIncrement()和 Token.startOffset() 以及 Token.endOffset() 信息。利用 Lucene 中新增加的 Token 信息保存结果以后，就不需要为了高亮显示而在运行时解析每篇文档。为了实现一列的高亮显示，索引的时候通过 Field 对象保存该位置信息。

```
//增加文档时保存term位置信息
private void addDoc(IndexWriter writer, String text) throws IOException{
    Document d = new Document();

    Field f = new Field(FIELD_NAME, text ,
                Field.Store.YES, Field.Index.TOKENIZED,
                Field.TermVector.WITH_POSITIONS_OFFSETS);
    d.add(f);
    writer.addDocument(d);
}
//利用term位置信息节省Highlight时间
void doStandardHighlights() throws Exception{
    Highlighter highlighter =new Highlighter(this,new QueryScorer(query));
    highlighter.setTextFragmenter(new SimpleFragmenter(20));
    for (int i = 0; i < hits.length(); i++) {
        String text = hits.doc(i).get(FIELD_NAME);
        int maxNumFragmentsRequired = 2;
        String fragmentSeparator = "...";
        TermPositionVector tpv = 
          (TermPositionVector)reader.getTermFreqVector(hits.id(i),FIELD_NAME);
        TokenStream tokenStream=TokenSources.getTokenStream(tpv);

        String result = highlighter.getBestFragments(
                        tokenStream,
                        text,
                        maxNumFragmentsRequired,
                        fragmentSeparator);

        System.out.println("\t" + result);
    }
}
```

最后把 highlight 包中的一个额外的判断去掉。对于中文来说没有明显的单词界限，所以下面这个判断是错误的：

```
tokenGroup.isDistinct(token)
```

注意上面代码中的 highlighter.setTextFragmenter(new SimpleFragmenter(20))，SimpleFragmenter 是一个最简单的段落分割器，它把文章按 20 个字分成一个段落。这种方式简单易行，但显得比较初步。有时会有一些没意义的符号出现在摘要的起始部分，例如逗号出现在摘要的开始位置。

RegexFragmenter 是一个改进版本的段落分割器。它通过一个正则表达式匹配可能的热点区域。但它是为英文定制的。我们可以让它认识中文的字符段。

```
protected static final Pattern textRE = Pattern.compile("[\\w\u4e00-\u9fa5]+");
```

这样使用 highlighter 就变成了：

```
highlighter.setTextFragmenter(new RegexFragmenter(descLenth));
```

比如"我的妈妈"，Google 搜索是这样："<em>我的</em> <em>妈妈</em>"。实际貌似 Lucene 都会变成"<em>我</em><em>的</em><em>妈妈</em>"，这样对 SEO(搜索引擎优化)很不好。<em>标签算是权重很高的标签，这样分使得页面会降很低，因为词都是分开的。另外，合并到一起，也省流量，对 SEO 有利。

不要高亮显示太长的文本，因为这样会影响搜索速度。

3.3.14 列合并

可以把鱼头和鱼尾分开做成不同的菜。假设要开发一个新闻检索系统，文章索引可能不会经常改变，但是评论索引会经常改变。所以可以把文章和评论独立建立索引，但可以同时查询。评论索引中包含文章索引的 id 列，如图 3-13 所示。

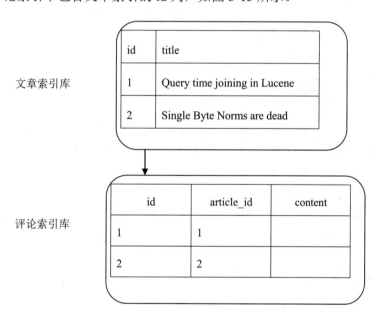

图 3-13 列合并查询

通过合并查询，可以将不同的部分分别按文档来存储，这将带来更多的灵活性，但是会增加更多的执行时间。

查询时合并封装在 JoinUtil.createJoinQuery()方法中，它需要下面几个参数。

- fromField：连接的来源列。

- toField：连接的目的列。
- fromQuery：指定查询的词。通常是用户指定的查询。
- fromSearcher. fromQuery：执行查询的搜索器找那个索引。
- multipleValuesPerDocument：指定 fromField 是否为多值列，也就是是否每个文档有不止一个值。

静态合并方法返回一个查询，能够在 IndexSearcher 上执行这个查询，用于检索所有在 to 列和找到的 from 词相匹配的文档。只有用于连接的条目展现给用户，而完全隐藏实际的实现方法，在保证 API 向后兼容性的同时允许 Lucene 实现者改变实现方法。

通过两步搜索实现在经过索引的词之上做查询时间合并：第一步从 from 域内(在我们的例子中，为文章标记域)搜集所有与 from 查询匹配的术语；第二步返回所有在 to 域(在我们的例子中，为评论文档中的文章标记域)中有匹配术语的文档给第一步从搜集到的术语。

返回给静态合并方法的查询能够作为一个静态合并方法的参数在不同的 IndexSearcher 上执行。这种灵活性允许合并来自不同索引中的数据，只要 toField 确实存在于索引之中。在这个例子中，意味着文章和评论数据能够位于两个不同的索引之中。文章索引可能不会经常改变，但是评论索引会经常改变。这样就可以去微调这些索引来满足各自特定的需求。

例如，搜索文章标题是 byte norms 的评论：

```
IndexSearcher articleSearcher = ... //打开文章索引
IndexSearcher commentSearcher = ... //打开评论索引
String fromField = "id";
boolean multipleValuesPerDocument = false;
String toField = "article_id";
// 这个查询产生 id 是 2 的文章作为结果
BooleanQuery fromQuery = new BooleanQuery();
fromQuery.add(new TermQuery(new Term("title", "byte")),
        BooleanClause.Occur.MUST); // byte 这个词必须出现
fromQuery.add(new TermQuery(new Term("title", "norms")),
        BooleanClause.Occur.MUST); // norms 这个词也必须出现
Query joinQuery = JoinUtil.createJoinQuery(fromField,
multipleValuesPerDocument,
                    toField, fromQuery, articleSearcher);
TopDocs topDocs = commentSearcher.search(joinQuery, 10);
```

3.3.15 关联内容(BlockJoinQuery)

一个简单的例子：假如有个网店卖衬衫，每件衬衫有一些常见的字段，如名称、描述、织物、价格等。对于每件衬衫，有许多单独的库存单元，它具有自己的列，如大小、颜色、库存数量等。库存单元叫作 SKU。这些型号是你实际出售的，或在你的库存中，因为当有人买衬衫时，他们买的是特定的 SKU 产品(大小和颜色)。

也许你卖的是袋鼠牌短袖 T 恤，有下列 SKU(大小, 颜色)：

- 小号, 蓝色
- 小号, 黑色

- 中号，黑色
- 大号，灰色

可能用户首先搜索"袋鼠 衬衫"，得到一个结果分支。然后下钻到一个特定的型号和颜色。产生这个查询：

```
name:袋鼠 AND size=小号 AND color=蓝色
```

这应该匹配名称是"衬衫"列，而大小和颜色是 SKU 列的衬衫。

但是，如果用户下钻到一个小号的灰色衬衫：

```
name:袋鼠 AND size=小号 AND color=灰色
```

这件衬衫查询不会返回任何结果，因为小号衬衫只有蓝色和黑色。

如何使用 BlockJoinQuery 运行这个查询？从把每件衬衫(父亲)和它所有的 SKU(孩子)当作独立的文档索引开始。

使用新的 IndexWriter.addDocuments API 增加一件 T 恤和它所有的 SKU 作为一个独立的文档块。这个方法自动增加一个文档块成为一个有相邻的文档编号的独立的段。这是 BlockJoinQuery 起作用的基础。

也要加一个标志列到每个衬衫文档，例如，type = shirt。因为 BlockJoinQuery 需要一个 Filter 标识出父文档。

要在搜索阶段运行 BlockJoinQuery，首先要创建一个父过滤器匹配 T 恤。注意，过滤器在底层必须使用 FixedBitSet，就像 CachingWrapperFilter：

```
Filter shirts = new CachingWrapperFilter(
             new QueryWrapperFilter(
                new TermQuery(
                   new Term("type", "shirt"))));
```

一旦创建好这个过滤器，任何时间需要执行 Join 的时候，都可以重用这个过滤器。

然后每个查询都需要 join，因为涉及 SKU 和衬衫列。从孩子查询开始，只匹配 SKU 字段：

```
BooleanQuery skuQuery = new BooleanQuery();
skuQuery.add(new TermQuery(new Term("size", "small")), Occur.MUST);
skuQuery.add(new TermQuery(new Term("color", "blue")), Occur.MUST);
```

接下来，使用 BlockJoinQuery 转换命中结果，把它们从 SKU 文档空间上升到衬衫文档空间：

```
BlockJoinQuery skuJoinQuery = new BlockJoinQuery(
   skuQuery,
   shirts,
   ScoreMode.None);
```

这里的 ScoreMode enum 类型决定应该如何为多个 SKU 命中结果打分。在这个查询中，不需要 SKU 匹配出来的分数，但是如果需要的话，可以用平均数、最大值或者求和的方法打分。

最后，可以使用 skuJoinQuery 作为子句，构建任意的衬衫查询：

```
BooleanQuery query = new BooleanQuery();
query.add(new TermQuery(new Term("name", "wolf")), Occur.MUST);
query.add(skuJoinQuery, Occur.MUST);
```

这个连接是一对多(父亲对孩子)的内连接。

3.3.16 查询大容量索引

一般情况下，索引达到 100GB 以上，搜索的响应时间可能会增加到秒级。可以使用并发多索引查询来改进性能。

```
/**
 * 并发多索引查询
 * @param index 索引目录
 * @param q 查询词
 * @return 查询结果
 */
public static Hits Multisearch(String[] index,String q) {
  int length = index.length;
  IndexSearcher[] is = new IndexSearcher[length];
  for ( int i = 0 ; i < length ; i ++){
     is[i] = new IndexSearcher(index[i]);
  }
  Searcher searcher = new ParallelMultiSearcher(is);
  Query query = QueryParser.parse(q, "temp", new StandardAnalyzer());
  Hits hits = searcher.search(query);
  return hits;
}
```

ParallelMultiSearcher()把查询任务委托给多个 IndexSearcher 执行，也就是 Searcher 类把任务交给多个同类执行。

分布式计算参考 http://www.spark-project.org。RDD 是一个由分布在各个计算节点上的数据组成的不可变的数据集合，简单来说就是一个数据集。(参考 https://github.com/zouzias/spark-lucenerdd)

固态硬盘读写速度比机械硬盘快，能够提高查找大容量索引的速度。机械硬盘在 RAID 0 阵列环境下使用也可以提高读写性能。通过玻璃中晶体和非晶体之间的相变来实现 0 和 1 之间的存储转变。

需要限制每个索引段最多只能存储 21 亿个独立的单词。超过此限制将会产生 ArrayIndexOutOfBounds 异常。

有些老师往往选学习成绩好的同学回答问题。把学习成绩好的同学调到前排，这样能更加方便地提问。分层索引就是为优质的文档开小灶，放在一个小索引中，其他的都放在大索引。

早期结束策略：如果完全执行查询耗时太长，则可以结束查询并估计结果。结束条件可以根据文档的数量或者增加时间来限制。为了让搜索结果的质量不会下降太多，可以对索引排序，把质量好的文档放在前面。

可以把 Lucene 的索引存到 HBase。首先使用内存缓存，然后再同步缓存到 HBase 后端。在 Hbase 中创建两个表用来存储索引：Index 表存储倒排索引，而 Document 表存储正排索引。HBase 中存储数据的格式为 Avro。

3.4 读写并发

在一个时刻只能够有一个线程修改索引库。Lucene 通过锁文件控制并发访问。在 Lucene 2.1 版本以后，控制写入的锁文件 write.lock 默认存储在 index 路径。

如果出现多余的锁文件，则有可能会抛出 Lock obtain timed out 异常。如果确定没有线程在修改索引，可以手工删除 write.lock 文件。

例如，当全文索引放在只读光盘中时，需要设置这个索引只读。只读索引的初始化设置如下：

```
Directory indexDir =
    FSDirectory.getDirectory(indexPath,NoLockFactory.getNoLockFactory());
```

如果一台机器用来索引的同时也用来执行搜索，就不会预热 reader，这样的后果会是灾难性的。当搜索的时候，用户会突然经历长时间的延时。因为一个大的合并能花费数小时，这样就意味着突然搜索性能变差达数小时。

因为 Java 没有暴露底层的 API，例如文件咨询信息(posix_fadvise)、内存咨询信息和对齐控制(posix_madvise)，所以要使用一个小的 JNI 扩展来改进性能。操作系统级别的功能应该能修复这个问题。

要仅仅在合并索引的时候完全忽略所有的操作系统缓存，可以通过使用 Linux 特定的 O_DIRECT 标志实现。合并索引的性能会变差，因为操作系统不再做预读也不写缓存，每个 I/O 请求都会接触到硬盘。但这可能是一个很好的折中。

有些类似的应用，例如视频解码，可能不需要缓存。因此创建了一个原型 Directory 实现，一个 DirectNIOFSDirectory 的变体。LUCENE-2056 使用 O_DIRECT 打开所有的输入和输出文件，这是通过 JNI 实现的。因为所有的 I/O 都必须用某种规则对齐，所以实现代码有点乱。

由于按顺序读，Linux 倾向于赶出已经加载的页面。

这个方法工作得很好。在执行索引优化时，搜索性能没有改变。

但是，优化阶段的时间从 1336 秒延长到了 1680 秒(慢了 26%)。可以通过增加缓存来减少优化所需的时间，或者创建自己的预读/写缓存模式。

3.5 Lucene 深入介绍

经过十多年的发展，Lucene 拥有了大量的用户和活跃的开发团队。Eclipse 软件和 Twitter 网站等都在使用 Lucene。如果说 Google 是拥有最多用户访问的搜索引擎网站，那么拥有最多开发人员支持的搜索软件项目也许是 Lucene。下面从 Lucene 开发者的角度理解 Lucene。

分析 Lucene 源代码，可以使用引用引擎，例如 http://ctags.sourceforge.net/。

3.5.1 整体结构

在 Lucene 中，org.apache.lucene.index.TermInfosReader 类的 getIndexOffset()方法实现了一个类似的折半查找。对于特别大的顺序集合可以通过采用插值法提高查找速度。

Lucene 的整体结构如图 3-14 所示。

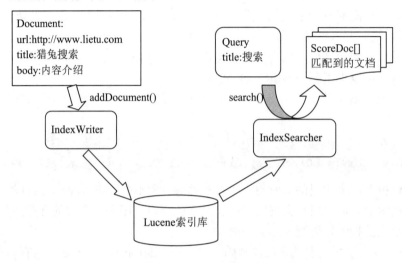

图 3-14　Lucene 原理图

可以使用 SVN 客户端从 https://svn.apache.org/repos/asf/lucene/dev/trunk/lucene/下载正在开发中的 Lucene 源代码。

Lucene 源码分为核心包和外围功能包：核心包实现搜索功能；外围功能包实现高亮显示等辅助功能。Lucene 源码的核心包中共包括 7 个子包，每个子包完成特定的功能。

最基本的是索引管理包 (org.apache.lucene.index) 和检索管理包 (org.apache.lucene.search)。索引管理包实现索引的建立、删除等。检索管理包根据查询条件，检索得到结果。

索引管理包调用数据存储管理包(org.apache.lucene.store)，主要包括一些底层的 I/O 操作。同时也会调用一些公用的算法类 (org.apache.lucene.util)。编码管理包 (org.apache.lucene.codecs) 用于方便自定义索引的编码和结构。文档结构包 (org.apache.lucene.document)用于描述索引存储时的文档结构管理，类似于关系型数据库的表结构。

查询分析器包(org.apache.lucene.queryParser)实现查询语法，支持关键词间的运算，如与、或、非等。语言分析器(org.apache.lucene.analysis)主要用于对放入索引的文档和查询词切词，支持中文主要是扩展此类。

下面先介绍索引是如何生成的，然后介绍查询原理。

3.5.2 索引原理

为了方便索引大量的文档,可以将 Lucene 中的一个索引分成若干个子索引,称为段(segment)。段中包含一些可搜索的文档。在给定的段中可以快速遍历任何给定索引词在所有文档中出现的频率和位置。IndexWriter 收集在内存中的多个文档,然后在某个时间点把这些文档写入一个新的段,写入点可以通过 Lucene 内部的配置类或者外部程序控制。这些文档组成的段将保持不动,直到 Lucene 把它合并进大的段。MergePolicy 控制 Lucene 如何合并段。

索引文档时,首先对文档分词后建立正排索引,然后建立倒排索引。在索引优化阶段,会把小的段合并成大的段。其中可能用到的算法有:合并两个排好序的数组成为一个大的排好序的数组。

为了提高性能,索引首先缓存在内存中,如果缓存达到预定的内存数量,就会写入硬盘。然而,即使 IndexWriter 从缓存中把这些文档的索引写入硬盘,在没有提交之前 IndexReader 也不能看到这些新加入的文档。如果频繁地调用 IndexWriter.commit 就会降低索引的通量。所以不要过于频繁地提交索引。可以通过测试来决定具体加入多少篇文档后再提交索引。索引文件的逻辑视图如图 3-15 所示。

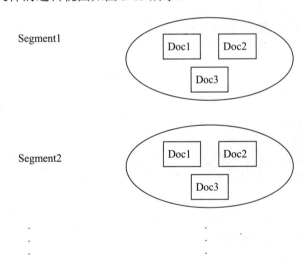

图 3-15 索引文件的逻辑视图

Lucene 把一个文档写入索引时,首先生成这个文档的倒排索引,然后再把文档的倒排索引合并到段的倒排索引中。先看一下最简单的统计一个文档中所有单词出现频次的例子:

```
//单词和对应的次数
HashMap<String, Integer> postingTable = new HashMap<String, Integer>();

StringTokenizer st = new StringTokenizer(inputStr);
while(st.hasMoreTokens()){ //按空格切分输入串
    String word = st.nextToken();
```

```
//检查单词是否在 HashMap 中
if (postingTable.containsKey(word)) {
    //取得单词出现次数, 加1后放回去
    postingTable.put(word, postingTable.get(word) + 1);
} else {
    //第一次看到这个单词, 设置次数为1次
    postingTable.put(word, 1);
}
```

描述单个文档的 Posting 表由 Posting 类组成的数组构成:

```
final class Posting {        // 一个文档中的一个词相关的信息
  Term term;                 // 词
  int freq;                  // 这个词在文档中的频率
  int[] positions;           // 词出现的位置
}
```

生成文档的倒排索引的简略代码如下:

```
final void addDocument(Document doc)
      throws IOException {
  //使用散列表缓存词和位置对应关系
  Hashtable<Term,Posting> postingTable = new Hashtable<Term,Posting>();

  //invertDocument 对文档倒排, 具体执行的工作有:
  //对文档内容分词, 然后把每个词放入 postingTable
  //postingTable.put(term, new Posting(term, position, offset));
  invertDocument(doc);

  //对 postingTable 排序后放入数组
  Posting[] postings = sortPostingTable();
}
```

首先按列读入每个文档, 然后处理同名的列。之后, 把每个新列中的数据写到缓存。

textStart 也是用来区分词的唯一值, 但是这个值太大了, 不方便存储, 所以要重新生成一个连续编号的 termID。

postingsArray 用来记录一个词之前有没有看到过。如果以前没看到过, 就调用 newTerm(termID)方法, 否则就调用 addTerm(termID)方法。

```
if(!postingsArray.textStarts.contains(textStart)){
  int termID = numPostings++;
  postingsArray.textStarts[termID] = textStart;
  consumer.newTerm(termID);
}else{
  consumer.addTerm(termID);
}
```

postingsHash 数组用来加快这个判断过程。可以把它看成散列表中的数组表。并行数组

postingsArray 中的 textStarts 则用来判断是否包含当前词。

在 reset()方法中，把 postingsHash 中的初始值设置成-1。

```
Arrays.fill(postingsHash, 0, numPostings, -1);
```

初始化的原理代码如下：

```
int postingsHashSize = 4;
int[] postingsHash = new int[postingsHashSize];
int postingsHashMask = postingsHashSize-1;
```

根据词所在的位置来找槽的位置。值是词编号。使用二次探测再散列法解决冲突。

```
int code = textStart;

int hashPos = code & postingsHashMask;

// 定位散列表中的 RawPostingList
int termID = postingsHash[hashPos];

if (termID != -1 && postingsArray.textStarts[termID] != textStart) {
  // Conflict: keep searching different locations in
  // the hash table.
  final int inc = ((code>>8)+code)|1;
  do {
    code += inc;
    hashPos = code & postingsHashMask;
    termID = postingsHash[hashPos];
  } while (termID != -1 && postingsArray.textStarts[termID] != textStart);
}
```

如果发现一个词没有在 postingsHash 存过，就在 postingsHash 中记录这个词的编号。

```
int numPostings = 0;
int termID = numPostings++;

int code = 100;  //词所在的位置
int hashPos = code & postingsHashMask;
postingsHash[hashPos] = termID;
```

然后压缩掉没有用到的位置。

```
private void compactPostings() {
  int upto = 0;  //记录已经填实的位置
  for(int i=0;i<postingsHashSize;i++) { //从前往后遍历postingsHash中的每一个值
    if (postingsHash[i] != -1) {  //如果不是初始值
      if (upto < i) {  //而且中间有空位
        postingsHash[upto] = postingsHash[i];  //压缩
        postingsHash[i] = -1;  //标志这个位置已经空了
      }
      upto++;  //已经填实的位置标志加1
```

```
    }
  }
}
```

得到的是一个词编号序列。

倒排索引在 FreqProxTermsWriter 类的 appendPostings()方法中创建，基于从文档中统计出的信息，例如词频、文档频率、词位置等。

DocumentsWriter 中包含一个索引链。

Consumer 是一个定义接口的抽象类。索引时，在不同的层次，有不同的 Consumer 类。例如 DocConsumer 处理整个文档。DocFieldConsumer 处理不同的列，每次处理一个。

InvertedDocConsumer 消耗产生的词序列。

TermsHashConsumer 写和它自己相关的字节到内存中的 posting lists(投递列表)。posting lists 以字节切片存储，按词索引。

DocumentsWriter*.java 更简单了，它只和 DocConsumer 打交道，并不知道 Consumer 在做什么。

DocConsumer 下面是做实际工作的索引链。

- NormsWriter 在内存中记录归一化信息，然后把这些信息写入_X.nrm。
- FreqProxTermsWriter 在内存中记录 postings 数据，然后写入_X.frq/prx。
- StoredFieldsWriter 负责写入列存储的值信息，把内存中的值信息写入_X.fdx/fdt。
- TermVectorsTermsWriter 把内存中的词向量信息写入_X.tvx/tvf/tvd。

DocumentsWriter 虽然没有做具体的事情，但是仍然管理一些事情，例如清空一个段，关闭文档储存，缓存和使删除生效，释放内存，当需要的时候中止整个过程，等等。

DocInverterPerField 得到位置信息。

FieldInvertState 记录要加入索引的词数量。通过 FieldInvertState 的 attributeSource 属性取得最后坐标，也就是字段长度。

为了让 posting list 中的文档序列压缩后更小，可以把相似的文档聚类，让前后两个文档的编号值尽可能小。

3.5.3 文档值

FieldCache 是索引的正向视图，也就是文档编号→值的视图。FieldCache 需要从倒排索引中解析出值，如图 3-16 所示。

可以轻松地使用 FieldCache 加载某一特定列的本地值数组。例如，如果每个文档有一个字段为 weight，也就是 PageRank 值列。

```
Field f = new Field("weight", doc.pageRank ,
    Field.Store.YES, Field.Index.UN_TOKENIZED,
    Field.TermVector.NO);
```

则可以这样得到所有文档的权重：

```
float[] weights = FieldCache.DEFAULT.getFloats(reader, "weight");
```

第 3 章　Lucene 的原理与应用

然后，每当需要知道一个文档的权重值时，只需参考 weights[docID]。

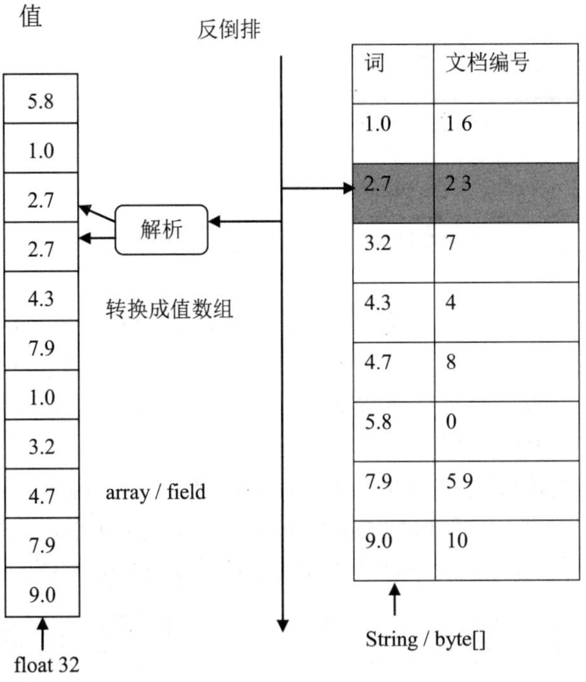

图 3-16　从倒排索引解析出值

FieldCache 支持许多本地类型：字节、短整型、长整型、浮点数、双精度和 StringIndex 类，其中包括字符串值的排序顺序。

第一次访问给定的 reader 和列的 field cache 时，将访问所有文档的值，并作为一个独立的大数组加载到内存中。值记录到一个以 reader 实例和列名作为键的内部缓存中。因为要从倒排索引中解析出值，所以对大的索引来说，加载过程很耗时。

IndexDocValues 不需要从倒排索引中解析出值，因为在建立索引时提供了一个文档到值的映射。IndexDocValues 允许在文档索引时，可以更多地控制数据。每一个普通的 Lucene 列接收一个类型的值，以列的方式存储。通过 ValueType 指定某个列的 IndexDocValues 类型，例如长整型、浮点型或者字节数组。

对整数类型，IndexDocValues 提供字节对齐的变体，如 8, 16, 32 和 64 位，以及压缩的 PackedInts。对于浮点值，当前只提供 float 32 和 float 64。但是对于字节数组值，IndexDocValues 提供更灵活的存取方式。

可以指定值是否有固定或可变的长度，这些值是否应直接存储，或以取消引用的方式存储，这样在不同值的数很少的情况下可以得到良好的压缩。

因为不需要从倒排索引中解倒排或者解析值，所以 IndexDocValues 的加载速度很快。

FieldCache 只能完全位于内存中，而 IndexDocValues 可以完全在内存或者位于硬盘。IndexDocValues 提供同样的随机读出接口。

读出数据的性能依赖于值的类型。如果选择使用 PackedInts 压缩整数，就要付出代价。

但是如果选择 32 位对齐的变体，则就好像 Java NIO ByteBuffers 一样访问底层数组。通过 Source.getArray()方法访问数据。

```
IndexReader r = IndexReader.open(w, true);
DocValues docValues = MultiDocValues.getDocValues(r, field);
Source source = docValues.getSource();

byte[] values = (byte[]) source.getArray();
```

有很多应用可以受益于 IndexDocValues，例如灵活的评分、分类统计、排序、合并和结果分组这样的搜索功能。

一般来说，当有像排序、分类统计和分组结果这样的功能要实现时，通常需要额外的专用列。以作者列为例，可以为了全文检索而对这个列分词，但这可能会导致分类统计或分组结果出现意外。然而，只需简单地添加一个非分词列就能解决这个问题。非分词列往往用 facet_author 这样的前缀命名。

IndexDocValues 是不分词的，所以可以简单地增加一个同名的 IndexDocValuesField 来达到和额外列同样的效果。使用 IndexDocValues 后，就能用更少的逻辑列实现同样的功能。

默认 Lucene 使用 FieldCache，除非声明 SortField.setUseIndexValues(boolean)。

```
IndexSearcher searcher = new IndexSearcher(...);
SortField field = new SortField("myField", SortField.Type.STRING);
field.setUseIndexValues(true);  //使用 IndexValues 实现排序
Sort sort = new Sort(field);
TopDocs topDocs = searcher.search(query, sort);
```

3.5.4　FST

FST 指有限状态转换机，根据输入的字符串返回文档列表存储的位置。CharsRef 表示一个字符引用。作为 Map 来用的 FST 示例代码如下：

```
String base1 = "fast";
String syn1 = "rapid";
String base2 = "slow";
String syn2 = "sluggish";

SynonymMap.Builder sb = new SynonymMap.Builder(true);
sb.add(new CharsRef(base1), new CharsRef(syn1), true);
sb.add(new CharsRef(base2), new CharsRef(syn2), true);
SynonymMap smap = sb.build();
```

3.6　查询语法与解析

Lucene 不仅提供了 API 来自己创建查询，而且还通过查询分析器提供了丰富的查询语言。

第 3 章 Lucene 的原理与应用

- 一个短语查询可以用双引号括起来，这样只有精确匹配该短语的文档才会匹配查询条件。比如，搜索 "上海世博会" 只会出现包含连续出现 "上海" 和 "世博会" 的文档。如果 "上海" 和 "世博会" 之间有其他词，则不会匹配这样的情况。所以搜索 "上海世博会" 比 "上海 AND 世博会" 这样的查询返回的结果更少。
- 使用^表示加权。例如搜索 "solr^4 lucene"。
- 修饰符 "+" "-" "NOT"。例如搜索 "+solr lucene"。
- 布尔操作符 "OR" "AND"。例如搜索 "(solr OR lucene) AND user"。注意这里的 AND 必须是大写的，如果是小写，即 and，就会被当作普通查询词看待了。
- 按域查询。一个字段名后面跟冒号，再加上要搜索的词语或短句，就可以把搜索条件限制在该字段。例如搜索 "title:NBA"，匹配标题包含 NBA 的文档。

QueryParser 将输入查询字串解析为 Lucene 的 Query 对象，如图 3-17 所示。

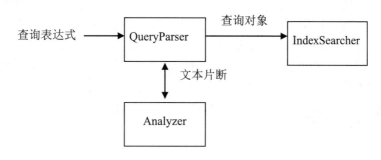

图 3-17 使用查询表达式搜索

下面两个写法在功能上是等价的：

```
QueryParser parser = new QueryParser("name", wrapper);
Query query = parser.parse("size:small AND color:blue");
```

与

```
BooleanQuery skuQuery = new BooleanQuery();
skuQuery.add(new TermQuery(new Term("size", "small")), Occur.MUST);
skuQuery.add(new TermQuery(new Term("color", "blue")), Occur.MUST);
```

可以检查返回的结果：

```
QueryParser parser = new QueryParser("title",
                analyzer);
Query query = parser.parse("Monitor");
System.out.println(query.getClass().getName());
//org.apache.lucene.search.TermQuery
```

这表示 QueryParser.parse()方法返回的是一个简单的 TermQuery。
直接这样写可能找不到结果：

```
Query query = new QueryParser("url", analyzer).parse(url);
```

需要把 URL 地址中的特殊字符当作普通字符看待。

```
Query query = new QueryParser("url",
                    analyzer).newTermQuery(
                        new Term("url", url)).parse(url);
```

有一些特殊字符，如单引号、转义符等符号，怎么让 Lucene 忽略它们，而把它们当作一般的字符串去查询？方法是使用"\\"转义或者直接调用 org.apache.lucene.queryparser.classic.QueryParserBase.escape(String)方法。

3.6.1 JavaCC

查询解析器用到的词法分析不深入处理普通查询串，而只是处理像:和*这样的特殊字符，以及像 AND 和 OR 这样的特殊单词。

查询分析一般用两步实现：词法分析和语法分析。词法分析阶段根据用户的输入返回单词符号序列，而语法分析阶段则根据单词符号序列返回需要的查询串。

词法分析的功能是从左到右扫描用户输入的查询串，从而识别出查询词、运算符、数值等单词符号，把识别结果返回到语法分析器，以供语法分析器使用。输入的是用户查询串，输出的是单词符号串的识别结果。例如，定义词的类型如下：

```
public enum TokenType{
    AND,    //与
    OR,     //或
    NOT,    //非
    PLUS,   //加
    MINUS,  //减
    LPAREN, //左括号
    RPAREN, //右括号
    COLON,  //冒号
    TREM    //词
}
```

对于如下的输入片断：

```
title:car site:http://www.lietu.com
```

词法分析的输出可能是：

```
TREM title
COLON :
TREM car
TREM site
COLON :
TREM http://www.lietu.com
```

可以把词法分析的结果进一步转换成语法树，如图 3-18 所示。

第3章 Lucene 的原理与应用

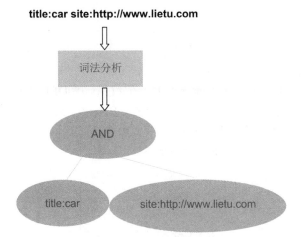

图 3-18 从词法分析到语法树

如果只满足于用空格分开多个词,那么 Java 的 StringTokenizer 对于解析任务就绰绰有余了。因为直接写 Lucene 这样的查询语法的分析器代码很麻烦,所以可以使用 JavaCC 生成查询解析器源代码。

QueryParser 是使用 JavaCC 生成的词法和语法解析器,它把字符串解释成一个 Query 对象。文本文件 QueryParser.jj 中定义了如何做查询串的词法分析和语法解析。JavaCC(Java Compiler Compiler)是一个软件工具,用于把 QueryParser.jj 转换成 Java 源代码。JavaCC 的下载地址为 http://javacc.java.net/。解压后执行 javacc.jar 中的 javacc 类。

```
java -cp javacc.jar javacc QueryParser.jj
```

JavaCC 首先把用户输入定义的 Token 转换成为与正规文法等价的形式,然后把正规文法转换成 NFA,再把 NFA 转换成 DFA,最后生成代码模拟 DFA,如图 3-19 所示。

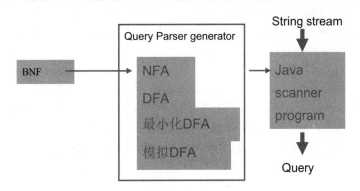

图 3-19 JavaCC 的原理

词法分析器采用有限状态机实现。有限状态机的状态在这里称为词法状态。至少会生成一个 DEFAULT 状态,还可以自己定义一些词法状态。

当初始化词法分析器时,它从 DEFAULT 状态开始。当构造一个 TokenManager 对象时,也可以通过一个参数指定开始状态。

```
public QueryParserTokenManager(CharStream stream, int lexState){
  this(stream);
  SwitchTo(lexState);
}
```

例如，匹配权重，也就是一个上箭头加上一个数字。用 DEFAULT 和 Boost 两个状态实现。在 DEFAULT 状态接收一个上箭头以后就进入 Boost 状态，在 Boost 状态接收数值串后回到 DEFAULT 状态，对应的有限状态机如图 3-20 所示。

例如，输入"^19"，词法分析的输出如下：

```
CARAT ^
NUMBER 19
```

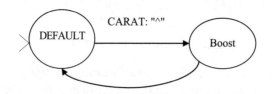

图 3-20 有限状态机

用若干个正则表达式产生式定义词法分析。有 4 种类型的产生式：SKIP、MORE、TOKEN 和 SPECIAL_TOKEN。产生式的分类依据是：成功匹配一个正则表达式后，接下来做什么。4 种产生式的动作分别说明如下。

- SKIP：执行指定的词法动作后，扔掉这个匹配的字符串。
- MORE：无论接下来的状态是什么，都继续拿着匹配的字符串，当前匹配的字符串将会成一个新的匹配的字符串的前缀。
- TOKEN：用匹配的字符串创建一个新的 Token，并且返回给解析器或者调用者。
- SPECIAL_TOKEN：创建一个不参与分析的特殊 Token。

例如，SKIP 类型的产生式跳过空格/制表符/新行符：

```
SKIP : { " " | "\t" | "\n" }
```

TOKEN 类型的正则表达式表示会用匹配的字符串创建一个新的 Token 并且返回给解析器。查询解析器只用到了 SKIP 和 TOKEN 两种类型的正则表达式。

例如，用两个 TOKEN 类型的正则表达式产生式实现这个词法分析功能：

```
<DEFAULT> TOKEN : {
  <CARAT:      "^" > : Boost
}

<Boost> TOKEN : {
  <NUMBER:    (<_NUM_CHAR>)+ ( "." (<_NUM_CHAR>)+ )? > : DEFAULT
}
```

状态名称成为 QueryParserConstants 类中的整数：

```
public interface QueryParserConstants {
 /** 词法状态 */
 int Boost = 0; //加权
 int Range = 1; //区间
 int DEFAULT = 2; //默认状态
}
```

但是 QueryParser 并没有用到 QueryParserConstants 类中的这些常量，还存在其他类似这样的无用代码。

词法状态的列表描述了相应的正则表达式产生式适用的词法状态集合。如果写成"<*>"表示适用于所有的词法状态。例如，一个 TOKEN 类型的产生式：

```
<*> TOKEN : {
 <#_NUM_CHAR:   ["0"-"9"] >
| <#_ESCAPED_CHAR: "\\" ~[] >
| <#_TERM_START_CHAR: ( ~[ " ", "\t", "\n", "\r", "\u3000", "+", "-", "!",
"(", ")", ":", "^",
                       "[", "]", "\"", "{", "}", "~", "*", "?", "\\", "/" ]
                     | <_ESCAPED_CHAR> ) >
| <#_TERM_CHAR: ( <_TERM_START_CHAR> | <_ESCAPED_CHAR> | "-" | "+" ) >
| <#_WHITESPACE: ( " " | "\t" | "\n" | "\r" | "\u3000") >
| <#_QUOTED_CHAR: ( ~[ "\"", "\\" ] | <_ESCAPED_CHAR> ) >
}
```

在标签 _WHITESPACE 之前的"#"表示它存在只是为了定义其他的 Token。实际上，执行查询解析器的时候不会进入这个<*>状态。

当在<Range>状态看到"]"时就切换到 DEFAULT 状态。

```
<Range> TOKEN : {
<RANGE_TO: "TO">
| <RANGEIN_END:  "]"> : DEFAULT
| <RANGEEX_END:  "}"> : DEFAULT
| <RANGE_QUOTED: "\"" (~["\""] | "\\\"")+ "\"">
| <RANGE_GOOP: (~[ " ", "]", "}" ])+ >
}
```

文法由若干个产生式组成，每个产生式对应若干个词法状态，词法状态的集合唯一定义这个产生式。例如，<DEFAULT, Range>产生式包含 DEFAULT 和 Range 两个词法状态。可以接收的字符串用正则表达式定义。每个正则表达式都可以有一个名字，例如一个叫作 AND 的正则表达式匹配"AND"或者"&&"：

```
<AND:        ("AND" | "&&") >
```

这样，词法分析返回的 Token 的类型就是 QueryParserConstants.AND。

如果可以接收多个正则表达式，则用"|"隔开。例如，匹配布尔逻辑表达式：

```
<DEFAULT> TOKEN : {
 <AND:        ("AND" | "&&") >
```

```
| <OR:      ("OR" | "||") >
| <NOT:     ("NOT" | "!") >
| <PLUS:    "+" >
| <MINUS:   "-" >
}
```

这里是一个 TOKEN 类型的产生式。

正则表达式产生式首先声明词法状态名称集合，然后是正则表达式类型，最后是正则表达式本身。例如：

```
<DEFAULT, Range> SKIP : {
  < <_WHITESPACE>>
}
```

表示在 DEFAULT 和 Range 状态扔掉匹配的空格类型的字符串。

如果把查询看成是一个简单的语言，则可以用巴科斯范式(BNF)定义查询语言的文法。

```
Query ::= ( Clause )*
Clause ::= ["+", "-"] [<TERM> ":"] ( <TERM> | "(" Query ")")
```

这里的中括号是可选的意思。Clause 中的"+""-"和<TERM>":"可以出现，也可以不出现。

- ?：是指操作符左边的符号(或括号中的一组符号)是可选项(可以出现 0 到 1 次)。
- *：是指可以出现 0 到多次。
- +：是指可以出现 1 到多次。

例如，定义一个网站域名：

```
(<_TERM_CHAR>)+ ( "." (<_TERM_CHAR>)+ )+
```

增加一个 Website 词法状态：

```
<Website> TOKEN:{
<SITEURL:   ("http://")? <_TERM_START_CHAR> (<_TERM_CHAR>)* ( "." (<_TERM_CHAR>)+ )+> : DEFAULT
}
```

增加从 DEFAULT 词法状态转移到 Website 词法状态：

```
<DEFAULT> TOKEN : {
  <SITE:    ("SITE" | "site") <COLON>> : Website
}
```

这个文法中的每个规则都是一个产生式。每个产生式由左边的一个符号和右边的多个符号组成，用右边的多个符号来描述左边的一个符号。符号有终结符和非终结符两种。这里的终结符是"+"和"-"等，非终结符是子句和 Query。例如，对于查询表达式："site:lietu.com"，可以表示成<TERM>":"<TERM>的形式，最终归约成一个有效的 Query。但是这里的 site 是一个保留列名。

JavaCC 把非终结符对应成 Java 中的方法。例如，非终结符 Clause 对应 Clause()方法。

JavaCC 的.jj 文件文法和标准的巴科斯范式的区别是：在 JavaCC 中，可以在每步推导的过程中执行一些相关的 Java 语句，在编译原理中，把这些语句叫作动作。动作写在{}中。例如，下面是QueryParser.jj 中定义的 BNF 类型的产生式：

```
int Conjunction() : {
  int ret = CONJ_NONE;
}
{
  [
    <AND> { ret = CONJ_AND; }
    | <OR> { ret = CONJ_OR; }
  ]
  { return ret; }
}
```

等价于：

```
Conjunction ::= [<AND> | <OR>]
```

这里的中括号是可选的意思。如果 AND 和 OR 都不出现，则会有个默认的布尔逻辑。

可以在产生式中指定在选择点做选择前向前看的 Token 数量，这个值称为 LOOKAHEAD。LOOKAHEAD 的默认值是 1。例如，通过 LOOKAHEAD(2)指定值是 2。

为了同时匹配"car"和"title:car"以及"*:car"这样的查询，选择前向前看 2 个 Token。只有看到第二个 Token 是否为冒号，才能知道第一个 Token 是不是列名。

```
Clause ::= [ LOOKAHEAD(2) ( <TERM> <COLON> | <STAR> <COLON> ) ] Term
```

LOOKAHEAD 的值越小，则解析器的速度越快。

JavaCC 是一个自顶向下的解析器生成器。JavaCC 可以用于词法分析和语法分析。JavaCC 根据输入文件中定义的语法生成实际用于词法分析和语法分析的程序源代码，然后 Lucene 会调用其中的功能，如图 3-21 所示。

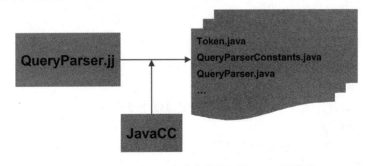

图 3-21　用 JavaCC 生成查询解析器 Java 源代码

JavaCC 生成的词法分析器叫作 TokenManager，其中的 TokenManager. getNextToken()方法返回下一个 Token。QueryParser.jj 生成的类 QueryParserTokenManager 就是词法分析器。而 QueryParser 则是语法分析。

和词法分析相关的几个类介绍如下。

- CharStream：字符流接口；
- FastCharStream：Lucene 按照 JavaCC 的要求提供的 CharStream 实现类；
- QueryParserConstants：定义 Token 类型常量，也就是正则表达式编号；
- QueryParserTokenManager：词法分析器；
- Token：词和类型；
- TokenMgrError：词法分析中出现的错误。

Token 类用于说明词对应的字符串和词的类型。

```
public class Token {
  public int kind; //Token 类型
  public String image; //Token 对应的字符串
}
```

测试词法分析器：

```
String strQuery = " title:car site:http://www.lietu.com";  //查询串
QueryParserTokenManager tokenManager = new QueryParserTokenManager(
        new FastCharStream(new StringReader(strQuery)));

for (Token next = tokenManager.getNextToken(); !next.toString().equals("");
        next = tokenManager.getNextToken())
  System.out.println("'" + next.image + "' "+ next.kind); //输出词本身和类别
```

输出结果：

```
'title' 20
':' 16
'car' 20
'site' 20
':' 16
'http' 20
':' 16
'//' 24
'www.lietu.com' 20
```

输出结果说明 QueryParserTokenManager 没有识别网址的功能，但可以处理"car AND price"这样的查询串中的逻辑运算符。

要在 QueryParser.jj 文件中定义语言，需要做以下 5 件事：

(1) 定义解析环境。
(2) 定义"空白"：例如在中文搜索中，全角空格也可以算空白。
(3) 定义 Token。
(4) 按照 Token 定义语言本身的语法。
(5) 定义每个解析阶段将发生的行为。

STATIC 是一个布尔选项，默认值是真。如果是真，则在生成的解析器和 Token 管理器中，所有的方法和类变量都声明成静态的。这样仅仅允许一个解析对象存在，但是查询分析器应该可以同时有很多个实例，所以这个值应该设成假。

JavaCC 的输入文法文件.jj 的格式：

```
javacc_options
PARSER_BEGIN ( <IDENTIFIER>1 )
    java 编译单元
PARSER_END ( <IDENTIFIER>2 )
( 产生式 )*
```

产生式有如下 4 种。

(1) Java 代码产生式：这类产生式以保留字 JAVACODE 开始。

(2) 正则表达式产生式：QueryParser.jj 定义了 5 个正则表达式产生式。

(3) BNF 产生式：写非终结符就像定义一个 Java 方法。因为会把每个非终结符都翻译成一个方法。QueryParser.jj 定义了 6 个 BNF 产生式，分别是 Conjunction、Modifiers、TopLevelQuery、Query、Clause 和 Term。

(4) 词法声明产生式：词法分析声明以保留字 TOKEN_MGR_DECLS 开始，后跟一个":"，然后是一组 Java 的声明和语句(Java 块)。把这些声明和语句写入生成的 TokenManager，并且在词法动作内可以访问到。QueryParser.jj 并没有定义这类产生式。

QueryParser.jj 的开始处是选项：

```
options {
  STATIC=false; //非静态
  //把 QueryParser.jj 文件中包含的 UNICODE 转义字符\u3000 替换成实际的 unicode 字符
  JAVA_UNICODE_ESCAPE=true;
  //为 Lucene 自己实现的 FastCharStream 生成 CharStream.java 接口
  USER_CHAR_STREAM=true;
}
```

然后是一些嵌入的 Java 代码，位于 PARSER_BEGIN(QueryParser)和 PARSER_END(QueryParser)之间。最后是交给 JavaCC 处理的文法定义。

import 声明复制到 QueryParser.java 和 QueryParserTokenManager.java 文件。

FastCharStream 需要实现的 JavaCC 生成的 CharStream 中的主要方法有：

```
public interface CharStream {
  // 返回下一个字符
  char readChar() throws java.io.IOException;

  //标志 Token 开始
  char BeginToken() throws java.io.IOException;

  //返回从 Token 开始处到当前位置的字符串
  String GetImage();
}
```

JavaCC 会生成 QueryParser(CharStream stream)和 ReInit(CharStream stream)方法：

```
public class QueryParser extends QueryParserBase {
  public QueryParserTokenManager token_source; // 生成的词法分析器
```

```
public Token token; // 当前 Token
public Token jj_nt; // 下一个 Token

// 根据用户提供的输入串构造解析器
protected QueryParser(CharStream stream) {
  //根据查询串构造词法分析器
  token_source = new QueryParserTokenManager(stream);
  token = new Token();
  //内部状态处理
}

// 重新初始化
public void ReInit(CharStream stream) {
  token_source.ReInit(stream);
  token = new Token();
  //内部状态处理
}
}
```

匹配的字符串存放在一个叫作 image 的变量中，匹配以后执行的动作可以访问到这个变量。例如：

```
<DEFAULT> TOKEN : {
  <AND:       ("AND" | "&&") >
| <OR:        ("OR" | "||") >
| <NOT:       ("NOT" | "!") >
| <PLUS:      "+" >
| <MINUS:     "-" >
| <BAREOPER:  ("+"|"-"|"!") <_WHITESPACE> >
| <LPAREN:    "(" >
| <RPAREN:    ")" >
| <COLON:     ":" >
| <STAR:      "*" >
| <CARAT:     "^" > : Boost
| <QUOTED:    "\"" (<_QUOTED_CHAR>)* "\"">
| <TERM:      <_TERM_START_CHAR> (<_TERM_CHAR>)* >
| <FUZZY_SLOP:    "~" ( (<_NUM_CHAR>)+ ( "." (<_NUM_CHAR>)+ )? )? >
| <PREFIXTERM: ("*") | ( <_TERM_START_CHAR> (<_TERM_CHAR>)* "*" ) >
| <WILDTERM: (<_TERM_START_CHAR> | [ "*", "?" ]) (<_TERM_CHAR> | ( [ "*",
"?" ] ))* >
| <REGEXPTERM: "/" (~[ "/" ] | "\\/" )* "/" >
| <RANGEIN_START: "[" > : Range
| <RANGEEX_START: "{" > : Range
}
```

匹配以后的动作是：取匹配的字符串中的第一个字符。

```
term=<BAREOPER> { term.image = term.image.substring(0,1); }
```

第 3 章 Lucene 的原理与应用

6 个 BNF 表达式如下：

```
Conjunction ::= [<AND> | <OR>]
Modifiers ::= [<PLUS> | <MINUS> | <NOT>]
TopLevelQuery ::= Query <EOF>
Query ::= Modifiers Clause ( Conjunction Modifiers Clause )*
Clause ::= [ LOOKAHEAD(2) ( <TERM> <COLON> | <STAR> <COLON> ) ] ( Term |
<LPAREN> Query <RPAREN> (<CARAT> <NUMBER>)? )
Term ::=
  (
    ( <TERM> | <STAR> | <PREFIXTERM> | <WILDTERM> | term=<REGEXPTERM> |
<NUMBER> | <BAREOPER> )
    [ <FUZZY_SLOP> ]
    [ <CARAT> <NUMBER> [ <FUZZY_SLOP> ] ]
    | ( ( <RANGEIN_START> | <RANGEEX_START> )
      ( <RANGE_GOOP>|<RANGE_QUOTED> )
      [ <RANGE_TO> ]
      ( <RANGE_GOOP>|<RANGE_QUOTED> )
      ( <RANGEIN_END> | <RANGEEX_END>))
    [ <CARAT> <NUMBER> ]
    | <QUOTED>
    [ <FUZZY_SLOP> ]
    [ <CARAT> <NUMBER> ]
  )
```

根据 QueryParser.jj 生成代码，在命令行运行：

```
>javacc QueryParser.jj
```

jj_consume_token()方法接收一个 Token 类型作为参数，试着从词法分析器得到一个指定类型的 Token。JavaCC 总是使用 0 编码 EOF 类别的 Token，所以 jj_consume_token(0) 从词法分析器得到一个 EOF 类别的 Token。JavaCC 把 QueryParser.jj 中的代码：

```
Query TopLevelQuery(String field) :
{
   Query q;
}
{
   q=Query(field) <EOF>
   {
      return q;
   }
}
```

转变成：

```
final public Query TopLevelQuery(String field) throws ParseException {
    Query q;
  q = Query(field); //自动生成的代码
  jj_consume_token(0); //自动生成的代码
```

```
            {if (true) return q;}
    throw new Error("Missing return statement in function");
  }
```

QueryParser 调用 QueryParserBase 中实现的 handleBareTokenQuery() 和 handleQuotedTerm()等方法生成查询对象。这些调用写在 QueryParser.jj 中。

3.6.2 生成一个查询解析器

根据现有的代码生成一个 QueryParser。首先找到 src\org\apache\lucene\queryparser\surround 目录下的 QueryParser.jj，然后生成 Java 源代码。把 lucene-core-6.3.0.jar 导入这个项目，src\org\apache\lucene\queryparser\surround\query 下的代码也导入这个项目。最后测试。

```
String query="A OR B ";
SrndQuery q = QueryParser.parse(query);
System.out.println(q);
```

3.6.3 简单的查询解析器

QueryParser 使用 Analyzer 对"title:car"中的查询词 car 再次处理成为 Query 对象。

如果要修改 QueryParser 的行为，可以继承 QueryParser 类，重写其中的 getFieldQuery 方法。例如，同时查询多个列的 MultiFieldQueryParser。

要改变查询分析器的某一部分，例如查询实例化，可以通过继承解析器类来实现，改变实际的查询语法需要深入地了解 JavaCC 解析器生成器。

使用 MultiFieldQueryParser 查询多个列：

```
MultiFieldQueryParser mq
    = new MultiFieldQueryParser(new String[]{"title","body"},
                analyzer);
```

在 QueryParser 传入一个字符串的地方 MultiFieldQueryParser 传入一个字符串数组。

3.6.4 灵活的查询解析器

为了实现更好的查询分析功能，可以分离一个查询的语法和语义。例如，"a AND b"，"+a +b"，"AND(a,b)"是同一个查询的不同的语法。区分不同查询组件的语义，例如，是否符号化/原型化/正规化不同的词，并且如何做到这些。或者对于词用哪个查询对象来创建。需要能够尽可能快地用新的语法编写一个解析器，重用底层的语义。

为了让查询解析器的实现更加灵活，把文本解析和查询构建独立出来。文本解析阶段返回一个 QueryNodeTree，它是一棵最初表示原始查询的语法的树。例如，"a AND b"的 QueryNodeTree 如图 3-22 所示。

这样查询解析至少有两个阶段：文本解析阶段和查询构建阶段。解析器需要实现 SyntaxParser 接口：

```
public interface SyntaxParser {
  public QueryNode parse(CharSequence query, CharSequence field);
}
```

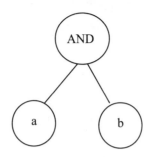

图 3-22　QueryNodeTree

这里的 QueryNode 对象包含分析出来的语法树。

+DisNey WOrld 的解析过程以及生成的语法树如图 3-23 所示。

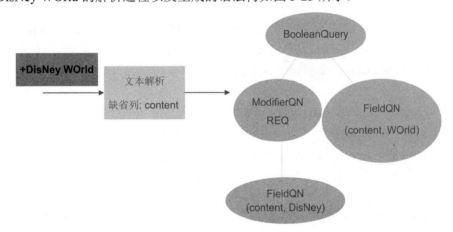

图 3-23　解析查询

查询解析器完整的三层具体介绍如下。

（1）QueryParser：最上层是解析层，简单地把查询字符串转换成一个 QueryNodeTree。当前使用 JavaCC 实现这一层。

（2）QueryNodeProcessor：查询节点处理器做了大部分工作。它实际上是一个可配置的处理器链。每一个处理器可以遍历树和修改节点甚至修改树的结构。这样就有可能在查询执行以前做查询优化或者把字符串分隔成词。

（3）QueryBuilder：第 3 层也是一个构造器的可配置的链，把 QueryNodeTree 转换成 Lucene 的 Query 对象。

使用 StandardQueryParser 得到查询对象：

```
StandardQueryParser queryParserHelper = new StandardQueryParser();
Query query = queryParserHelper.parse("a AND b", "defaultField");
```

得到解析树：

```
SyntaxParser sp = new StandardSyntaxParser();
QueryNode qTree = sp.parse("a AND b", "defaultField");
System.out.println(qTree);
```

多列查询：

```
StandardQueryParser queryParserHelper = new StandardQueryParser();
String[] fields = {"title","body"};
queryParserHelper.setMultiFields(fields);
Query query = queryParserHelper.parse("a AND b", null);
System.out.println(query);
```

StandardQueryParser 中的实现代码展示了对这 3 个阶段的调用。

```
QueryNode queryTree = this.syntaxParser.parse(query, getField());
queryTree = this.processorPipeline.process(queryTree);
return (Query) this.builder.build(queryTree);
```

灵活的查询解析器的主要实现代码包括接口定义和具体实现两部分：在包 org.apache.lucene.queryParser.core 中定义查询解析器 API 类，需要通过查询解析器实现扩展；在包 org.apache.lucene.queryParser.standard 中包含使用新的查询解析器 API 的查询解析器实现。

JavaCC 根据 StandardSyntaxParser.jj 转换出源代码：

```
Token.java
StandardSyntaxParserConstants.java
StandardSyntaxParser.java
...
```

StandardSyntaxParser.jj 中定义的 parse()方法定义了对用户查询串的词法分析功能，并完成初步的语法分析。例如，StandardSyntaxParser.jj 中对子句的定义：

```
QueryNode Clause(CharSequence field) : {
  QueryNode q; //定义局域变量
  Token fieldToken=null, boost=null;
  boolean group = false;
}
{
  [
    LOOKAHEAD(2) //向前看两个 Token
    (
    fieldToken=<TERM> <COLON>    //修改列名
{field=EscapeQuerySyntaxImpl.discardEscapeChar(fieldToken.image);}
    )
  ]

  (
  q=Term(field)
```

```
    | <LPAREN> q=Query(field) <RPAREN> (<CARAT> boost=<NUMBER>)? {group=true;}
//如果碰到 ^ 字符则对查询加权
    )
    {
      if (boost != null) {
          float f = Float.valueOf(boost.image).floatValue();
          q = new BoostQueryNode(q, f);
      }
      if (group) { q = new GroupQueryNode(q);}
      return q;  //返回生成的 QueryNode
    }
}
```

QueryBuilder 把 QueryNodeTree 转换成 Lucene 的 Query 对象。QueryTreeBuilder 实现了 QueryBuilder 接口。可以在 QueryTreeBuilder 中为每类节点指定一个 QueryBuilder。

```
public class QueryTreeBuilder implements QueryBuilder {
  private HashMap<Class<? extends QueryNode>, QueryBuilder>
queryNodeBuilders;
}
```

QueryConfigHandler 用来保留查询配置信息。例如设置默认操作符为 AND。

```
QueryConfigHandler config = new StandardQueryConfigHandler();
config.addAttribute(DefaultOperatorAttribute.class).setOperator(
            DefaultOperatorAttribute.Operator.AND);
```

使用 StandardQueryParser 的代码：

```
StandardQueryParser qpHelper = new StandardQueryParser();
StandardQueryConfigHandler config = qpHelper.getQueryConfigHandler();
config.setAllowLeadingWildcard(true); //允许通配符出现在开头
config.setAnalyzer(new WhitespaceAnalyzer());
Query query = qpHelper.parse("apache AND lucene", "defaultField");
```

使用 AnalyzerAttribute 属性确认 QueryConfigHandler 需要定义 AnalyzerQueryNodeProcessor 属性。

StandardSyntaxParser 知道优先级规则，但是 QueryParser 不知道。例如，QueryParser 把 A AND B OR C AND D 处理成和+A +B +C +D 一样。

FieldQueryNode 表示一个包含<列, 查询文本>元组的元素。如果已经定义了 AnalyzerQueryNodeProcessor 属性，而且 analyzer 存在，就把 analyzer 应用到查询节点树中的每个 FieldQueryNode 对象，条件是 FieldQueryNode 对象不是 WildcardQueryNode、FuzzyQueryNode 或者 ParametricQueryNode。

StandardQueryBuilder 接口只定义了一个返回 Query 对象的 build(QueryNode)方法。要想从 QueryNode 对象构建 Query 对象的类，都要实现 StandardQueryBuilder 接口。

```
public static class PayloadQueryNodeBuilder implements StandardQueryBuilder
{
```

```
    private PayloadFunction function; //如何利用分值
    private boolean includeSpanScore; //包含 Span 分值

    /**
     * 用默认的选项创建一个 PayloadQueryNodeBuilder
     * PayloadFunction 默认是 AveragePayloadFunction
     * Including span scores is disabled by default.
     */
    PayloadQueryNodeBuilder() {
        this(new AveragePayloadFunction());
    }

    /**
     * 用提供的 PayloadFunction 创建一个新的 PayloadQueryNodeBuilder
     */
    PayloadQueryNodeBuilder(PayloadFunction function) {
        this(function, false);
    }

    @Override
    public Query build(QueryNode queryNode) throws QueryNodeException {
        FieldQueryNode node = (FieldQueryNode) queryNode;
        return new PayloadTermQuery(new Term(node.getFieldAsString(),
node.getTextAsString()), function, includeSpanScore);
    }
}
```

输出查询对象的类型：

```
Query q = qp.parse(qtxt);
System.out.println(q.getClass().getName());
```

如果是 BooleanQuery，则输出每个子句是否是必需的：

```
BooleanQuery booleanQuery = (BooleanQuery)query;

for(BooleanClause bc : booleanQuery){
    System.out.println(bc.getQuery() + bc.getOccur());
}
```

QueryVisitor 用来遍历查询对象的内部结构。

查询"NBA 视频"当作查询"NBA AND 视频"来处理。

```
queryParser.setDefaultOperator(QueryParser.Operator.AND);
```

ModifierQueryNode 修饰查询。

QueryNodeProcessorImpl 是 QueryNodeProcessor 接口的默认实现。它是一个抽象类，处理 QueryNode 树的类要继承 QueryNodeProcessorImpl。这个类从左到右处理树中的 QueryNode。它沿着树向下走，对每个节点调用 preProcessNode(QueryNode)。当一个节点的

孩子处理完以后，用该节点作为参数，调用 postProcessNode(QueryNode)。

如果节点有至少一个孩子，就在 postProcessNode(QueryNode) 之前调用 setChildrenOrder(List)。在 setChildrenOrder(List)里面，实现者可以重定义孩子顺序，或者从孩子列表删除任何孩子。

如图 3-24 所示是一个演示 QueryNodeProcessorImpl 如何处理节点的例子。

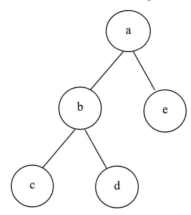

图 3-24　QueryNode 树

对上面的树，调用方法的顺序如下：

```
preProcessNode( a );
preProcessNode( b );
preProcessNode( c );
postProcessNode( c );
preProcessNode( d );
postProcessNode( d );
setChildrenOrder( bChildrenList );
postProcessNode( b );
preProcessNode( e );
postProcessNode( e );
setChildrenOrder( aChildrenList );
postProcessNode( a );
```

3.7　检索模型

查询返回的是一个相关度从高到低的文档列表。检索模型也就是文档和查询词的相关度的评分方法，给每个文档分配一个 0～1 的相关度值。

根据主题特征和质量特征得到文档分值。特征函数给出一个特征的量化值。文档和查询中的特征都有自己的特征函数。文档分值是文档和查询对应特征值的乘积之和。写成公式是：

R(Q, D) = Sum(gi(Q)fi(D))

其中，i是每一个特征，如每个词、每个文档质量得分；f是对文档计算的得分；g是对查询计算的得分。

需要把Lucene评分归一化到0~1。例如，一个查询返回下列评分：

```
8.864665
2.792687
2.792687
2.792687
2.792687
0.49009037
0.33730242
0.33730242
0.33730242
0.33730242
```

这是Lucene评分归一化问题。可以把所有的评分除以最大评分来让所有的评分值都为0~1。

但是，请注意归一化的评分应该仅用来比较单个查询中的文档分值。来自两个不同的查询中的分值计算结果不能正确地比较评分。不能比较来自两个不同查询的分值，无论评分是归一化的还是没有归一化的。

相关度评分在Lucene中由Similarity的子类实现。这些类都位于org.apache.lucene.search.similarities中。Lucene中有很多检索模型可供选择：

- TFIDFSimilarity 实现向量空间模型。
- BM25Similarity 实现BM25概率模型。
- IBSimilarity 实现基于信息的模型。
- LMSimilarity 实现语言模型。
- DFRSimilarity 实现随机性差异模型，也是一个概率模型，根据它产生更复杂的BM25概率模型。
- MultiSimilarity 从多个相似性值合并凭证。

首先看一下如何定制Similarity，然后再详细介绍常用的几种检索模型。

可以扩展Similarity，下面是扩展DefaultSimilarity的一个例子：

```
private class SimilarityOne extends DefaultSimilarity {
  public float lengthNorm(String fieldName, int numTerms) {
    return 1;
  }
}
```

然后通过IndexSearcher.setSimilarity让IndexSearcher对象使用这个SimilarityOne。

```
Similarity similarity = new SimilarityOne();
searcher.setSimilarity(similarity );
```

通过org.apache.lucene.search.Explanation中提供的方法可以查看某个文档的得分的具体构成。

```
TopDocs docs = searcher.search(query, 2);
for (ScoreDoc doc : docs.scoreDocs) {
    Explanation explanation = searcher.explain(query, doc.doc);
    System.out.println("explain: "+explanation.toString());   //输出得分原因解释
}
```

3.7.1 向量空间模型

如果只查询一个词，词频(也就是 TF)高的文档相关度高即可。如果查询多个词，可以根据这几个词构成一个 n 维空间。

向量空间模型假设在向量空间中相似文档说的是同样的事情，如图 3-25 所示。

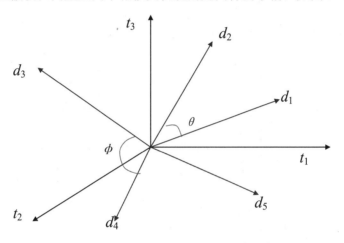

图 3-25 向量空间

Lucene 使用布尔模型来确定哪些文档与查询词匹配，使用向量空间模型(VSM)来对这些文档评分。评分算法中的向量空间模型使用 TF-IDF 计算权重。

TF(Term Frequence)代表词频，IDF(Invert Document Frequence)代表文档频率的倒数。例如，"的"在 100 篇索引文档中的 40 篇文档中出现过，则文档频率 DF(Document Frequence)是 40，IDF 是 1/40。"的"在第 1 篇文档中出现了 15 次，则 TF*IDF(的) = 15 * 1/40=0.375。另外一个词"户口"在这 100 篇文档中的 5 篇文档中出现过，则 DF 是 5，IDF 是 1/5。"户口"在第 1 篇文档中出现了 5 次，则 TF*IDF(户口)=5*1/5=1。结果是：TF×IDF(户口)>TF×IDF(的)。对给定的词 t 和文档(或者查询)x，TF(t,x)的值和词 t 在文档 x 中出现的次数正相关，而 IDF(t)的值和索引文档集合中包含词 t 的次数负相关。

文档 d 对于查询词 q 的 VSM 分值是带权重的查询向量 $V(q)$ 和文档向量 $V(d)$ 的夹角余弦(cos)相似度：

$$\text{cosine-similarity}(q,d) = \frac{V(q) \bullet V(d)}{\|V(q)\| \bullet \|V(d)\|} = \frac{V(q)}{\|V(q)\|} \bullet \frac{V(d)}{\|V(d)\|}$$

其中：

- 查询向量 $V(q) = <w(t_1, q), w(t_2, q), \cdots, w(t_n, q)>$。

- 文档向量 $V(d) = <w(t_1, d), w(t_2, d), \cdots, w(t_n, d)>$。
- $V(q) \cdot V(d)$ 是两个带权重的向量的点积,计算方法是 $V(q) \cdot V(d) = w(t_1, q) * w(t_1, d) + w(t_2, q) * w(t_2, d) + \cdots + w(t_n, q) * w(t_n, d)$。
- $\|V(q)\|$ 和 $\|V(d)\|$ 表示欧几里得范数。例如 $\|V(d)\| = \sqrt{t_1^2 + t_2^2 + \cdots + t_n^2}$,这里的 t 是文档 d 中出现的词的权重。

分母涉及向量的长度,用来归一化。向量长度归一化的方法是:每个分量除以它的长度。这里使用 L_2 范数计算向量长度:

$$\|\vec{x}\|_2 = \sqrt{\sum_i x_i^2}$$

这里使用 L_2 范数把文档归一化为单位向量。$\cos(q,d)$ 是 q 的单位向量和 d 的单位向量的点乘积。

如图 3-26 所示是一个简单的例子。

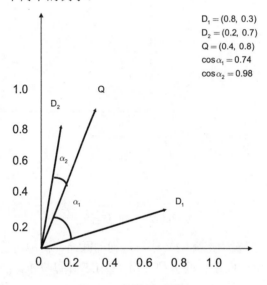

图 3-26 查询相似度

举个例子,共有 11 个词,分别是 t_1, t_2, …, t_{11}。共有 3 篇文档搜索出来,分别是 D_1、D_2、D_3。用户查询 Q 包含 t_6、t_9 和 t_{11} 3 个词。其中各自的权重(Term weight)见表 3-6。

表 3-6 词项—文档矩阵表

	t_1	t_2	t_3	t_4	t_5	t_6	t_7	t_8	t_9	t_{10}	t_{11}
D_1	0	0	0.477	0	0.477	0.176	0	0	0	0.176	0
D_2	0	0.176	0	0.477	0	0	0	0	0.954	0	0.176
D_3	0	0.176	0	0	0	0.176	0	0	0	0.176	0.176
Q	0	0	0	0	0	0.176	0	0	0.477	0	0.176

计算 3 篇文档与查询语句的相关性分值:

$$SCORE(D_1,Q) = \frac{0.176^2}{\sqrt{0.477^2+0.477^2+0.176^2+0.176^2}*\sqrt{0.176^2+0.477^2+0.176^2}} \approx 0.08$$

$$SCORE(D_2,Q) = \frac{0.954*0.477+0.176^2}{\sqrt{0.176^2+0.477^2+0.954^2+0.176^2}*\sqrt{0.176^2+0.477^2+0.176^2}} \approx 0.825$$

$$SCORE(D_3,Q) = \frac{0.176^2+0.176^2}{\sqrt{0.176^2+0.176^2+0.176^2+0.176^2}*\sqrt{0.176^2+0.477^2+0.176^2}} \approx 0.327$$

计算结果是文档 D_2 的相关性最高,其次是 D_3,最后是 D_1。

不同长度之间的文档的分值可以比较,称为长度归一化,此外还有查询归一化,让不同查询之间的分值可以比较大小。

Lucene 通过以下方法改进了 VSM 评分公式来提高搜索结果相关性和实际可用性。

- 归一化向量 $V(d)$ 成单位向量是有问题的,因为去除了所有的文档长度信息。对一些特定的文档可以归一化为单位向量,比如某些含有 10 次重复段落的文档,尤其是段落是由不同索引词组成的文档。但是对包含不重复的段落的文档,应当归一化成为大于单位向量的向量。为了避免这个问题,对不同的文档使用不同的文档长度归一化因子:doc-len-norm(d)。这样把向量归一化成等于或者大于单位向量。为了节省计算时间,在索引文档语料的时候计算了 lengthNorm(d),因此 lengthNorm(d)并没有包含文档语料长度的统计信息。后面介绍的 BM25 算法在这方面有改进。

- 在索引阶段,用户可以通过给某个文档加权(boost)来调整文档的重要程度。所以,每个文档的分值要乘以它的加权值 doc-boost(d)。下面的代码设置整个文档的加权:

```
doc.setBoost(1.5F);
```

- Lucene 是按列搜索的,因此每个查询词应用于一个独立的列,文档长度归一化是按索引列的长度来计算的,而且除了可以对整个文档加权外,还可以对文档中不同的列设置不同的加权。下面的代码可以对标题列加权:

```
Field bodyField = new Field("body", body,
        Field.Store.YES,
        Field.Index.ANALYZED);
Field titleField = new Field("title", title,
        Field.Store.YES,
        Field.Index.ANALYZED);
titleField.setBoost(1.2F);
```

- 同样的列可以很多次地加入文档,因此对一个列的加权是对文档中该列的多个值设置为其倍数。

- 在搜索时,用户可以对每个查询、子查询和每个查询项说明加权值,因此可以用查询词的加权值 query-boost(q)倍增查询词对文档分值的贡献度。

- 一个文档可能匹配多词查询,但是可能不包含查询中所有的词。用户可以通过协调因子 coord-factor(q,d)进一步奖励匹配更多查询词的文档,当匹配上更多的词时,

coord-factor(q,d)的值就更大。如果要修改这个值，则不用重建索引就可以生效。

Lucene 的概念评分公式：

$$score(q,d) = coord\text{-}factor(q,d) \bullet query\text{-}boost(q) \bullet \frac{V(q)\ ?\ V\ d}{|V(q)|\ |V(d)|} \bullet doc\text{-}len\text{-}norm(d) \bullet doc\text{-}boost(d)$$

Similarity 类的实际计算公式：

$$score(q,d) = coord(q,d) \bullet queryNorm(q) \bullet \sum_{t\ in\ q}(TF(t\ in\ d) \bullet IDF(t)^2 \bullet t.getBoost() \bullet norm(t,d))$$

其中：TF(t in D)是 t 在文档 d 中的词频，默认实现是：$\sqrt{t在文档d中的原始词频}$；
IDF(t)是词 t 在整个文档库中的倒文档频率，默认实现是：$\ln[N_{Docs}/(docFreq+1)]+1$。

Lucene 的打分算法涉及的因素如下：

- TF = 词频：度量文档里词出现的频率；
- IDF = 反向文档频率：度量词在索引中出现的频率；
- coord：文档中发现的查询词的频率；
- lengthNorm：根据索引列中的词总数来衡量一个词的重要度；
- queryNorm：归一化的参数便于比较查询；
- boost (index)：索引时的域的加权；
- boost (query)：查询时的域的加权。

前 4 个因素的实现、含义和原理说明如下。

1. TF

实现函数：sqrt(freq)。

含义：一个词在文档里出现的频率越高，则该文档的分值越高。

原理：多次包含同一个词的文档的相关度更高。

2. IDF

实现函数：log(numDocs/(docFreq+1)) + 1。

含义：一个词越多出现在不同的文档里，则它的分值越低。

原理：常见词的重要性要低于不常出现的词。

3. coord

实现函数：overlap / maxOverlap。

含义：对于查询中的词，文档中包含这些词越多分值就越高。

原理：分值高的文档要更多地覆盖查询中的词，也就是说查询和文档中的词重合越多越好。

4. lengthNorm

实现：1/sqrt(numTerms)。

含义：如果与包含较少词的索引列中的词匹配，则这个文档有较高的权重。

原理：如果一个词在含有少量的词的列中，则它比在包含较多词的列中的词更重要。

queryNorm 和文档的相关性没关，不会改变一个具体查询中的任何文档的相对排序，是为了让不同的查询之间的分值有可比性。queryNorm 仅仅用来保证相关度分值不是高到浮点数都无法保证精度的程度。queryNorm 的实现函数是：1/sqrt(sumOfSquaredWeights)。

因此，大致上说：

- 包含所有搜索词的文档是好的；
- 匹配很少出现的词比常用词更好；
- 长文档不如短文档好；
- 多次提及搜索词的文档更好。

DefaultSimilarity 继承 TFIDFSimilarity。

存在的一个问题是，Lucene 默认的长度归一化公式对长的文档打分太低。想象在打羽毛球或网球，球拍面中有个最佳击球区，叫作甜区(Sweet Spots)，只要在甜区范围内的回球都同样好。SweetSpotSimilarity 根据这个思想来改进长度归一化公式。SweetSpotSimilarity 实现了一个 lengthNorm，把一段区间内的文档长度看成同样好。可以定义一个全局的 min/max，在这段区间内的 lengthNorm 都是 1.0。低于最小值或大于最大值的 lengthNorm 以平方根下降。这样的结果是，在此区间内稍微长点的文档就不会有罚分了。

也可以对每列设置不同的 min/max，这样它们有不同的甜区。选择 min/max 时，可以根据文档的平均长度。须提前知道和推测出这个平均长度，并且通过 SweetSpotSimilarity.setLengthNormFactors(yourAvg,yourAvg,steepness)方法把它明确地设定为优选区。比如要做个论文的搜索系统，经过统计发现大多数论文的长度在 8000～10000 个词，因而 min 值设为 8000，max 值设为 10000。

做好这个 SweetSpotSimilarity 之后，用它来替换默认的 Similarity。在 org.apache.lucene.search.Searcher 中有个方法可以指定用哪个 Similarity。

```
public void setSimilarity(Similarity similarity) {
   this.similarity = similarity;
}
```

Nutch 中的 NutchSimilarity 各类是一个计算网页相关性的例子。

向量空间模型没有考虑搜索词在文档中的相对位置。Scorer 包含评分公共功能的抽象类，提供了评分和解释评分结果的功能。需要让 Scorer 包括位置信息，这样就可以让 SpanQuery 中的匹配位置直接影响相关性排名。

Scorer 以 docId 增序的方式遍历匹配查询的文档集合。其中的 score()方法返回当前文档的分值。Scorer 中的 positions()方法返回用 PositionIntervalIterator 表示的与位置相关的信息。

3.7.2 DFR

向量空间模型中经常用 TF*IDF 来计算词的重要度，这样的计算方法没有理论依据。认为太常见的词没用，把它们当作停用词去掉也没有理论依据，都只是经验。这里根据词频分布的规律来估计词的重要度。

把某个词在某一个文档中出现的次数看成是一次试验。试验的结果是一个非负整数。

某个词在文档 0 中出现 1 次，在文档 1 中出现 3 次。词在文档集合中出现的次数构建一个直方图，称为词频直方图，如图 3-27 所示。

图 3-27　词频直方图

要计算词在文档中出现 TF 次的概率，这个值写作 $P(\text{TF})$。词频分布是一种概率分布，所有可能的词频概率加起来是 1。即满足如下公式：

$$\sum p(\text{TF}) = 1$$

把词在文档中的分布看成是随机的，则可以用各种概率分布来近似计算 $P(\text{TF})$。把无差别的小球看成词，一个盒子看成一个文档，把词在文档中的每次出现看作一次试验，则每次试验就是往一个盒子中放入一个小球。

把一个词放入一个文档可以看成是泊松试验，因为符合泊松试验的如下特征：

- 试验的输出结果可以分成成功和失败两类。把词放入观察的那个文档就是成功，否则就是失败。
- 成功放入一个指定文档的平均次数是已知的。
- 如果文档数量增多，则词成功放入这些文档的概率成比例地增大。
- 在很短的文档内成功放入词的概率接近于 0。

如果文档集合中的词 t 平均在每个文档中出现的次数是已知的，则可以用泊松分布来计算 $P(\text{TF})$：

$$P(\text{TF}) = \frac{\lambda^{\text{TF}}}{\text{TF}! e^{\lambda}}$$

其中，λ 是文档中出现的词的平均数。例如，如果词平均出现 2 次，则词在文档中出现 3 次的概率是：

$$P(3) = \frac{2^3}{3! e^2} \approx \frac{8}{6 \times 7.3891} = 0.1804$$

当 $\lambda=2$ 时的概率表见表 3-7。

表 3-7　$\lambda=2$ 时泊松分布概率表

TF	P(TF)
0	0.1353
1	0.2707
2	0.2707
3	0.1804

续表

TF	P(TF)
4	0.0902
5	0.0361
6	0.0120
7	0.0034
8	0.0009
9	0.0002

计算泊松分布的代码：

```
/**计算泊松概率
 * @param rate - lambda，文档中出现词 t 的平均数
 * @param observed - 实际观察到的词次数
 * @return 泊松概率 */
public static double poissonProbablility (double rate, int observed){
    double a = Math.pow(rate, observed);
    double b = Math.pow(Math.E, -1*rate);
    double c = Num.factorial(observed); //计算阶乘
    return (a*b)/c;
}
```

对特定的文档集合，每次都对其建立索引，并同时统计词在文档中平均出现次数等相关信息。

使用 Excel 可以生成泊松分布和直方图。另外，搜索系统在一小时内到的用户发送的请求次数也是一个泊松分布。

换一种方式考虑词在一个文档中出现的次数。假设已知词在整个文档集合中出现 n 次，也就是要把词 n 次放入文档集合，在这 n 次放入中，成功放入考察的文档中的次数设为 x。n 次试验中，成功 x 次的概率是一个二项分布。

二项分布的基本事件是：一个词在一个文档中的一次出现。

只有两种可能结果的单次随机试验称为伯努利试验。如果集合中的文档数量是 N，则每次试验成功的概率是

$$p = 1/N$$

一列独立同分布的伯努利试验称为伯努利过程。

假设在一次试验中，词在任何一个文档中出现的概率是 p，则一共出现过 n 次的词在某个文档中出现 TF 次的概率是：

$$P(\text{TF}) = \binom{n}{\text{TF}} p^{\text{TF}} (1-p)^{n-\text{TF}}$$

这就是二项分布。这里

$$\binom{n}{\text{TF}} = \frac{n!}{\text{TF}!(n-\text{TF})!}$$

例如，假设共有 1024 个文档，即 $N=1024$。一个词在集合中总共出现 10 次，也就是说集合频率是 10，即 $n=10$。词在一个文档中出现 4 次的概率是 $TF=4$，则有

$$P(4) = \binom{n}{TF} p^{TF}(1-p)^{n-TF} = \binom{10}{4}(\frac{1}{1024})^4(1-\frac{1}{1024})^6 = 0.00000000019$$

用二项分布计算概率的代码如下：

```
public double probability(double tf) {
    double binomial = fact(n) / (fact(n- tf) * fact(tf));
    return binomial*Math.pow(p,tf)*Math.pow(1.0-p,n-tf);
}
```

当 $x=0,1,2,\cdots,n$ 时，用上面的公式计算概率。当 x 取其他的值时，$P(X=x)=0$。泊松分布给所有非负整数一个非零的概率，但是二项分布在文档长度后就停止分配概率。因为一个词的出现次数只是文档长度的一小部分，所以泊松分布用在不可行的频率值上的密度是微乎其微的。

可以把二项分布推广之后，得到泊松分布。当 n 很大，而 p 很小的时候，$\lambda \approx n \cdot p$。假设集合中的文档数量是 N，则 $p=(1/N)$。可以用泊松分布作为二项分布的近似。注意，泊松分布并没有完全去掉 n，而是用参数 λ 代替了。

例如：总频率是 50，而词在指定文档中出现的概率是 0.004。如果使用二项分布，则 $n=50$，$p=0.004$。如果使用泊松分布，则 $\lambda=50\times 0.004=0.2$。

一般假设词在文档中随机出现，且符合泊松分布。实际上，一个词并不可能是完全随机出现的，实际词频值比简单的模型预测的变化更大。

在一个特定的词和一个特定的文档有一个显著的关系的情况下，不会纯属偶然地发生关系时，玻色—爱因斯坦统计可能是有用的方程。

在 N 个文档中，随机地把一个词放入 F 次。事件可以用词在每个文档中的出现次数完全描述。每个出现是等概率的。出现次数用 n 个值表示：$TF_1, \cdots, TF_N$。其中，TF_k 表示在第 k 个文档中的词频。

关于出现问题，所有 N 元组满足方程：$TF_1+\cdots+TF_N=F$

根据这个方程找到所有可能的解的个数：

$$s_1 = \binom{N+F-1}{F} = \frac{(N+F-1)!}{(N-1)!F!}$$

理解为 F 个小球用 $N-1$ 个隔板隔开。假设有 2 个文档，某个词出现 6 次，则有 7 种可能的出现形式：$(6,0), (5,1), (4,2), (3,3), (2,4), (1,5)$ 和 $(0,6)$。

$$s_1 = \frac{(N+F-1)!}{(N-1)!F!} = \frac{(2+6-1)!}{(2-1)!6!} = 7$$

第 k 个文档词频是 TF。随机分配剩下的 $(F-TF)$ 个出现次数到其他的 $N-1$ 个文档。

$$TF_1+\cdots+TF_{k-1}+TF_{k+1}+\cdots+TF_N = F-TF$$

共有如下几种可能的解：

$$s_2 = \binom{N-1+(F-TF)-1}{F-TF} = \frac{(N+F-TF-2)!}{(N-2)!(F-TF)!}$$

某个文档词频是 TF 的概率是:

$$\frac{s_2}{s_1} = \frac{\binom{N+F-TF-2}{F-TF}}{\binom{N+F-1}{F}} = \frac{(N+F-TF-2)!F!(N-1)!}{(F-TF)!(N-2)!(N+F-1)!}$$

玻色—爱因斯坦分布可以用几何分布的概率近似计算。

可以统计词在索引库中的每个文档中出现的次数,然后看它的概率分布最符合哪个分布。

随机性差异模型基于简单的想法:单个文档内的词频和文档集合内的词频会有差异,这个差异越大,则在文档 d 中的词 t 蕴含更多的信息。

有些词不拥有代表性的文档,这样的词的频率遵循随机分布,例如遵循泊松分布或者二项分布,而且包含的信息较少。

非特长词的词频倾向于随机分布,而特长词倾向于泊松分布,这就是 2-泊松模型。

一个优秀的文档集合见证了富含信息的词。词越多地出现在精英文档集合,则词频越不太可能是由于随机性产生的。例如,"有木有"这个词出现在咆哮体文章这样的精英文档中。

词的重要度和从随机模型中得到的词频概率负相关。

$$weight(t \mid d) \propto -\log Prob_M(t \in d \mid Collection)$$

如果随机模型选择二项分布,则基本模型是 P。

$$-\log Prob_P(t \in d \mid Collection) = -\log \binom{F}{TF} p^{TF}(1-p)^{F-TF}$$

这里的 F 是词 t 在文档集合中的总频率,TF 是词 t 在文档 d 中的频率。假设集合中的文档数量是 N,则 $P=(1/N)$。

词不达意的情况并不少见,所以需要平滑词的重要度。即使计算 weight(t|d) 的公式给出了一个高的值,但是一个小的风险系数值也有负面影响,可能导致最终产生一个小的信息增益。平滑词的方法是乘一个风险概率。把这个平滑后的值叫作这个词给文档增加的信息。

$$P_{risk}[-\log Prob_P(t \in d \mid Collection)] = GAIN(t|d)$$

当一个罕见的词没有在一个文档中出现时,则它对文档有益的可能性几乎为零。相反,如果一个罕见的词在一个文档中出现很多次,则它有非常高的概率(几乎肯定)对文档所描述的主题信息有益。如果文档中的词频高,则这个词是没有益处的风险很小。

可以采用拉普拉斯方法计算 P_{risk}。

$$P_{risk} = \frac{1}{TF+1}$$

这是拉普拉斯模型,还可以采用两个伯努利过程的比例计算 P_{risk}:

$$P_{risk} = \frac{F}{DF(TF+1)}$$

这里的 DF 是包含词的文档数。

$$w = Inf1 \times Inf2$$

这里的 Inf1 表示词在文档集合上的分布模型(玻色—爱因斯坦或者伯努利模型)；Inf2 表示一个词在精英集合中的某个文档的多个出现(包含词的文档集合)的模型。

DFR 的打分公式由 3 个独立的部分组成：基本模型、后效和额外归一化组件。对应 DFRSimilarity 类用到 3 个类：BasicModel、AfterEffect 和 Normalization 类。这个模型基本上不需要参数，但是归一化部分接收浮点参数。

AfterEffect 类实现信息增益的首次归一化可以采用拉普拉斯方法或者伯努利方法。

Normalization 类实现长度归一化。

```
protected float score(BasicStats stats, float freq, float docLen) {
  float tfn = normalization.tfn(stats, freq, docLen);
  return stats.getTotalBoost() *
      basicModel.score(stats, tfn) * afterEffect.score(stats, tfn);
}
```

使用相关反馈的概率模型：对于一个给定的查询，如果我们知道一些文档是相关的，则在寻找其他相关文档时，这些已知的相关文档中出现的词，应给予更大的权重。

概率模型的想法很好，但性能不行。但是后来随着硬件的升级，情况不一样了。

使用 DFRSimilarity：

```
Similarity similarity = new DFRSimilarity(new BasicModelP(), new AfterEffectL(), new NormalizationH2());
searcher.setSimilarity(similarity );
```

3.7.3 BM25 概率模型

可以认为打上了和查询词同样标签的文档是相关文档。但很多时候，猜测文档是否有相关内容是没有把握的，所以要用概率来量化这种不确定性。把信息检索作为分类问题，一类是相关文档 R，还有一类是无关的文档 NR。根据贝叶斯判别规则，如果 $P(R|D)>P(NR|D)$，则 D 是相关的文档。如果 $P(R|D)<P(NR|D)$，则 D 是不相关的文档。例如，$P(R|D)=0.8$，$P(NR|D)=0.2$，则 D 是和用户查询相关的文档。如图 3-28 所示把"新生婴儿入户须知"这个索引库中的文档分成相关文档或者不相关文档。

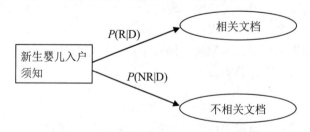

图 3-28　把信息检索看成分类问题

假设知道相关文档集合就能够计算出 $P(D|R)$。例如，如果知道某个词在相关文档集合中频繁出现，然后，给定一个新文档，就能直接计算出这个文档中词的组合有多大可能性出现在相关文档集合中。

第3章　Lucene 的原理与应用

使用贝叶斯公式估计概率：

$$P(R|D) = \frac{P(D|R)P(R)}{P(D)}$$

比较 $P(R|D)$ 和 $P(NR|D)$ 的值。如果满足 $P(D|R)\ P(R) > P(D|NR)\ P(NR)$，就把文档分到相关类。把一个文档分类成相关类的条件可以写成：

$$\frac{P(D|R)}{P(D|NR)} > \frac{P(NR)}{P(R)}$$

左边的公式叫作似然比，需要计算 $P(D|R)$ 和 $P(D|NR)$。为了简化计算，把文档表示成词的组合，用词概率估计 $P(D|R)$ 和 $P(D|NR)$。

用一个二值特征的向量表示文档的特征，表示文档中出现或者不出现某个词。把文档表示成二项特征组成的向量。$D=(d_1, d_2, \cdots, d_t)$，如果词 i 出现在文档中，则 $d_i=1$，否则就是 0。如果假设词都是独立出现的，则 $P(D|R)$ 可以用词概率的乘积 $\prod_{i=1}^{t} P(d_i|R)$ 计算。因为这个模型假设词独立出现，而且使用文档的二项特征，所以叫作二项独立模型。

假设索引库包含 5 个词，某文档 D 根据二元假设，表示为 $\{1,0,1,0\}$，其含义是这个文档出现了第 1 个、第 3 个和第 5 个词，但不包含第 2 个和第 4 个词。

用 P_i 来代表第 i 个词在相关文档集合内出现的概率，于是在已知相关文档集合的情况下，文档 D 相关的概率为：

$$P(R|D)=P_1*(1-P_2)*P_3*(1-P_4)*P_5$$

其中的 $1-P_2$ 代表第 2 个词不出现在相关文档的概率，因为 $P(t_2|R) + P(\neg t_2|R)=1$。

为了计算 $P(D|NR)$，假设用 S_i 代表第 i 个词语或单词在不相关文档集合内出现的概率，于是在已知不相关文档集合的情况下，观察到文档 D 的概率为：

$$P(D|NR)=S_1*(1-S_2)*S_3*(1-S_4)*S_5$$

例如，查询为"信息 检索 教程"，所有词项在相关、不相关情况下的概率 P_i、S_i 分别为见表 3-8。

表 3-8　P_i、S_i 值

词项	信息	检索	教材	教程	课件
R 中的概率 p_i	0.8	0.9	0.3	0.32	0.15
NR 中的概率 s_i	0.3	0.1	0.35	0.33	0.10

假设文档 D1 中只有 2 个词：检索 课件，则：

$P(D|R)=(1-0.8)*0.9*(1-0.3)*(1-0.32)*0.15$

$P(D|NR)= (1-0.3)*0.1*(1-0.35)*(1-0.33)*0.10$

$P(D|R)/P(D|NR)=4.216$

返回到似然比。使用 P_i 和 S_i 得到分值：

$$\frac{P(D|R)}{P(D|NR)} = \prod_{i:d_i=1} \frac{p_i}{s_i} \prod_{i:d_i=0} \frac{1-p_i}{1-s_i}$$

其中，$\prod_{i:d_i=1}$ 表示在文档向量中值为 1 的词对应的乘积。把上面的公式转换如下：

$$\prod_{i:d_i=1} \frac{p_i}{s_i} \prod_{i:d_i=0} \frac{1-p_i}{1-s_i} = \prod_{i:d_i=1} \frac{p_i}{s_i} (\prod_{i:d_i=1} \frac{1-s_i}{1-p_i} \prod_{i:d_i=1} \frac{1-p_i}{1-s_i}) \prod_{i:d_i=0} \frac{1-p_i}{1-s_i}$$
$$= \prod_{i:d_i=1} \frac{p_i(1-s_i)}{s_i(1-p_i)} \prod_{i} \frac{1-p_i}{1-s_i}$$

第 2 项在所有定义向量维度的词上运算，因此对任何文档来说，值都是一样的。对文档评分时可以忽略该项。

因为多个很小的数乘起来可能导致精度丢失或者向下溢出，成为 0，所以对计算公式取 log。这样评分公式变成了：

$$\sum_{i:d_i=1} \log \frac{p_i(1-s_i)}{s_i(1-p_i)}$$

如果存在相关性反馈，可以得到相关文档和无关文档集合。也就是说，给定用户查询，如果可以确定哪些文档构成了相关文档集合，哪些文档构成了不相关文档集合，可以利用表 3-9 所列出的数据来估算单词概率。

表 3-9 某个查询的词出现情况的相依表

	相关文档	不相关文档	文档总数
$d_i=1$	r_i	n_i-r_i	n_i
$d_i=0$	$R-r_i$	$N-n_i-R+r_i$	$N-n_i$
文档总数	R	$N-R$	N

表中第 3 行的 N 为文档集合总共包含的文档个数，R 为相关文档的个数，于是 $N-R$ 就是不相关文档的个数。对于某个词语或单词 d_i 来说，假设包含这个词语的文档共有 n_i 个，而其中相关文档有 r_i 个，那么不相关文档中包含这个单词的文档则为 n_i-r_i。再考虑表中第 2 列，因为相关文档个数是 R，而其中出现过单词 d_i 的有 r_i 个，那么相关文档中没有出现过这个单词的文档为 $R-r_i$ 个，同理，不相关文档中没有出现过这个单词的文档为 $(N-R) - (n_i-r_i)$ 个。从表中可以看出，如果我们假设已经知道 N、R、n_i、r_i，则其他参数可以靠这 4 个值推导出来。

采用最大似然估计，计算 $p_i = \frac{r_i}{R}$，$s_i = \frac{n_i-r_i}{N-R}$。为了避免 r_i 为 0 导致 log 0 无法计算的问题，采用相关文档和不相关文档都加 0.5 的平滑方法。这样得到 $p_i = \frac{r_i+0.5}{R+1}$，$s_i = \frac{n_i-r_i+0.5}{N-R+1}$。把这些值代入打分公式，得到：

$$\sum_{i:d_i=q_i=1} \log \frac{(r_i+0.5)/(R-r_i+0.5)}{(n_i-r_i+0.5)/(N-n-R+r_i+0.5)}$$

这个打分公式没有考虑词频，相关度比考虑词频的公式低 50%。

Okapi BM25(简称 BM25)是一种相关性排序函数,适用于搜索引擎根据给定搜索查询的相关性对匹配文档进行排序。

1. 排名函数

BM25 是一个基于单词集合的检索函数,它依据出现在每个文档中的查询词对匹配文档集合排序,而不管查询词在文档内相互之间的联系。它不是一个单一的函数,而实际上是有略微不同的组件和参数变化的一群函数的集合。一个最典型的具体函数如下。

假定有一个查询词组 Q,含有关键词 $q_i, \cdots, q_n$,用 BM25 给文档 D 评分的公式如下:

$$SCORE(D,Q) = \sum_{i=1}^{n} IDF(q_i) \cdot \frac{f(q_i, D) \cdot (k_1 + 1)}{f(q_i, D) + k_1(1 - b + b \cdot \frac{|D|}{\text{avg}d1})}$$

其中,$f(q_i, D)$ 是检索词 q_i 在文档 D 中的频率,$|D|$ 是文档 D 以单词为单位的长度,$\text{avg}d_1$ 是从抽取出的文档的文本集合的平均文档长度。k_1 和 b 是自由参数,k_1 通常选择 $= 2.0$ 和 $b = 0.75$。$IDF(q_i)$ 是检索词 q_i 的 IDF(文档频率倒数)权重。$IDF(q_i)$ 的通常计算公式如下:

$$IDF(q_i) = \log \frac{N - n(q_i) + 0.5}{n(q_i) + 0.5}$$

这里,N 是集合中文档的总数,$n(q_i)$ 是包含 q_i 的文档个数。

对 IDF 的解释较多,例如有人用信息论来解释 IDF。使用 IDF 的公式也有一些变化。在最初的 BM25 发展过程中,IDF 组件来源于二元独立模型。

请注意:当把它应用于在一半以上的语料文档中出现的检索词(term)时,以上的 IDF 公式存在潜在的重大缺陷。这些检索词的 IDF 是负数,所以对于两个几乎相同的文件——一个包含检索词而另一个不包含它——后者可能得到更高的分数,这意味着检索词出现在一半以上的语料文档中时将对最后文档的相关性得分提供负贡献。通常,并不期望这样,所以许多实际的应用程序会以不同的方式处理 IDF 公式:

- 每个加数规定一个最小数 0,为了去除掉常见的检索词。
- 为了避免常见的检索词被完全忽略,给 IDF 方程一个常量 ε。
- 为了避免检索词被完全忽略,IDF 方程被包含一些非负的数或者严格正数的类似形式方程代替。

2. IDF 的信息理论性解释

下面是一个来源于信息理论的解释。假设查询词 q 出现在 $n(q)$ 个文档中,那么一份随机文档 D 包含查询词的概率为 $\frac{n(q)}{N}$(这里的 N 还是指文本集合包含的文档的总数)。因此,"D 包含 q"的信息量是:

$$-\log \frac{n(q)}{N} = \log \frac{N}{n(q)}$$

现在假设我们有两个检索词 q_1 和 q_2。如果两个检索词出现在完全不相关的文档中,那么随机文档 D 中 q_1 和 q_2 都出现的概率是:

$$\frac{n(q_1)}{N} \cdot \frac{n(q_2)}{N}$$

相应的信息量是：

$$\sum_{i=1}^{2} \log \frac{N}{n(q_i)}$$

加入一个小的变化，这恰好是 BM25 的 *IDF* 组件的公式表达。

Lucene 默认使用的检索模型是向量空间模型。为方便在 Lucene 中使用概率模型中著名的 BM25 模型，必须定制 Lucene 的评分系统。

为了将 BM25 概率模型排序集成到 Lucene，需要开发一个新的 Query、Weight 和若干 Scorer。其主要功能在于实现 Scorer 级别，因为 Query 和 Weight 的主要目的是为 Scorer 准备必要的参数及在调用搜索方法时创建 Scorer 实例。

在 Lucene 中，采用插件方式实现 similarity。

Lucene 的评分系统是一个非常复杂的机制，主要由下面 3 个类来实现：

(1) Query —— 表示用户需要的信息的抽象类。
(2) Weight —— 用户的 Query 的内部接口表示，以便能够重用 Query 对象。
(3) Scorer —— 包含评分公共功能的抽象类。提供了评分和解释评分结果的功能。

查询的执行可以分成两个部分：布尔过滤和排序评估。布尔过滤依赖于逻辑运算符的 ShouldBooleanScorer、MustBooleanScorer 和 NotBooleanScorer 实现，而排序评估则是通过 BM25TermScorer 和 BM25FTermScorer 的 score()方法来实现。

BM25BooleanScorer 创建 BM25TermScorer 的实例，构造函数如下：

```
public BM25BooleanQuery(String query, String field, Analyzer analyzer)
throws ParseException
```

IOException 使用 BM25 排序函数。

BM25BooleanScorer 将忽略任何 Lucene QueryParser 处理的和域有关的信息，因此将只在构造函数中作为参数传入的域中执行搜索。另外，它只支持布尔查询，其他查询类型将被分割成检索词然后作为布尔查询执行。

应当指出的是，这两个排名函数没有用到查询权重，因此，所有的计算是在 scorer 水平上执行的。

几乎所有为了计算 BM25 相关的必要信息都能通过 Lucene expert API 获得，如 termdocs、numdocs、docfreq 等。但文档的平均长度无法通过 Lucene 内部的 API 获得。可以在做索引的时候得到文档的平均长度，可以在初始化的时候通过执行一个特定文件比较工具，计算文件长度并存储文件来获得。取得文档的平均长度的主要代码如下：

```
public class CollectionSimilarityIndexer extends DefaultSimilarity{
    private static Map<String,Long> length = new HashMap<String, Long>();

    @Override
    public float lengthNorm(String fieldName, int numTokens) {
        Long aux = CollectionSimilarityIndexer.length.get(fieldName);
        if (aux==null)
```

```
                aux = new Long(0);
            aux+=numTokens;
            CollectionSimilarityIndexer.length.put(fieldName,aux);
             return super.lengthNorm(fieldName, numTokens);
        }
        
        public static long getLength(String field){
            return CollectionSimilarityIndexer.length.get(field);
        }
}
```

在索引完成后，可以检索到某个域的长度。把 CollectionSimilarityIndexer.length/numdocs 的值保存到一个文件中，可以在打开搜索器的时候读取这个文件。BM25Parameters 的 load(String filePath)方法用来载入文档平均长度。

可以在 BM25Parameters 类中找到相关的 BM25 参数，默认设置 $k_1 = 2$，$b = 0.75$。BM25F 的情况更为复杂，因为 BM25F 需要更多的参数，主要是一个包含被搜索域的字符串数组。可以在 BM25FParameters 中找到所有的参数，同样道理，k_1 也是一个数组。对于 b 来说，每个域均设置为 0.75。但是构建 Query 对象时，可以使用更合适的参数来代替默认值。可以用浮点数组来设置这些参数。boost 的情况有些相似，这些值已经初始化为 1，但却可以提供一个浮点型数组来修改初始值。必须有序提供所有基于 BM25F 的数组参数，如 $boost_{field}$ 和 b_{field}。这意味着对于域数组中的第 i 个域的参数 boost 和 b 在各自的数组中也处于第 i 个的位置。

在两个模型中，IDF 是在 BM25Similarity 中计算的，并且必须用到文档频率和 numdocs 这两个值在文档级进行计算。Lucene 在域级别返回文档频率，对于 BM25 非常合适，因为在 BM25 中只对唯一的一个域搜索。对于 BM25F 的情况却是一个问题，因为 IDF 只有在包含所有被索引的词的一个新的域中才能进行文档级别的计算。而提供的实现则在一个最长的平均长度的域中计算文档频率。

这个实现的用法和用 Lucene 搜索的方法类似，但是在查询执行前必须给 BM25Parameters 或 BM25FParameters 赋值，这样做是为了设置列的平均长度，而其他参数则可以忽略，因为它们被设置为默认值。使用 BM25 的例子如下：

```
IndexSearcher searcher = new IndexSearcher("Index//Path");
//加载平均长度
BM25Parameters.load(avgLengthPath);
BM25BooleanQuery query = new BM25BooleanQuery("This is my Query",
                                              "Search-Field",
            AnalyzerUtil.getPorterStemmerAnalyzer(new StandardAnalyzer()));

//取得归一化之后的评分值
Hits hits = searcher.search(query);

//打印结果
for (int i = 0; i < 10; i++)
    System.out.println(hits.id(i) + ":"+hits.score(i));
```

org.apache.lucene.search.similarities.BM25Similarity 实现了 BM25 相似性。为了使用 BM25Similarity，只需要在 IndexSearcher 中设置相似性：

```
searcher.setSimilarity(new BM25Similarity());
```

IndexWriterConfig 中也设置相似性：

```
config.setSimilarity(new BM25Similarity());
```

完整的代码如下：

```
Directory directory = new RAMDirectory();

Analyzer analyzer = new StandardAnalyzer();
IndexWriterConfig config = new IndexWriterConfig(analyzer);
config.setSimilarity(new BM25Similarity());
IndexWriter writer = new IndexWriter(directory, config);
//增加一些文档到索引...

IndexReader reader = DirectoryReader.open(directory);
IndexSearcher searcher = new IndexSearcher(reader);
searcher.setSimilarity(new BM25Similarity());
```

使用它对文档评分：

```
//得到不归一化的评分值
TopDocs top = searcher.search(queryF, 10);
ScoreDoc[] docs = top.scoreDocs;

//打印结果
for (int i = 0; i < top.scoreDocs.length; i++) {
    System.out.println(docs[i].doc + ":"+docs[i].score);
}
```

b 的取值范围为 0~1。BM25 模型没有提供如何设置这些参数的指导。实际应用时，k 取值范围为 0.75~2。默认的参数值是：$k_1 = 1.2$, $b = 0.75$。可以通过构造方法设置 k_1 参数和 b 参数。

```
float k1= 1.2f;
float b = 0.75f;

Similarity s = new BM25Similarity(k1, b);
```

当 b=0 时，去掉文档长度的影响。

3.7.4　BM25F 概率模型

Robertson 等人在 BM25 模型的基础上提出了词频加权的权重计算公式 BM25F。首先获得检索词在所有域上累计的权重：

$$\text{weight}(t,d) = \sum_{c \text{ in } d} \frac{\text{occurs}_{t,s}^d \cdot \text{boost}_c}{\left((1-b_c) + b_c \cdot \dfrac{l_c}{\text{avg}l_c}\right)}$$

其中，l_c 是域的长度；$\text{avg}l_c$ 是域 c 的平均长度；b_c 是一个与域长度有关的常数，b_c 与 BM25 中的 b 相似；boost_c 是运用于域 c 的加权因子。然后，再用一个非线性饱和函数 $\dfrac{\text{weight}}{k_1 + \text{weight}}$，饱和函数就是说输入达到一定的值以后输出就不再变化了。

$$R(q,d) = \sum_{t \text{ in } q} IDF(t) \cdot \frac{\text{weight}(t,d)}{k_1 + \text{weight}(t,d)}$$

$$IDF(t) = \log \frac{N - \text{df}(t) + 0.5}{\text{df}(t) + 0.5}$$

这里的 N 是文档集合中文档的个数；df 是出现过检索词 t 的文档的个数。

BM25F 的概率模型比较适合基于关键词来源位置进行加权的词频统计方法。它保证了词频的作用不会成为衡量关键词在文档中重要性的唯一标准；该模型综合考虑了词频、文档频、文档长度、文档集合平均长度等多种因素，尤其在 BM25F 模型中添加了赋予不同域的关键词不同权重的衡量参数，进一步保证了关键词的加权词频能够反映该关键词对文档主要内容的贡献程度。

使用混合词查询(Blended Term Query)实现 BM25F 的例子：

```
public class BM25FDemo {
    private static void addDoc(IndexWriter w, String title, String description) {
        Document doc = new Document();
        doc.add(new TextField("title", title, Field.Store.YES));
        doc.add(new TextField("description", description, Field.Store.YES));
        w.addDocument(doc);
    }

    static Similarity perFieldSimilarities = new PerFieldSimilarityWrapper() {
        @Override
        public Similarity get(String name) {
            if (name.equals("title")) {
                return new BM25FSimilarity(/*k1*/1.2f, /*b*/0.8f);
            } else if (name.equals("description")) {
                return new BM25FSimilarity(/*k1*/1.4f, /*b*/0.9f);
            }
            return new BM25FSimilarity();
        }
    };

    public static void main() throws IOException {
        StandardAnalyzer analyzer = new StandardAnalyzer();
        Directory index = new RAMDirectory();

        IndexWriterConfig config = new IndexWriterConfig(analyzer);
```

```
config.setSimilarity(perFieldSimilarities);

IndexWriter w = new IndexWriter(index, config);

addDoc(w, "Moby Dick", "Moby Dick was a pretty cool whale");
addDoc(w, "The moby Letter", "I listen to moby!");

IndexReader reader = DirectoryReader.open(index);
IndexSearcher searcher = new IndexSearcher(reader);

BlendedTermQuery bm25fQuery = new BlendedTermQuery.Builder()
    .add(new Term("title", "moby"), 2.0f)
    .add(new Term("description", "moby"), 4.0f)
    .setRewriteMethod(BlendedTermQuery.BOOLEAN_REWRITE).build();

TopDocs docs = searcher.search(bm25fQuery, 10);
ScoreDoc[] hits = docs.scoreDocs;

System.out.println("发现 " + hits.length + " 个结果");
for(int i=0;i<hits.length;++i) {
    int docId = hits[i].doc;
    Document d = searcher.doc(docId);
    System.out.println((i + 1) + ". " + d.get("isbn") + "\t" + d.get("title"));
}
}
}
```

3.7.5 统计语言模型

语言模型最初用来估计一段文本的生成概率，现在也可以用语言模型来估计一篇文档和某个查询词的相关程度。例如，n 元模型就是一种常用的语言模型。语言模型已经应用在机器翻译、语音识别和手写体识别等与自然语言处理相关的领域。

信息检索中的语言模型用于计算每个文档生成查询项的概率，基本公式是：$P(Q|M_d)$。这里，Q 是查询字符串，其中包含查询词 $t_1,\cdots,t_n$。M_d 是文档 d 的语言模型。文档 $d_1,\cdots,d_n$ 可以根据 $P(Q|M_{d_i})$ 评分。

例如，有 6 篇文档，计算文档生成查询项的概率如下：

$P(Q|M_{d_1})$= 0.3；$P(Q|M_{d_2})$= 0.1；$P(Q|M_{d_3})$= 0.08；$P(Q|M_{d_4})$= 0.06；$P(Q|M_{d_5})$= 0.18；$P(Q|M_{d_6})$= 0.28

这 6 篇文档的排名如下：

Doc 1   0.3
Doc 6   0.28
Doc 5   0.18
Doc 2   0.1

Doc 3　0.08

Doc 4　0.06

在像语音识别这样的应用中,使用了能够基于长序列预测单词的 n 元(n-gram)语言模型。n 元模型基于前面的 $n-1$ 个词来预测下一个词。最常用的 n 元模型是二元(bigram)模型和三元(trigram)模型。二元模型在前一个词的基础上预测下一个词,三元模型在前两个词的基础上预测下一个词。尽管二元模型已经用在信息检索中来表示两个词组成的短语,但是这里专门讨论一元模型,因为一元模型更简单,并且作为搜索结果排序算法的基础很有效。

关于搜索的应用,我们使用语言模型表示文档的主题内容,关于信息检索的讨论中很少给主题下一个定义。在这个方法中,我们把主题定义成在词汇上的概率分布,也就是语言模型。例如,如果一个文档是关于在嘉陵江钓鱼,我们会注意到在语言模型中和"钓鱼""位于嘉陵江"相关的词语有高概率。如果文档是关于在千岛湖钓鱼,有些高概率的词是一样的,但是会有更多的高概率词与位于"千岛湖"有关,如果这个文档是一个关于计算机的钓鱼游戏,很多高频词会与游戏制造和计算机使用相关,尽管这里仍然会有一些重要的词是关于钓鱼的。话题语言模型(简称话题模型)包含所有词的概率,不仅是最重要的词。大多数词有个默认概率,这个值在任何文档中都是一样的,但是对主题重要的词将有不同寻常的高概率。

可以把语言模型看成是朴素贝叶斯方法在文档检索领域的应用。把索引库中的 n 个文档中的每个文档都看成一个仅仅包含文档自身一个训练样本的类别。使用这个 n 类的分类器对查询分类。基于类别的后验概率对文档评分。

$$P(d\,|\,q) \propto \prod_{t \in q} P(t\,|\,D)$$

采用最大似然估计计算 $P(t|D)$:

$$P(t\,|\,D) = \frac{f_{t,D}}{|D|}$$

其中,$f_{t,D}$ 是词 t 在文档 D 中出现的次数。如果采用最大似然估计,则不在文档 D 中的词概率是 0。如果有任何查询词不在文档中出现,则 $P(Q|D)$ 的值是 0。所以需要平滑 $P(t|D)$ 的值。

$$P(d\,|\,q) \propto \prod_{t \in q} (\lambda P(t\,|\,D) + (1-\lambda) P(t))$$

这里使用了线性插值来平滑概率。

其中,$P(t_i\,|\,D) = \lambda \hat{P}(t_i\,|\,D) + (1-\lambda)\hat{P}(t_i)$

这里的 $P(t_i\,|\,D)$ 是从文档 D 生成查询词 t_i 的概率;$P(t_i)$ 是从整个语料库生成查询词 t_i 的概率。

λ 值控制平滑量。λ 的值越小,则越平滑,因为这时候查询词无条件的生成概率$(1-\lambda)$更大了。

设置合适的 λ 对于文档评分有重要的影响。可以根据一个查询项的集合,人工或者自动设置最大性能。较低的 λ 对于长的查询更好,较高的 λ 对于短的查询更好。为了提高计算效率,不要计算不包括任何查询词的文档概率。

用文档长度来决定 λ 值。

$$\lambda = \frac{\mu}{|D|+\mu}$$

这叫作狄利克雷平滑(Dirichlet smoothing)。μ是一个经验参数,往往设置成1000～2000的值。Lucene中默认的μ值是2000。

```
Similarity similarity = new LMDirichletSimilarity();
searcher.setSimilarity(similarity );
```

把相关文档信息整合进评分算法。KL偏离值衡量文档模型和相关模型。文档模型和相关模型偏离越小,则文档的分值越高。

3.7.6 相关性反馈

显式反馈方法需要评判人员或者用户显式地参与,对给定查询返回的文档集进行相关性判断,然后反馈给检索系统,这样系统就可以根据这些反馈信息优化查询结果。对于给定查询返回文档的相关性判断,可以是二元相关性或者多等级相关性。二元相关性仅指出文档对于给定查询是相关(Relevant)还是不相关(Irrelevant),而多等级相关性则需要对相关程度进行判定,譬如不相关(not relevant)、有些相关(somewhat relevant)、相关(relevant)和非常相关(very relevant)。显式相关反馈算法的主要优点是:

(1) 用户只需要判断文档的相关性,而不必理解查询扩展技术的细节;
(2) 把检索任务分成一系列小的步骤,更易于理解;
(3) 提供了一个可控的过程,即强调相关的词项(term),降低不相关词项的权重。

相对显式相关反馈,隐式相关反馈不需要用户直接参与。隐式相关反馈的主要目标是通过对用户行为的研究,如用户的点击数据(click-through data),浏览了哪些文档,不同文档使用的时间长度和鼠标滚动行为等,判断可能的相关文档,从而优化最终的查询结果。

一种简单的实现方法:把每个关键词和对应的文档及点击量存入BerkeleyDB这样的内存数据库。

一种简单的处理方法是:索引中有一列记录用户点击次数,把这列中的值作为文档提高权重的依据。与通过列中的值影响相关度(函数式查询)不同,该处理方法中,和用户点击过的文档很相似的文档分值也会得到提升。这有点类似于相关推荐的效果。

3.7.7 隐含语义索引

隐含语义索引(Latent Semantic Indexing,LSI)对于较小的文档集合很有效。LSI的基础是SVD(奇异值分解),SVD是一种分解矩阵的数值方法。SVD提供了一个美妙的方式将文档—词矩阵分解成一个向量。这允许我们通过建立一个降低等级的近似来查询文档数据库。可以用SVD计算查询词和文档的相关度。

SVD的速度慢,而且很耗内存。把词库缩小到2万个,都得用6GB左右内存。http://mahout.apache.org/ 有个分布式的SVD实现。

Mahout自身带了很多算法。Mahout看起来更像是一群方法的集合,而不像是某种可编程的接口,在实际项目中应该怎么用它呢?

用 Mahout 只是填一下 input 参数就可以了。例如：

```
<MAHOUT_HOME>/bin/mahout svd \
  --input (-i) <Path to input matrix> \
  --output (-o) <The directory pathname for output> \
  --numRows (-nr) <Number of rows of the input matrix> \
  --numCols (-nc) <Number of columns of the input matrix> \
  --rank (-r) <Desired decomposition rank> \
  --symmetric (-sym) <Is the input matrix square and symmetric>
```

3.7.8 学习评分

PageRank 导致了 Google 炸弹。合并多个特征比一个特征更准确。PageRank 仅仅是连接结构特征。

汇集搜索用户的智慧。利用用户的点击日志分析出哪些文档和查询词最相关。根据用户搜索行为调整搜索结果排序。此外，还可以通过社交网络判断文档相关度。

学习评分(Learning to Rank)采用机器学习方法训练出采用多种特征的模型用来对文档相关度进行评分。相关度评分是一种有监督或半监督的机器学习问题，其目标是从训练数据自动构建评分模型。这样恶意用户更难以操作排名。有人说学习评分是一种超越 PageRank 的算法。

可以使用流行度数据进行训练。例如，访问量，用户在页面停留了多久。

可以人工标注出一个理想的文档相关性排序。然后采用一种学习评分算法学习出模型。LambdaMART 是当前流行的一种学习评分算法，由微软的 Chris Burges 提出。目前，LambdaMART 在工业界被广泛使用，包括 Bing、Facebook 都在实际业务中使用了该算法。

LambdaMART 算法是从 Pairwise 方法逐渐发展起来。Pairwise 方法的主要思想是将排序问题形式化为二元分类问题。Pairwise 方法通过考虑文档两两之间的相对相关度进行排序。例如，文档 X 比文档 Y 更相关还是更不相关这是一个二元分类问题，其目标是最小化反转的排名，也就是让损失最小。RankNet 算法是一个 Pairwise 方法，它使用交叉熵作为损失函数。损失函数的值越低说明机器学得的当前排序越趋近于理想排序。RankNet 算法可以使用神经网络模型，也可以使用渐进梯度回归树(Gradient Boost Regression Tree，GBRT)模型。

如果 GDBT 模型求解过程使用求梯度的 Lambda 方法，就是 LambdaMART 算法。这里的 MART(Multiple Additive Regression Tree)，也就是 GBDT(Gradient Boost Decision Tree)。Lambda 的含义是一个待排序的文档下一次迭代应该排序的方向(向上或者向下)和强度。

机器学习评分包 RankLib(https://sourceforge.net/p/lemur/wiki/RankLib/)用于 Lemur 搜索引擎，但也可以修改后和 Lucene 一起使用。

可以直接在命令行运行 RankLib.jar：

```
> java -jar RankLib.jar
```

具体的使用可以参考 Solr 和 Elasticsearch 中的相关实现。

3.7.9 查询与相关度

ExactSimScorer 用来给准确匹配的查询打分，例如 TermQuery 和准确的 PhraseQuery。SloppySimScorer 用来给松弛匹配的查询打分，例如 SpanQuery 和松弛的 PhraseQuery。ExactSimScorer 根据词频返回一个文档的分值：

```
public static abstract class ExactSimScorer {
 /**
  * 对一个文档打分
  * @param doc 文档编号
  * @param freq 词频
  * @return 文档的分值
  */
  public abstract float score(int doc, int freq);
}
```

3.7.10 使用 Payload 调整相关性

搜索"烧鸡"，包含"烧鸡"这个词的记录应该排在只包含单个字"烧"和"鸡"的记录前面。在 Payload 中，说明词匹配"烧鸡"有更好的相关性，而字匹配"烧"和"鸡"相关性则低很多。

让词的 Payload 值比字的 Payload 值高。计算文档相关度时，取所有匹配 Term 的平均值。

首先在分析文本期间生成 Payload 信息。可以使用 PayloadAttribute 达到这一点。只需在分析过程中将该属性添加到 Token 属性中。使用 PayloadHelper 编码数字成为 Payload，然后就可以设置到 PayloadAttribute。例如，编码浮点数：

```
Payload p = new Payload(PayloadHelper.encodeFloat(42));
```

注意：这里的 PayloadHelper 类不在核心包中，而在 contrib/common/lucene-analyzers-3.x 里面。PayloadHelper 中的 decodeInt()方法从字节数组中得到一个整数。

```
public static final int decodeInt(byte [] bytes, int offset){
    return ((bytes[offset] & 0xFF) << 24) | ((bytes[offset + 1] & 0xFF) << 16)
        | ((bytes[offset + 2] & 0xFF) << 8) | (bytes[offset + 3] & 0xFF);
}
```

这里的 bytes[offset] & 0xFF 是为了得到整数结果，然后参与后续的位移运算。

在字词混合的 TokenFilter 中增加 PayLoad 信息：

```
public class SingleFilterPayload extends TokenFilter {
    private CharTermAttribute termAtt; // 词属性
    private OffsetAttribute offsetAtt; // 位置属性
    private PositionIncrementAttribute posIncr; // 位置增量属性
    private char[] curTermBuffer; // 要再次切分开的词缓存
```

```
    private int curPos; // 在词中的位置
    private int curTermLength; // 长度缓存
    private int i = 0; // 单字所在总的偏移量
    private PayloadAttribute attr;

    public SingleFilterPayload(TokenStream in) {
        super(in);
        this.termAtt = addAttribute(CharTermAttribute.class);
        this.offsetAtt = addAttribute(OffsetAttribute.class);
        this.posIncr = addAttribute(PositionIncrementAttribute.class);
        this.attr = addAttribute(PayloadAttribute.class);
    }

    public boolean incrementToken() throws IOException {
        // 把词分成单个字
        if (curPos < curTermLength && curTermLength > 1) {
            Payload p = new Payload(PayloadHelper.encodeFloat(1));
            attr.setPayload(p);
            offsetAtt.setOffset(i, i + 1);
            termAtt.copyBuffer(curTermBuffer, curPos , 1);
            posIncr.setPositionIncrement(0);
            curPos++;
            i++;
            return true;
        }
        if (input.incrementToken()) {
            Payload p = new Payload(PayloadHelper.encodeFloat(42));
            attr.setPayload(p);
            curTermBuffer = termAtt.buffer();
            curTermLength = termAtt.length();
            curPos = 0;
            i = offsetAtt.startOffset();
            return true;
        }
        return false;
    }
}
```

测试 TokenStream：

```
TokenStream ts = tokenStream("field", new StringReader("value"), a,
docValue);
CharTermAttribute cta = ts.getAttribute(CharTermAttribute.class);
PayloadAttribute payload = ts.getAttribute(PayloadAttribute.class);
while(ts.incrementToken()) {
    System.out.println("Term = " + cta.toString());
    System.out.println("Payload = " + new
String(payload.getPayload().getData()));
```

```
}
ts.reset();
```

搜索时通过遍历 TermPositions 得到索引中的 Payload 信息。

```
IndexReader reader = IndexReader.open(directory);

Term t = new Term("field", "java");
TermPositions tps = reader.termPositions(t);
int doc = 0; //文档编号
if (tps.skipTo(doc)) {
    byte[] buffer = new byte[4];
    tps.getPayload(buffer, 0); //得到Payload信息
}
```

在搜索阶段，应该使用特殊的 query 类 PayloadTermQuery。这个类像 SpanTermQuery，但是会提取索引中的 Payload。

```
private SpanNearQuery spanNearQuery(String fieldName, String words) {
    String[] wordList = words.split("[\\s]+");
    SpanQuery clauses[] = new SpanQuery[wordList.length];
    for (int i = 0; i < clauses.length; i++) {
        clauses[i] = new PayloadTermQuery(new Term(fieldName, wordList[i]),
                new AveragePayloadFunction());
    }
    return new SpanNearQuery(clauses, 10000, false);
}
```

PayloadTermQuery 可以选择包含或不包含 spanScore。

Lucene 对字节数组中的内容不做任何假设，重写 Similarity 中的 scorePayload()方法来解释有什么在 Payload 里面。DefaultSimilarity 中的 scorePayload()默认实现返回 1。使用自定义的 Similarity 实现可以为文档中出现的每个 Payload 打分。

```
public class PayloadSimilarity extends DefaultSimilarity {

  public float scorePayload(int docID, String fieldName,
                    int start, int end, byte[] payload,
                    int offset, int length) {
    if (payload != null) {
      return PayloadHelper.decodeFloat(payload, offset);
    } else {
      return 1.0f;
    }
  }
}
```

把 PayloadSimilarity 作为默认的 Similarity。

```
PayloadSimilarity mySimilarity = new PayloadSimilarity();
Similarity.setDefault(mySimilarity); // IndexSearcher在评分时会使用它
```

最后，使用 PayloadFunction 汇总文档上的 Payload 分值来产生最后的文档分值。AveragePayloadFunction 把看到的所有 Payload 的平均得分算作最后得分。

```
IndexSearcher searcher = new IndexSearcher(dir, true);

//只是对文档中的查询词汇总 Payload 平均值，而不是匹配文档中所有的词
PayloadTermQuery btq = new PayloadTermQuery(new Term("body", "fox"),
                                new AveragePayloadFunction());
TopDocs topDocs = searcher.search(btq, 10);
```

PayloadNearQuery 结合了 SpanNearQuery 和 PayloadTermQuery 的功能，如果下面两个条件满足，则相关性就高：
(1) 词有 Payload 加权它们的分值；
(2) 词连续出现。

PayloadTermQuery 是 SpanTermQuery 的子类，而 SpanTermQuery 又是 SpanQuery 的子类，所以 PayloadTermQuery 是 SpanQuery 的子类。因此可以传递 PayloadTermQuery 的实例组成的数组给 PayloadNearQuery。

```
SpanQuery q2 = new SpanTermQuery ("companyDesc", "烧", false,
         new AveragePayloadFunction());
SpanQuery q3 = new SpanTermQuery ("companyDesc", "鸡", false,
         new AveragePayloadFunction());
Query   query = new PayloadNearQuery(new SpanQuery[] {q2,q3}, 2, false);
//上面的 2 就是字的个数
TopDocs topDocs = searcher.search(query, 10);
```

把 PayloadNearQuery 和 SpanTermQuery 搭配在一起用：

```
String[] words = phrase.split("[\\s]+");
SpanQuery clauses[] = new SpanQuery[words.length];
for (int i=0;i<clauses.length;i++) {
  clauses[i] = new SpanTermQuery(new Term(fieldName, words[i]));
}
return new PayloadNearQuery(clauses, 0, inOrder, function);
```

如果要查询多个不同的词，则 PayloadNearQuery 包含字和词的 PayloadTermQuery，然后再用 BooleanQuery 连接 PayloadNearQuery。

```
BooleanQuery booleanQuery = new BooleanQuery();
booleanQuery.add(chickenQuery, BooleanClause.Occur.SHOULD);
booleanQuery.add(duckQuery, BooleanClause.Occur.SHOULD);
```

目前只有两个查询类型在分值计算中考虑 Payload，即 PayloadTermQuery 和 PayloadNearQuery。Lucene 的默认 QueryParser 不利用这些查询。因此要覆盖 QueryParser 类的 newTermQuery()方法，让它返回 PayloadTermQuery 而不是 TermQuery。

```
public class PayloadQueryParser extends QueryParser {
    // 构造方法
```

```
    protected Query newTermQuery(Term term) {
        return new PayloadTermQuery(term);
    }
}
```

使用 PayloadQueryParser：

```
Analyzer analyzer = new CnAnalyzer(); //支持字词混合切分的分析器
QueryParser qp = new PayloadQueryParser(fields, analyzer);
query = qp.parse(queryString);
```

一个词的同义词也可以放入这个词的 Payload 中。

查询词：烧鸡
测试文档："烧肉"，
"烤鸡"，
"鸡腿饭烧青菜，"，
"愤怒的小鸡"，
"非主流烧鸡"，
"烧秸秆"，
"正宗烧鸡"，
"烧鱼"，
"烧鸡玩具"，
"百岁鸡"
结果正常：text=非主流烧鸡
text=正宗烧鸡
text=烧鸡玩具
text=鸡腿饭烧青菜
text=烧肉
text=烤鸡
text=愤怒的小鸡
text=烧秸秆
text=烧鱼
text=百岁鸡

3.8 查询原理

在查询过程中首先找到相关文档集合，然后要计算每个文档和查询对象的相似度。早期版本的 Scorer 类同时负责两件事情：匹配和打分。后来把与打分相关的代码都分离到 Similarity 类。Lucene 5 以后的 Similarity 类负责所有和打分相关的代码，Scorer 类只负责匹配文档和调用 Similarity 类中的打分方法。

Query 对象表示用户需要的信息的抽象类。为了能够重用 Query 对象，把计算某一次查询所需要的一些全局信息放入 Weight 对象。Weight 类是一个抽象类。每种不同的 Query 对象有不同的 Weight 对象与之对应。例如，基本的词查询 TermQuery 有对应的 TermWeight，短语查询 PhraseQuery 有对应的 PhraseWeight。TermWeight 作为 TermQuery 的内部类实现。

PhraseWeight 也作为 PhraseQuery 的内部类。

每个词都有个文档频率 DF。TermContext 对象中包含这些信息，这些信息最终都放在 Weight 类的子类中。TermContext 部分代码如下：

```
public final class TermContext {
  private int docFreq;           //文档频率
  private long totalTermFreq;   //索引中总的词频率，也就是该词在索引中总共出现的次数
}
```

用户往往只看搜索结果中和查询词最相关的前面几条，所以可以在 IndexSearcher.search()方法中指定返回多少条结果，并把查询结果封装成一个 TopDocs 对象，但是这样就没办法通过 hits.length()知道总共有多少条和关键词相关的结果。因此要通过 TopDocs.totalHits 属性知道一共找到了多少条结果，这样方便翻页处理。可以把搜索结果看成是一个优先队列，相关度分值高的文档排在前面。

3.8.1 布尔匹配

查询两个词"NBA"和"AND 视频"，则对"NBA"这个词对应的文档编号列表和"视频"这个词对应的文档编号列表做交集(Intersection)运算后返回。例如在倒排索引表中检索出包含"NBA"一词的文档列表为 docList("NBA")=(1, 5, 9, 12)，表示这 4 个文档编号的文档含有"NBA"这个词汇。包含"视频"的文档列表为 docList("视频")= (5, 7, 9, 11)，这样同时包含"NBA"和"视频"这两个关键词的文档为 docList("NBA")∩docList("视频")= (5, 9)。这里的"∩"表示文档列表集合求交运算。

这样的查询也叫作合取查询(conjunctive query)。Lucene 中的 ConjunctionScorer 类包含类似的实现。

如果文档列表的长度是 m 和 n，则合并时间复杂度是 $O(m+n)$。

在索引阶段使用跳跃指针。这样可以进行跳跃式查找。两个文档列表如图 3-29 所示。

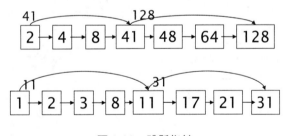

图 3-29　跳跃指针

假设每个列表都已经遍历到 8，然后前进，第一个列表是 41，第二个列表是 11。第二个列表的 11 更小，而且 11 的跳跃后继是 31，仍然比 41 小，所以可以完全跳过这段。

例如，把跳跃间隔设置成 16。如果想要跳过 100 个文档，没有必要从文档列表中读取 100 个条目，而只需要 100/16 = 6 个跳表条目，外加文档列表中的 100%16 = 4 个条目。

Lucene 使用了跳表。这显著地加快了合取查询(AND)和短语查询的速度。

Lucene Boolean Query search 使用接口 DocsEnum 中的两个方法：

(1) boolean.advance(int target):当前文档跳表前进到当前文档编号等于或者大于目标文档编号的位置。

(2) boolean next():当前文档编号序列前进一个文档位置。

使用 MultiFields.getTermDocsEnum 得到 DocsEnum 对象。例子如下:

```
private static void addDoc(IndexWriter writer, String content) throws
IOException {
        FieldType fieldType = new FieldType();
        fieldType.setStoreTermVectors(true);
        fieldType.setStoreTermVectorPositions(true);
        fieldType.setIndexed(true);
        fieldType.setIndexOptions(IndexOptions.DOCS_AND_FREQS);
        fieldType.setStored(true);
        Document doc = new Document();
        doc.add(new Field("content", content, fieldType));
        writer.addDocument(doc);
}

public static void main(String[] args) throws IOException, ParseException
{
    Directory directory = new RAMDirectory();
    Analyzer analyzer = new WhitespaceAnalyzer();
    IndexWriterConfig config = new IndexWriterConfig(analyzer);
    IndexWriter writer = new IndexWriter(directory, config);
    addDoc(writer, "bla bla bla bleu bleu");
    addDoc(writer, "bla bla bla bla");
    writer.close();
    DirectoryReader reader = DirectoryReader.open(directory);
    DocsEnum de =
        MultiFields.getTermDocsEnum(reader,
            MultiFields.getLiveDocs(reader), "content", new
BytesRef("bla"));
    int doc;
    while((doc = de.nextDoc()) != DocsEnum.NO_MORE_DOCS) {
        System.out.println(de.freq());
    }
    reader.close();
}
```

合取查询原理。例如,包含 bla 这个词的文档序列是 t1,包含 bleu 这个词的文档序列是 t2。输出 t1 和 t2 两个文档序列都有的文档编号的方法如下:

```
public static void DocsEnumAnd(DocsEnum t1, DocsEnum t2) throws IOException
{
    t1.nextDoc();
    t2.nextDoc();
    while (t1.docID() != DocsEnum.NO_MORE_DOCS
```

第 3 章 Lucene 的原理与应用

```
                && t2.docID() != DocsEnum.NO_MORE_DOCS) {
            if (t1.docID() < t2.docID()) {
                if (t1.advance(t2.docID()) == DocsEnum.NO_MORE_DOCS)
                    return;
            }
            if (t2.docID() < t1.docID()) {
                if (t2.advance(t1.docID()) == DocsEnum.NO_MORE_DOCS)
                    return;
            }
            if (t1.docID() == t2.docID()) {
                System.out.println("bla found doc:" + t1.docID() +"
freq:"+t1.freq());
                System.out.println("bleu found doc:" + t2.docID() +"
freq:"+t2.freq());
                t1.nextDoc();
            }
        }
    }
}
```

MultiDocsEnum 用来管理多个 DocsEnum。org.apache.lucene.codecs.MultiLevelSkipListWriter 写入多层跳表。org.apache.lucene.codecs.MultiLevelSkipListReader 读出多层跳表。

例如 skipInterval = 3：

```
                                              c            (跳跃层 2)
                          c              c    c            (跳跃层 1)
      x   x   x   x   x   x   x   x   x   x   x            (跳跃层 0)
      d d d d d d d d d d d d d d d d d d d d d d d d d d d d d d  (posting list)
      3       6       9       12      15      18      21      24      27      30   (DF)
```

这里的：
- d 表示文档；
- x 表示跳跃数据；
- c 表示带孩子指针的跳跃数据。

跳跃表的第 i 层包含跳跃表第 i-1 层的每 skipInterval 个条目。

因此第 i 层的条目数量是：floor(df / ((skipInterval ^ (i + 1)))。

每个 i>0 水平的跳过条目包含一个指向在 i-1 列表中相应的跳过条目的指针。这保证了找到目标文档只需要对数次的跳跃。

MultiLevelSkipListWriter 类用于实现不同的跳跃层次，子类必须定义跳跃数据的实际格式。

MultiLevelSkipListReader 中的 skipTo()方法的实现代码如下：

```
public int skipTo(int target) throws IOException {
    if (!haveSkipped) {
        // 第一次加载跳过层次
```

```
    loadSkipLevels();
    haveSkipped = true;
}

//走到最高层次直到这个目标
int level = 0;
while (level < numberOfSkipLevels - 1 && target > skipDoc[level + 1]) {
    level++;
}

while (level >= 0) {
    if (target > skipDoc[level]) {
        if (!loadNextSkip(level)) {
            continue;
        }
    } else {
        //在这个层面上没有更多的跳跃，往下一个层面
        if (level > 0 && lastChildPointer > skipStream[level - 1].getFilePointer()) {
            seekChild(level - 1);
        }
        level--;
    }
}

return numSkipped[0] - skipInterval[0] - 1;
}
```

3.8.2 短语查询

PhraseQuery 的实现，需要读取索引中词的位置信息。PhraseWeight 的 scorer()方法中有读取索引中词的位置信息的代码。

3.8.3 索引统计

给文档打分，需要索引统计信息。为了理解索引统计，假设索引如下两个文档，每个只有一个 title 列：

文档 1：The Lion, the Witch, and the Wardrobe
文档 2：The Da Vinci Code

假设按空格分词，去掉逗号，所有的词都小写，不去掉停用词。下面是分项统计说明：

- TermsEnum.docFreq()

有多少个文档至少包含 1 次这个词。对于词 lion 来说，docFreq 是 1，而对于词 the 来说，docFreq 是 2。

- Terms.getSumDocFreq()

投递列表的数量，例如，所有词的 TermsEnum.docFreq()的和。对于这个例子文档来说，

Terms.getSumDocFreq()的值是 9。

- TermsEnum.totalTermFreq()

这个词在该列出现的次数，包括所有的文档。对于词 the，totalTermFreq()的值是 4，对于词 vinci，这个值是 1。

- Terms.getSumTotalTermFreq()

这个列词的总数。包括所有的文档，这是包括该列所有不同词的 TermsEnum.totalTermFreq()值的和。对这个例子文档集合来说，这个值是 11。

- Terms.getDocCount()

有多少文档至少包含这个列的一个词。在例子文档中，Terms.getDocCount()的值是 2。但是，如果其中一个文档缺少 title 列，则 Terms.getDocCount()的值是 1。

- Terms.getUniqueTermCount()

这个列中有多少唯一的词。对于例子文档集合来说，Terms.getUniqueTermCount()是 8。注意这个值只是每段得到一个值，不能快速地计算索引中所有段的值，除非只有一个段。

- Fields.getUniqueTermCount()

所有列中唯一词的数量，它是 Terms.getUniqueTermCount()在所有列上的和。在例子文档集合中，这个值是 8。注意，这个值也只能按段得到。

- Fields.getUniqueFieldCount()

列的数量，对例子文档来说，这个值是 1。如果还有一个 body 列和 abstract 列，则这个值是 3。注意，这个值也只能按段得到。

3.x 版本的索引只存 TermsEnum.docFreq()，因此如果想要体验 Lucene 4.0 中新的评分模型，就要重新建索引，或者使用 IndexUpgrader 更新索引。

新的评分模型都使用单字节范数格式，因此可以在这些评分模型之间自由地切换，而不需要重新索引。

除了存储在索引中的，索引时还有按列和按文档的统计信息。这些信息由 FieldInvertState 传递给 Similarity.computeNorm()方法。FieldInvertState 记录有多少词加到索引。它用来计算列的归一化因子。

- Length：文档中有多少词。对于文档 1，length 值是 7；对于文档 2，length 值是 4。
- uniqueTermCount：对于该文档中的该列，统计有多少不同的词。对文档 1 来说，这个值是 5；对文档 2 来说，这个值是 4。
- maxTermFrequency：文档中最常出现的词的出现次数。对文档 1 来说，是 3(the 出现了 3 次)；对文档 2 来说则是 1。

从 4.0 版本开始，可以自由地编码这些统计信息，然后用自己定制的 Similarity 从中拿出数据。

从这些可用的统计方法，可以自由地获得其他常用的统计信息：

- 所有文档的平均列长度：Terms.getSumTotalTermFreq() 除以 maxDoc(或者 Terms.getDocCount()，如果不是所有的文档都有该列)。
- 文档内的平均词频：FieldInvertState.length 除以 FieldInvertState.uniqueTermCount。

- 每列中,所有文档中不同词的平均数量:Terms.getSumDocFreq()除以 maxDoc (或者 Terms.getDocCount(field),如果不是所有的文档都有该列)。

注意,统计数据不反映已删除的文档,直到把这些文档合并走。一般来说,这也意味着段的合并将改变分值!同样,如果该字段省略了词的频率,那么统计不会正确。

3.8.4 相关性

ScoreDoc 记录了文档编号和相关度分值。

```
public class ScoreDoc implements java.io.Serializable {
  public float score; //对这个查询来说,这个文档的相关度分值
  public int doc; //文档编号

  public ScoreDoc(int doc, float score) {   //根据文档编号和分值构造一个ScoreDoc
    this.doc = doc;
    this.score = score;
  }
}
```

Similarity 决定 Lucene 如何计算词的重要度,Lucene 在索引和查询的时候都会用到这个类。

在索引的时候,索引器调用 computeNorm(FieldInvertState),允许 Similarity 对每个文档设置一个值,以后可以通过 AtomicReader.getNormValues(String)方法访问这个值。Lucene 对于范数中有什么不做任何假设,但往往在范数中编码长度归一化信息。

SimScorer 根据词频对文档打分。SimScorer 的子类 ExactSimScorer 用来给准确匹配的查询打分。例如 TermQuery 和准确的 PhraseQuery。SimScorer 的另外一个子类 SloppySimScorer 用来给松弛匹配的查询打分。例如 SpanQuery 和松弛的 PhraseQuery。

ExactSimScorer 根据词频返回一个文档的分值:

```
public static abstract class ExactSimScorer {
  /**
   * 对一个文档打分
   * @param doc 文档编号
   * @param freq 词频
   * @return 文档的分值
   */
  public abstract float score(int doc, int freq);
}
```

NumericDocValues 是一个从文档编号到长整型数值的映射。

Scorer 以文档编号增序的方式遍历匹配查询的文档集合。其中的 score()方法返回当前文档的分值。它使用一个 Similarity 对象计算文档分值。

```
public abstract class Scorer extends DocIdSetIterator {
  // 返回匹配上查询的当前文档的分值
  public abstract float score() throws IOException;
```

```
/** 返回当前文档的匹配次数
 * This returns a float (not int) because
 * SloppyPhraseScorer discounts its freq according to how
 * "sloppy" the match was.
 */
public abstract float freq() throws IOException;
}
```

SpanScorer 是 Scorer 的子类，用来根据匹配区间打分。PhraseScorer 用来给 PhraseQuery 打分。

Weight 类的目的是保证搜索时不改变查询对象，方便重用查询实例。所以查询类实际调用 Weight 类，而不是对应的 Scorer 类。IndexSearcher 依赖的查询状态应该位于 Weight，例如：

```
private class PhraseWeight extends Weight {
  private final Similarity similarity;
  private final Similarity.SimWeight stats;
  private transient TermContext states[];
}
```

首先从最简单的 TermScorer 开始：

Query 创建一个在整个索引上都有效的 Weight 对象：它整合加权、IDF、queryNorm 以及外部的查询对象的加权，例如保持这个词的布尔查询。这个重要度计算一次。

Weight 针对每个索引段创建一个 Scorer，例如这里是 TermScorer。对一个词来说，这个 Scorer 有评分公式所需要的一切值，除了文档相关的：文档内部的词频 TF，必须从 postings 读出来，以及文档的长度归一化值，这个值在 norm 变量中。

这是 TermScorer 对一个文档用 weight * sqrt(tf) * norm 打分的原因。实际上，对于 TF 值小于 32 都有缓存，所以对于大多数文档来说，只是一个乘法。在缓存中存了 weight * sqrt(tf)。

sqrt(tf)的实现在 DefaultSimilarity 类中的 tf(float freq)方法中。

```
public class DefaultSimilarity extends TFIDFSimilarity {
  // 用 sqrt(freq) 实现
  @Override
  public float tf(float freq) {
    return (float)Math.sqrt(freq);
  }
}
```

TFIDFSimilarity 中的内部类 ExactTFIDFDocScorer 利用缓存打分的代码如下：

```
private final class ExactTFIDFDocScorer extends ExactSimScorer {
  private final float weightValue;
  private final byte[] norms;
  private static final int SCORE_CACHE_SIZE = 32; //缓存大小
  private float[] scoreCache = new float[SCORE_CACHE_SIZE];
```

```
@Override
public float score(int doc, int freq) {
  final float raw =                                   // 计算 tf(f)*weight
    freq < SCORE_CACHE_SIZE                           // 检查缓存
    ? scoreCache[freq]                                //命中缓存
    : tf(freq)*weightValue;                           // 没有命中缓存

  return norms == null ? raw : raw * decodeNormValue(norms[doc]); //列归一化
  }
}
```

跟踪 BooleanScorer 的执行很有挑战性,首先来理解 BooleanScorer 2,这个最容易,然后可以把 BooleanScorer 看成是 BooleanScorer 2 的优化实现。BooleanScorer 2 使用子评分器,例如 ConjunctionScorer 和 DisjunctionSumScorer。

BooleanQuery 并没做多少事情。它的打分器负责调用子打分器的 nextDoc() 和 advance() 方法。

如果有更多不同的查询词匹配,那么应该奖励这样的文档。例如对于 "A OR B OR C" 这样的查询,如果一个文档中 A、B、C 这 3 个不同的词各出现了一次,另一个文档中 A 词出现了 2 次,B 词出现 1 次,而 C 词并没有出现,则前一个文档的分值应该比后一个文档的分值高。

Coord()方法在 TFIDFSimilarity 类中定义成一个抽象的方法:

```
public abstract class TFIDFSimilarity extends Similarity {
 @Override
 public abstract float coord(int overlap, int maxOverlap);
}
```

它组合子打分器的分值,基于匹配多个子打分器应用协调工厂。协调因子奖励包含更多可选的子句的文档。如果所有的子查询都匹配,则协调因子是 1。如果 3 个子条件匹配 2 个,则协调因子是 2/3。在 DefaultSimilarity 类中实现成为: overlap / maxOverlap。

```
public class DefaultSimilarity extends TFIDFSimilarity {
 @Override
 public float coord(int overlap, int maxOverlap) {
    return overlap / (float)maxOverlap;
 }
}
```

测试协调因子的效果:

```
RAMDirectory directory = new RAMDirectory();
Analyzer analyzer = new StandardAnalyzer();
IndexWriterConfig config = new IndexWriterConfig(analyzer);
IndexWriter writer = new IndexWriter(directory, config);
Document doc1 = new Document();

Field f1 = new TextField("content", "common hello world", Store.YES);
```

```
doc1.add(f1);
writer.addDocument(doc1);
Document doc2 = new Document();
Field f2 = new TextField("content", "common common common", Store.YES);
doc2.add(f2);
writer.addDocument(doc2);
writer.close();
IndexReader reader = DirectoryReader.open(directory);

IndexSearcher searcher = new IndexSearcher(reader);
QueryParser parser = new QueryParser("content", new StandardAnalyzer());
Query query = parser.parse("common world");
TopDocs docs = searcher.search(query, 2);
for (ScoreDoc doc : docs.scoreDocs) {
    System.out.println("docid : " + doc.doc + " score : " + doc.score);
}
```

输出结果：

```
docid : 0 score : 0.581694
docid : 1 score : 0.13156208
```

输出结果显示：前一个文档分值比后一个文档高。
SloppySimScorer 对模糊匹配打分，返回 spanScore 的分值。

3.9 分析文本

如何表示文档和查询串中的文本？也就是说，索引什么样的字符串，按什么样的字符串查询。查询和索引英文的例子见表 3-10。

表 3-10　文档和查询串中的文本

文　　本	查　　询
… Official Michael Jackson website …	michael jackson
小写和大写	
… Michael Jackson's new video …	michael jackson
切分问题	
… Fender Music, the guitar company …	Fender guitars
词干化/原型化	
… Microsoft WindowsXP …	windows xp
词边界	
… the cat is on the table …	cat table
停用词	

3.9.1 Analyzer

全文索引是按词组织的，词是怎么来的？是 Analyzer 类分出来的。分析文本的工作交给了 Analyzer 类，这也是它唯一的工作。Lucene 把索引中的单词称为 Token。Analyzer 把接收的字符串流解析成单词序列，也就是 Token 流，如图 3-30 所示。

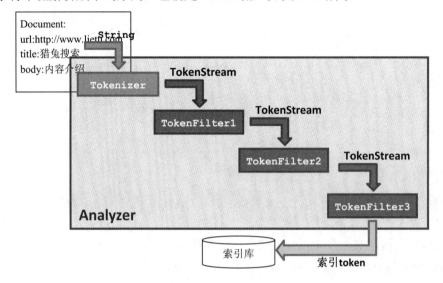

图 3-30　分析文档

Analyzer 可以使用两类组件完成功能。Tokenizer 切分词，Filter 对切分结果做后续处理。例如，处理英文的 EnAnalyzer：

```
public class EnAnalyzer extends Analyzer {
    @Override
    protected TokenStreamComponents createComponents(String fieldName,
                                                    Reader reader) {
        //以空格方式切分 Token
        final Tokenizer source = new WhitespaceTokenizer(reader);
        //小写化
        result = new LowercaseFilter(source);
        return new TokenStreamComponents(source, result);
    }
}
```

Lucene 把查询串中的单词叫作 Term，如图 3-31 所示。

Lucene 在两个地方使用 Analyzer：索引文档的时候和按关键词搜索的时候。索引文档的时候 Analyzer 分析出的词成为倒排索引中的词。

好的 Analyzer 可以把文本转换成更适合检索的形式。例如，搜索用户一般不关心单词是大写还是小写的，也就是说查找是大小写无关的。Analyzer 往往会做的一个简单的转换，即小写化单词。

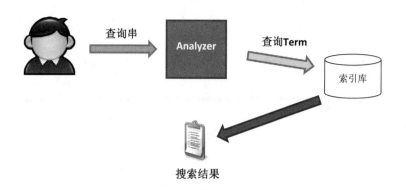

图 3-31　分析查询串

例如，对于执行小写化的 Analyzer 会把文档中的"CAT"转化成倒排索引中的"cat"。也就是说"cat"会关联这个文档。

源文本是：

The Cat in the Hat.

分析器的输出是：

[the]　[cat]　[in]　[the]　[hat.]

注意，这里用括号包装每一个 Token。

索引和搜索使用的 Analyzer 必须能够切出同样的 Token。如果索引的时候不使用 LowerCaseFilter，而搜索的时候使用 LowerCaseFilter，就会导致很多词搜索不出来。因为这时候用"cat"找不到"CAT"。

往往对不同语言的文档和查询使用不同的 Analyzer。例如，ClassicAnalyzer 专门用来处理英文。在 org.apache.lucene.analysis 中包含很多常用的 Analyzer。其中，TokenStream 类用来进行基本的分词工作；Analyzer 类是 TokenStream 的外围包装类，负责整个解析工作。有人把文本解析比喻成人体的消化过程，输入食物，分解出有用的氨基酸和葡萄糖等。Analyzer 类接收的是整段的文本，解析出有意义的词语。

TokenStream 是生产者，产生 Token。生成词索引的程序是消费者，调用 TokenStream 的 incrementToken()方法得到一个 Token。

很少有人会为了重复使用酱油瓶而上街打酱油，但是却有很多人用一个固定的杯子喝水，重用喝水的水杯。因为如果不重用水杯，喝水的容器每天都要换新的，而一瓶酱油可以用很久。处理一个文档会产生很多个词，如果不重用 Token，就会有大量的 Token 需要回收。incrementToken()方法通过 Attribute 对象可以实现在分析文档的时候少创建对象。

TokenStream.incrementToken()方法修改内部状态，然后通过相关的 Attribute 得到词本身，以及它所在的位置信息等。incrementToken()方法返回一个布尔值，如果返回 true 则表示后面还有 Token 可以取；如果返回 false，则表示后面没有更多的 Token 了。示例代码如下：

```
String text = "The door has been opened for us.";

StandardAnalyzer analyzer = new StandardAnalyzer();
TokenStream stream = analyzer.tokenStream("title", new StringReader(text));
```

```
stream.reset();          //重置 TokenStream
// 增加 Token 表示的字符串属性
CharTermAttribute term = stream.addAttribute(CharTermAttribute.class);

while (stream.incrementToken()) {
    System.out.println(term.toString()); // 逐个单词输出
}
analyzer.close();
```

对于一个 TokenStream 的实例来说,每个 Attribute 的实现类都只有一个,这个 TokenStream 返回的所有 Token 都重用这个 Attribute 实现类。例如,上面的 CharTermAttribute 实现类 term 只有一个。除了表示词本身的 CharTermAttribute,还有表示词所在位置的 OffsetAttribute,如下所示。

```
CharTermAttribute termAtt =
tokenStream.addAttribute(CharTermAttribute.class);  //词
OffsetAttribute offsetAtt =
tokenStream.addAttribute(OffsetAttribute.class);    //用于高亮显示
tokenStream.addAttribute(PositionIncrementAttribute.class); //用于处理同义词
```

TokenStream.reset()不是在内部调用的,是在外部重复使用 TokenStream 的时候调用的。例如:

```
WhitespaceTokenizer tokenizer =
    new WhitespaceTokenizer(Version.LUCENE_48, reader);
// 消耗 Token …
reader.reset();
tokenizer.reset();
// 可以再次消耗 …
```

在调用 stream.incrementToken()方法之前,必须调用 TokenStream.reset()方法,否则 stream.incrementToken()方法可能抛出异常。

自定义的 Tokenizer 往往需要重写 reset()方法。

TokenStream API 的流程如下:

(1) 初始化 TokenStream,TokenStream 增加属性到 AttributeSource;
(2) 消费者调用 reset();
(3) 消费者从 TokenStream 找属性,并把它想要得到的值存在本地变量中;
(4) 消费者调用 incrementToken()直到它返回 false,每次调用后,消费属性值;
(5) 消费者调用 end(),以便执行流结束操作;
(6) 当结束使用 TokenStream 后,消费者调用 close()方法来释放资源。

在 Lucene 中通过 WhitespaceTokenizer、WordDelimiterFilter、LowercaseFilter 处理英文字符串的流程,如图 3-32 所示。

第 3 章 Lucene 的原理与应用

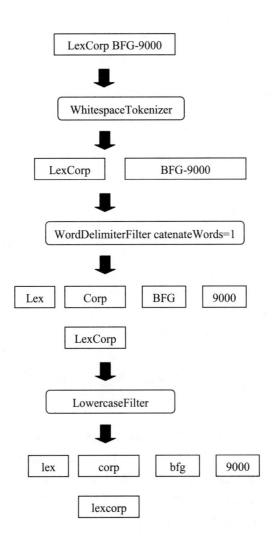

图 3-32 Lucene 处理英文字符串流程

Analyzer 内部包括一个对字符串的处理流程，先由 Tokenizer 做基本的切分，然后再由 Filter 做后续处理。Analyzer 的内部类 TokenStreamComponents 封装了 TokenStream 的输出部分。分析器需要在 createComponents(String, Reader) 方法中创建自己的 TokenStreamComponents。调用分析器的 tokenStream(String, Reader) 方法时，会重用 TokenStreamComponents。

```
//重用TokenStreamComponents
TokenStream stream = analyzer.tokenStream("", new StringReader(text));
```

tokenStream()方法重用 TokenStreamComponents 的代码如下：

```
public final TokenStream tokenStream(final String fieldName,
                        final Reader reader) throws IOException {
TokenStreamComponents components =
            reuseStrategy.getReusableComponents(fieldName);
   final Reader r = initReader(fieldName, reader);
```

```
    if (components == null) {
      components = createComponents(fieldName, r);
      reuseStrategy.setReusableComponents(fieldName, components);
    } else { //重用TokenStreamComponents
      components.setReader(r);
    }
    return components.getTokenStream();
}
```

一个简单的处理英文的分析器例子如下:

```
public class EnAnalyzer extends Analyzer {
    @Override
    protected TokenStreamComponents createComponents(String fieldName,
                                                    Reader reader) {
        //以空格方式切分Token
        final Tokenizer source = new WhitespaceTokenizer(reader);
        //处理特殊的分隔符
        TokenStream result = new WordDelimiterFilter(source, 1);
        //小写化
        result = new LowercaseFilter(result);
        return new TokenStreamComponents(source, result);
    }
}
```

Lucene 调用 EnAnalyzer 来做分词。当调用 EnAnalyzer.incrementToken()取得下一个 Token 相关信息的时候，EnAnalyzer 内部会调用 LowercaseFilter.incrementToken()方法，然后 LowercaseFilter.incrementToken()方法内部会调用 WordDelimiterFilter.incrementToken()方法，最终 WordDelimiterFilter.incrementToken()方法会调用 WhitespaceTokenizer.incrementToken()方法从输入的 StringReader 中分出一个词。

StopFilter 用于去除停用词。判断一个词是否为停用词的方法是查询一个固定的停用词表。去掉停用词的目的是去掉一些没多大意义的词。但有些英文短语完全是由停用词组成的。例如，"to be or not to be"是一句经典台词，意思是：生存还是毁灭。采用 StopFilter 之后，这句话就被完全去掉了，所以有时候不用 StopFilter。

英文有词性变换，例如名词的单复数形式或者动词的被动形式。PorterStemFilter 用于词干化，它使用 Porter stemming 算法。例如，通过执行后缀替换，把 falling 转换成 fall，把 troubled 转换成 trouble。Snowball 是一个专门用于写词干化算法的字符串处理语言。Porter stemming 算法最初是用 Snowball 实现的，Lucene.Analysis.PorterStemmer 是一个用 Java 重写的版本。

如果碰到单词 saw，词干化可能仅返回 s，而原型化(Lemmatization)则可能依据这个词是动词还是名词而决定返回 see(看见)或者 saw(锯子)。

TokenStream 相关类的层次结构如下：

```
* TokenStream
```

第 3 章　Lucene 的原理与应用

```
    * Tokenizer
        * KeywordTokenizer
        * CharTokenizer
            * WhitespaceTokenizer
            * LetterTokenizer
                * LowerCaseTokenizer
        * StandardTokenizer

    * TokenFilter
        * LowerCaseFilter
        * StopFilter
        * StandardFilter
        * PorterStemFilter
        * LengthFilter
        * ISOLatin1AccentFilter
```

LowerCaseFilter 和 StopFilter 位于 org.apache.lucene.analysis.core 包，而 StandardAnalyzer、StandardFilter 和 StandardTokenizer 则位于 org.apache.lucene.analysis.standard 包。这些类都位于 lucene-analyzers-common-6.3.0.jar 文件中。

一般在 Tokenizer 的子类实际执行词语的切分。需要设置的值有：和词相关的属性 termAtt，和位置相关的属性 offsetAtt。在搜索结果中高亮显示查询词时，需要用到和位置相关的属性。但是在切分用户查询词时，一般不需要和位置相关的属性。此外还有声明词类型的属性 TypeAttribute。Tokenizer 的子类需要重写 incrementToken()方法。通过 incrementToken()方法遍历 Tokenizer 分析出的词，当还有词可以获取时，incrementToken()方法返回 true，已经遍历完 Token 流时，返回 false。

基于属性的方法把无用的词特征和想要的词特征分隔开。每个 TokenStream 在构造时增加它想要的属性。在 TokenStream 的整个生命周期都保留一个属性的引用。这样在获取所有和 TokenStream 实例相关的属性时，可以保证属性的类型安全。例如，在 CnTokenStream 中增加词属性。

```
protected CnTokenStream(TokenStream input) {
    super(input);
    termAtt = addAttribute(CharTermAttribute.class);
}
```

这里的 addAttribute(CharTermAttribute.class)方法相当于创建一个 CharTermAttribute 的实例。

在 TokenStream.incrementToken()方法中，Token 流仅仅操作在构造方法中声明过的属性。例如，如果只要分词，则只需要 CharTermAttribute。其他的属性，例如 PositionIncrementAttribute 或者 PayloadAttribute 都被这个 TokenStream 忽略掉了，因为这时不需要其他的属性。

char[]比 String 性能更好，例如下面的 TokenStream。

```
public boolean incrementToken() throws IOException {
    if (input.incrementToken()) {
```

```
            final char[] termBuffer = termAtt.termBuffer();
            final int termLength = termAtt.termLength();
            if (replaceChar(termBuffer, termLength)) {
                termAtt.setTermBuffer(output, 0, outputPos);
            }
            return true;
        }
        return false;
    }
```

虽然也可以通过 termAtt 对象中的 term() 方法返回词，但这个方法返回的是字符串，直接返回字符数组的 termBuffer() 方法性能更好。

3.9.2 TokenStream

"wi-fi" 这个词可以用 WordDelimiterFilter 创建出如下 3 个 Token：

`wi(posinc=1), fi(posinc=1), wifi(posinc=0)`

这样的做法，"wifi" 简单地堆放在 "fi" 上是一个损失。PositionLengthAttribute 修正了这个错误，它允许一个词声明它的跨度，因此不会丢失任何信息。

Lucene 的 TokenStream 类产生文档字段中要索引的 Token 序列。API 是一个迭代器。调用 incrementToken 推进到下一个 Token，然后通过查询特定的属性来获得当前 Token 的细节。

例如，CharTermAttribute 持有 Token 的文本。为了能够突出显示这个词，OffsetAttribute 中有原始字符串对应此 Token 的字符开始和结束偏移量。

TokenStream 实际上是一个链。从一个 Tokenizer 开始，把输入字符串初步分解成 Token 序列，然后再用任意数量的 TokenFilter 修改或者插入已有的 Token 序列。TokenStream 是一个抽象类，Tokenizer 和 TokenFilter 是它的两个子类。Tokenizer 用 StringReader 构造，而 TokenFilter 则用另外一个 TokenStream 构造。

也可以使用 CharFilter 在分出 Token 前预处理字符，例如去掉 HTML 标记，或者根据正则表达式替换字符，同时保留合适的偏移量到原来的输入字符串。Analyzer 根据需要创建 TokenStream 的工厂类。

Lucene 包含支持 34 种语言的 Tokenizer 和 TokenFilter，包括英语、汉语、韩语、德语、法语、日语、阿拉伯语、俄语、西班牙语、葡萄牙语等。没有专门处理维吾尔语的 Analyzer，需要自己开发。维吾尔语文本中已经有空格，不需要分词，但是需要处理词的原型化。

StandardAnalyzer 可以用来分析 Unicode 文本。ClassicAnalyzer 专门用来处理英文，所以速度比 StandardAnalyzer 快。StandardAnalyzer 返回一个 TokenStream 的实例。

标记化一个简单的例子：fast wi fi network is down。假设保留停用词。当看成一个图时，Token 看起来像这样：每个节点是一个位置，而每个边是一个 Token。TokenStream 列举一个有向无环图，每次一个弧。TokenStream 本质上是根据要形成倒排索引的文本生成一个词图，如图 3-33 所示。

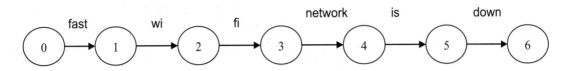

图 3-33　TokenStream 词图

接下来，增加 SynoynmFilter 到分析链，应用下面这些同义词：

```
fast → speedy
wi fi → wifi
wi fi network → hotspot
```

产生如图 3-34 所示的 TokenStream 词图。

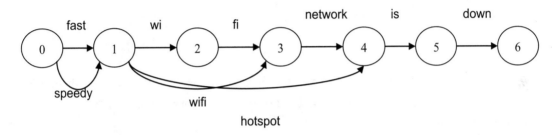

图 3-34　增加同义词后的 TokenStream 词图

PositionIncrementAttribute 告诉我们当前 Token 从边的开始位置前进了多少个位置。PositionIncrementAttribute 的值默认是 1。例如，对于 Token(fast) 来说，PositionIncrementAttribute 的值就是 1。

PositionLengthAttribute 说明边向前到达什么位置，也就是前进多少个节点。例如，Token(hotspot) 的 PositionLengthAttribute 的值是 3。

OffsetAttribute 中虽然保存了词的开始和结束位置的信息，但只是用于高亮显示，没有用于查询过程。OffsetAttribute 保存词在文档中的实际位置，而 PositionLengthAttribute 保存词的语义位置。

除了 SynonymFilter，还有其他几个分析组件可以产生词图。如 Kuromoji 的 JapaneseTokenizer 对于复合 Token 输出分解后的词。例如，ショッピングセンター(购物中心)这样的 Token 也会有一个可选的路径：ショッピング(购物)跟着センター(中心)。很多中文词是复合词，因此也需要用词图来描述词之间的关系。日文分词包 kuromoji 用到了 PositionLengthAttribute：

```
public final class JapaneseTokenizer extends Tokenizer {
  private final PositionLengthAttribute posLengthAtt =
      addAttribute(PositionLengthAttribute.class);

  @Override
  public boolean incrementToken() throws IOException {
    //…
```

```
            posLengthAtt.setPositionLength(token.getPositionLength());
            //…
        }
}
```

当 ShingleFilter 和 CommonGramsFilter 合并两个输入 Token 的时候,将设置位置长度为 2,也就是把 PositionLengthAttribute 设置成 2。

```
public final class ShingleFilter extends TokenFilter {
  private final PositionLengthAttribute posLenAtt =
                      addAttribute(PositionLengthAttribute.class);

  public final boolean incrementToken() throws IOException {
       //...
       posLenAtt.setPositionLength(builtGramSize);
       //...
   }
}
```

对于 CommonGramsFilter 来说,输入:the quick brown fox,输出|"the","the-quick"|"brown"|"fox"|。Token[the-quick]的 PositionIncrementAttribute 的值就是 0,而 PositionLengthAttribute 的值是 2。

此外,WordDelimiterFilter、DictionaryCompoundWordTokenFilter、HyphenationCompoundWordTokenFilter、NgramTokenFilter、EdgeNGramTokenFilter 等还需要增加 PositionLengthAttribute 的值。

但是,索引目前忽略 PositionLengthAttribute,它只注意 PositionIncrementAttribute。需要修改 Codec API,然后让与位置相关的查询用到这个值。

查询分析器也忽略了位置长度。如果增加对 PositionLengthAttribute 的支持以后,就可以在查询时运行图式分析器,用词图做查询扩展,更容易得到正确的结果。

Filter 需要考虑多词同义。例如:

```
(US | united states) gold medals
```

这里,US 的长度是 2,这样它就能跳到 gold,而 united 和 states 的长度都只是 1。AutoPhrasingTokenFilter(https://github.com/LucidWorks/auto-phrase-tokenfilter)实现了这个功能。

3.9.3 定制 Tokenizer

Lucene 通过 Tokenizer 类切分文本。通常使用的 Tokenizer 类是 StandardTokenizer,StandardTokenizer 虽然处理中文时仍然是单字的方式,但是在 Lucene 2.3 以后的版本中已经可以区分出像 lsdfsd@sina.com 这样的邮件地址,或者是 AT&T 这样的公司名字以及 U.S.A. 这样的缩写。下面用个简单的测试类验证:

```
StandardAnalyzer cna = new StandardAnalyzer();
String input = "公司:P&G 信箱:webmaster@ertt.com 网址: http://www.lietu.com";
```

```
TokenStream ts = cna.tokenStream("dummy", new StringReader(input));

CharTermAttribute termAtt = ts.addAttribute(CharTermAttribute.class);
OffsetAttribute offsetAtt = ts.addAttribute(OffsetAttribute.class);
TypeAttribute typeAtt = ts.addAttribute(TypeAttribute.class);
ts.reset();

while (ts.incrementToken())
    System.out.println(termAtt.toString() + " /"
            + offsetAtt.startOffset() + "," + offsetAtt.endOffset()
            + " " + typeAtt.type());
cna.close();
```

输出结果是：

```
公 0 1 <CJ>
司 1 2 <CJ>
p&g 3 6 <COMPANY>
信 7 8 <CJ>
箱 8 9 <CJ>
webmaster@ertt.com 10 28 <EMAIL>
网 29 30 <CJ>
址 30 31 <CJ>
http 33 37 <ALPHANUM>
www.lietu.com 40 53 <HOST>
```

我们可以根据需要修改 StandardTokenizer，它的实现是根据编译原理中的状态机理论使用工具自动生成的。早期版本采用 JavaCC 生成，Lucene 2.3 的版本采用 JFlex 生成。相对于 JavaCC，JFlex 生成的代码性能更好。JFlex 首先生成 NFA，然后把 NFA 转换成 DFA，并最小化这个 DFA。

JFlex 把语法声明规范转换成实际的 Java 源代码。可以从 http://www.jflex.de/ 下载 JFlex 的最新版本。解压后，执行 C:\JFLEX\bin\jflex.bat 批处理文件。输入声明规范 StandardTokenizerImpl.jflex，输出 StandardTokenizerImpl.java。

比如我们希望增加对电话号码的识别，就需要修改 StandardTokenizerImpl.jflex。StandardTokenizerImpl.jflex 由%%分开的 3 部分组成，分别是用户代码段、选项声明段和语法规则段。比如我们定义电话号码是由 3 位区号和 8 位号码以及中间的 "-" 组成的。

```
// 电话号码
TEL       = ({DIGIT}){3} "-" ({DIGIT}){8}
          | ({DIGIT}){4} "-" ({DIGIT}){7}
```

这样，当碰到 010-51667560 这样的电话号码时，就可以正确识别成<TEL>类型的 Token。用重复多次切分字符串的方法测试 Tokenizer 的 reset()方法能够正常工作：

```
String text = "包裹的退回方法";

QuestionAnalyzer analyzer = new QuestionAnalyzer();
```

```java
TokenStream stream = analyzer.tokenStream("title", new StringReader(
        text));
stream.reset(); // 重置 TokenStream
// 增加 Token 表示的字符串属性
CharTermAttribute termAtt = stream
        .addAttribute(CharTermAttribute.class);
OffsetAttribute oa = stream.addAttribute(OffsetAttribute.class);
TypeAttribute tp = stream.addAttribute(TypeAttribute.class);

while (stream.incrementToken()) {// 逐个单词输出
    System.out.print("词:" + termAtt + " " + oa.startOffset() + "~"
            + oa.endOffset());
    System.out.println(" 词性:" + tp.type());
}

text = "为什么邮件被海关扣留了? ";
stream = analyzer.tokenStream("title", new StringReader(text));
stream.reset(); // 重置 TokenStream
while (stream.incrementToken()) {// 逐个单词输出
    System.out.print("词:" + termAtt + " " + oa.startOffset() + "~"
            + oa.endOffset());
    System.out.println(" 词性:" + tp.type());
}

analyzer.close();
```

3.9.4 重用 Tokenizer

假设索引中有两列：关键词列和类别列。关键词列通过自动提取关键词产生，类别列通过自动分类产生。为了生成一段文本的关键词，需要先做分词，然后做关键词提取。为了自动分类，也要先做分词，然后做文本分类。重用 CnTokenizer 的代码如下：

```java
TeeSinkTokenFilter source1 = new TeeSinkTokenFilter(new
CnTokenizer(reader1));
TeeSinkTokenFilter.SinkTokenStream sink1 = source1.newSinkTokenStream();
TeeSinkTokenFilter.SinkTokenStream sink2 = source1.newSinkTokenStream();

TeeSinkTokenFilter source2 = new TeeSinkTokenFilter(new
CnTokenizer(reader2));
source2.addSinkTokenStream(sink1);
source2.addSinkTokenStream(sink2);

TokenStream final1 = new LowerCaseFilter(source1);
TokenStream final2 = source2;
TokenStream final3 = new KeyWordsExtract(sink1);// 提取关键词
TokenStream final4 = new Classify(sink2);// 分类
```

第 3 章 Lucene 的原理与应用

```
d.add(new Field("f1", final1));
d.add(new Field("f2", final2));
d.add(new Field("f3", final3));
d.add(new Field("f4", final4));
```

在这个例子中，sink1 和 sink2 都会得到 reader1 和 reader2 中的 Token。

3.9.5 有限状态转换

经常需要在内存中保存整个词典。有限状态转换(FST)使用的内存少，而且查找速度快。采用的实现算法是从已经排好序的输入递增地建立最小的无加权的 FST。这很适合 Lucene，因为索引已经按顺序存储了所有的词。

从本质上讲，FST 是一个 SortedMap<ByteSequence,SomeOutput>。如果边按顺序排好。而且使用正确的方法表示边，它比 SortedMap 的实现需要的 RAM 要少得多，但 FST 在查找过程中 CPU 计算量更大一些。低内存占用对 Lucene 是至关重要的，因为一个索引就可以轻松获取几百万(有时甚至数十亿)个不同的词。

先看一下如何使用 org.apache.lucene.util.fst 中的 FST 实现。FST 编码成 byte[]，也就是 BytesRef 对象。首先把输入字符串转换成 BytesRef，然后进一步转换成 IntsRef。

```
String inputValue = "cat"; // 词
BytesRef scratchBytes = new BytesRef();
scratchBytes.copyChars(inputValue);
IntsRef scratchInts = new IntsRef();
Util.toIntsRef(scratchBytes, scratchInts);
System.out.println(scratchInts); //输出 [63 61 74]，也就是 c、a、t 这 3 个字符的编码 Unicode
```

根据预先排好序的输入构建最小无环 FST。FST 只能一次性构建好。使用 org.apache.lucene.util.fst.Builder 构建 FST。调用 Builder.add()方法增加键到值的映射。调用 Builder.finish()方法得到一个 FST 的实例。

构建一个 FST，输入是词，输出是长整数。代码如下：

```
// 输入键必须按 Unicode 顺序排序
String inputValues[] = {"cat", "dog", "dogs"}; //词
long outputValues[] = {5, 7, 12}; //词对应的值

//FST 会共享输出
final PositiveIntOutputs outputs = PositiveIntOutputs.getSingleton();
//构建一个 FST 映射 BytesRef → Long
final Builder<Long> builder = new Builder<Long>(FST.INPUT_TYPE.BYTE1,
outputs);
//输入
BytesRef scratchBytes = new BytesRef();
IntsRef scratchInts = new IntsRef();   //创建可以重用的整数数组
for (int i = 0; i < inputValues.length; i++) {
    scratchBytes.copyChars(inputValues[i]);
    //从输入到输出
```

```
        builder.add(Util.toIntsRef(scratchBytes, scratchInts),
outputValues[i]);
}

//得到 fst
final FST<Long> fst = builder.finish();
```

采用一次一个边的方式遍历 FST，在查找的过程中解码 byte[]。通过键检索的代码如下：

```
Long value = Util.get(fst, new BytesRef("dog"));   //通过输入的词找对应的值
System.out.println(value); // 7
```

IntsRef 是 int[]。通过值检索的代码如下：

```
IntsRef key = Util.getByOutput(fst, 12);   //通过输出得到输入
System.out.println(Util.toBytesRef(key, scratchBytes).utf8ToString());
//输出：dogs
```

因为 org.apache.lucene.util.fst 使用泛型实现，所以代码难懂。

可以用 MappingCharFilter 转换字符。例如，把查找英文人名 Karoline 转换成搜索 Caroline，可以把其中的 K 替换成 C。MappingCharFilter 可以用 FST 实现。

FST 的寻找时间通常只是整个查询执行时间的一小部分，通常大部分时间都花在遍历文档列表上，一旦词典上的查找完成，只是特别依赖词典的查询(例如 FuzzyQuery 和 WildcardQuery)真正对词典有压力。

3.9.6 索引数值列

一个索引库类似一个数据库的表结构，但是在 Lucene 2.9 以前的版本中只能存储字符串，如果是日期或者数字，需要专门的方法转换成字符串后再索引。新的版本可以直接存储数字：

```
document.add(new NumericField(name).setIntValue(value));
```

高级搜索中可能会用到范围查询，例如查询价格区间范围。

Lucene 2.9 以前版本实现的区间查询，性能上有问题。RangeQuery 采用扩展成 TermQuery 来实现，如果查询区间范围太大，RangeQuery 会导致 TooManyClausesException。为了避免产生这个异常，ConstantScoreRangeQuery 没有采用扩展成布尔查询的方式实现，而是采用 Filter 来实现，但是当索引很大的时候，查询速度会变慢。

在 Lucene 2.9 以后的版本中，用 Trie 结构索引日期和数字等类型。例如，把 521 这个整数索引成为：百位是 5、十位是 52、个位是 521。这样重复索引的好处是可以用最低的精度搜索匹配区域的中心地带，用较高的精度匹配边界。这样就减少了要搜索的 Term 数量。例如，TrieRange:[423 TO 642]分解为 5 个子条件来执行：

```
handreds:5 OR tens:[43 TO 49] OR ones:[423 TO 429] OR tens:[60 TO 63] OR
ones:[640 TO 642]
```

TrieRange 的工作原理如图 3-35 所示。

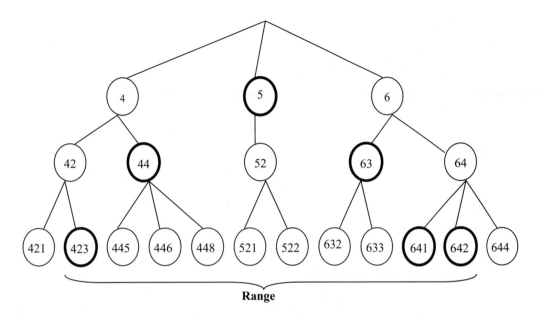

图 3-35　TrieRange 的工作原理

可以用 NumericField 来实现 Trie 结构索引数字。这样做的好处是更有效地按区间查找数字和排序。下面是增加一个整数列到索引的例子：

```
document.add(new NumericField(name).setIntValue(value));
```

为了优化索引性能，可以重用 NumericField 和 Document 实例：

```
NumericField field = new NumericField(name);
Document document = new Document();
document.add(field);

for(all documents) {
 ...
 field.setIntValue(value)
 writer.addDocument(document);
 ...
}
```

然后可以使用 NumericRangeQuery 来查询这样的数字列。例如：

```
Query q = NumericRangeQuery.newFloatRange("weight", //列名称
            new Float(0.3f), //最小值从它开始
            new Float(0.10f),//最大值到它结束
            true,   //是否包含最小值
            true);  //是否包含最大值
```

如果要索引日期列，若为精确到天的搜索，只需要把日期类型的值用 DateTools 转换成 "yyyyMMdd" 格式的字符串，然后再转换成整数来索引就可以了。例如：

```
DateTime d=DateTime.Now;
```

```
int n= Integer.valueOf(d.ToString("yyyyMMdd"));
doc.Add(new NumericField("num", Field.Store.YES, true).SetIntValue(if));
```

如果精确到秒,需要先用 Date.getTime()把日期转换成 long 类型。索引过程如下:

```
NumericField documentDateTimeField = new NumericField("Pub_Date",
                                    1,Field.Store.NO, true);
                                    Document document = new Document();
                                    document.add(documentDateTimeField);

while(hasDoc) {
    documentDateTimeField.setLongValue(scoreDetails.getDocumentDate().getTime());
    writer.addDocument(document);
}
```

查询过程如下:

```
Long begin = esq.getBeginDate().getTime(); //开始时间
Long end = esq.getEndDate().getTime(); //结束时间

NumericRangeQuery rangeQuery = NumericRangeQuery.newLongRange("Pub_Date",
                               1, begin, end,
                               esq.isBeginDateInclusive(),
                               esq.isEndDateInclusive());
```

如果精确到天,可以用 DateTools 把日期转换成 int 类型。搜索过程如下:

```
int int_Date_Begin = Integer.valueOf(DateTools.dateToString(date1,
                     Resolution.DAY));
int int_Date_End = Integer.valueOf(DateTools.dateToString(date2,
                   Resolution.DAY));

SortField sortField = new SortField("Pub_Date",SortField.INT,false);
Sort sort = new Sort(sortField);

//检索
NumericRangeQuery rangeQuery = NumericRangeQuery.newIntRange("Pub_Date",
                               int_Date_Begin,int_Date_End, true, true);
ScoreDoc[] hits = searcher.search(rangeQuery, null, 1000, sort).scoreDocs;
```

可能需要索引数值列并按数值排序。

```
NumericField priceField = new NumericField("price");
Document document = new Document();
document.add(priceField);

for(all documents) {
  ...
  priceField.setIntValue(value);
  writer.addDocument(document);
```

```
...
}
```

按价格升序排列：

```
Sort sort= new Sort(new SortField("price",SortField.INT,false));
ScoreDoc[] hits = searcher.search(query,null,1000,sort).scoreDocs;
```

3.9.7 检索结果排序

检索出来的文档一般按相关度排序后返回。虽然搜索结果中可能有很多文档，但是大部分用户使用搜索引擎查询时只关注搜索结果的首页，所以实际上需要返回最相关的前 n 项结果即可。这个计算过程叫作"top-n 查询"。

因此，IndexSearcher.search()方法设计成为 TopDocs search(Query query, int n)。

Lucene 用优先队列(Priority Queue)记录前 n 个评分最高的文档。通过遍历词计算前 n 个评分最高的文档的伪代码如下：

```
inverted_search(query){
    double[] scores = new double[N]; // 文档对应的评分初始化为 0
    PriorityQueue queue = new PriorityQueue(); //按评分排列的优先队列
    for(t∈query) { //遍历用户查询中的词
        ps = postings(t) //得到词 t 的 posting stream
        while (p = nextposting(ps) ) {//遍历 postings, 也就是所有包含词 t 的文档
            int id = p.id;//得到文档编号
            double weight = p.weight;
            scores[id] += qt * weight;
            if(queue.length()>n) queue.pop();
            queue.insert(id,scores[id]);
        }
    }
    return queue;
}
```

还可以采用并发合并(Parallel Merge)进一步加快计算前 n 个最相关的文档。

3.9.8 处理价格

Java 内部有一个与表示价格相关的类：java.util.Currency。

```
Number number = NumberFormat.getCurrencyInstance(Locale.US).parse("$123.45");
```

对于国际化的搜索网站，价格由值和货币类型组成。货币类型可能是美元或者英镑等。索引中存储两列。

价格一般是精确到小数点后两位。

```
public static String CurrencyToString(System.Double d){
    long l = (long)(d * 100);
    return LongToString(l);
```

```
}
public static Double StringToCurrency(System.String str){
    long l = StringToLong(str);
    System.Double d = (double)l / 100;
    return d;
}
```

查询单个价格，例如"price:4.00USD"。或者查找价格区间，例如"price:[$5.00 TO $10.00]"。

3.10 Lucene 中的压缩算法

倒排索引的大小可能和文档内容本身差不多大。当文档数量多时，倒排索引也会很大。为了节省存储空间，可以采用压缩格式存储倒排索引。从一个硬盘读入索引数据时，采用压缩存储后，能降低磁头需要移动的距离，从而提高性能。另外，为了在搜索系统内部快速地传输文档编号数组，可以压缩文档编号。为了让压缩效果更好，需要尽量做到使文档相似，文档越相似，则文档编号的差别越小。

因为信息存在冗余，所以可以压缩。例如：在生活中登记信息时，如果发现当前这行的信息和上面一行的信息相同，为了少写字，可以采用这种写法：同上。这样的压缩原理称为预测编码(Predictive Encoding)，可以对前后相似的内容压缩。

压缩算法存在两个过程：编码过程和解码过程。编码过程也就是压缩过程，而解码过程就是解压缩过程。编码过程可以时间稍长，而解码过程则需要速度快。这有点类似 ADSL 优化上网速度的机制：用户往往下载文件的时候多，而上传文件的时候少。所以 ADSL 设计成下载速度快，而上传速度慢。因为在索引数据阶段执行编码过程，而在搜索阶段执行解码过程。索引数据的速度可以稍慢，但是搜索速度不能慢。因为索引一般只需要在后台完成一次，而搜索则需要经常调用。

3.10.1 变长压缩

int 类型数据在 Java 里固定占 4 个字节，而全文索引中的文档编号和词频都是正整数，并且大部分其值比较小，所以会降低性能。Lucene 的内部实现考虑到了压缩存储较小的正整数问题。

压缩的原理是出现次数较多的数用较短的编码表示。这样一来，变短的数相对于变长的数更多，文件的总长度就会减少。在倒排索引中，小的数字出现概率大，如图 3-36 所示，说明较小的数值用较短的编码可以取得不错的压缩效果。

Lucene 采用了变长压缩方法(Variable byte encoding)。变长压缩算法的原理是：较小的数使用较短的编码，较大的数使用较长的编码。可以使用变长压缩方法压缩 int 或者 long 类型数据。如果使用一个字节流连续存储多个变长压缩的整数，就需要确定多少个字节对应一个整数。用每个字节的最高位表明是否有更多的字节在后面，如果是 0 表示这个字节是尾字节，1 表示还有后续字节。以字节流 10000010 00000001 00000001 为例，通过检查每个字节的最高位发现：前面两个字节对应一个整数，后面一个字节对应另外一个整数。

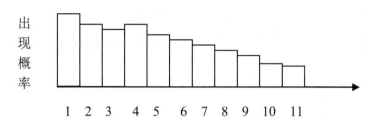

图 3-36　倒排索引中数值的出现频率

VInt 是一个变长的正整数表示格式，是一种整数的压缩格式表示方法。每字节分成两部分，最高位和剩下的低 7 位。最高位表明是否有更多的字节在后面，如果是 0 表示这个字节是尾字节，1 表示还有后续字节。低 7 位表示数值。按如下的规则编码正整数 x：

- if (x < 128)，则使用一个字节(最高位置 0，低 7 位表示数值)；
- if (x < 128×128)，则使用 2 个字节(第 1 个字节最高位置 1，低 7 位表示低位数值；第 2 个字节最高位置 0，低 7 位表示高位数值)；
- if (x < 128^3)，则使用 3 个字节，依次类推。

每字节的低 7 位表明整数的值，可以把 VInt 看成是 128 进制的表示方法，低位优先，也就是说随着数值的增大，向后面的字节进位，从 VInt 值表示可以看出，见表 3-11。

表 3-11　VInt 编码示例

值	二进制编码	十六进制编码
0	00000000	00
1	00000001	01
2	00000010	02
127	01111111	7F
128	10000000 00000001	00 01
129	10000001 00000001	01 01
130	10000010 00000001	02 01
16383	11111111 01111111	7F 7F
16384	10000000 10000000 00000001	00 00 01
16385	10000001 10000000 00000001	01 00 01

从 0～127 的值可以存储在一个字节中，128～128×128−1(= 16383)的值可以存储在 2 个字节中。这里 16383 = 127×128+127−1，二进制表示方式是 11111111 01111111。可以认为 Vint 能表示的整数值范围没有上限。因为小的数值更经常出现，相对于 int 的 4 个字节的表示方法而言，VInt 表示方法对小的数值使用的字节数更少。所以 VInt 提供了正整数的压缩表示，而且解码效率高。

为了方便调试，写一个方法把字节转换成二进制字符串。方法是：首先创建一个初始值为 8 个 0 的字符串缓存，然后把字节中对应位不为 0 的值设置为 1。

```java
public static String toBinaryString(byte n) {
    //创建一个字符串缓存初始值，并设置初始值为 00000000
    StringBuilder sb = new StringBuilder("00000000");
    for (int bit = 0; bit < 8; bit++) { //遍历每一位
        if (((n >> bit) & 1) > 0) { //n 右移 bit 位
            sb.setCharAt(7 - bit, '1'); //设置一个字符的值
        }
    }
    return sb.toString();
}
```

把一个整数的 VInt 编码写入字节缓存：

```java
int data = 824; //待编码的正整数
ByteBuffer buff = ByteBuffer.allocate(4); //写入缓存
while ((data & ~0x7F) != 0) { //如果大于 127
    buff.put((byte) ((data & 0x7f) | 0x80)); // 写入低位字节
    data >>>= 7; // 右移 7 位
}
buff.put((byte) data); //取低 8 位，并写入高位字节
for (int k=0;k<buff.position();++k) //遍历已经写入的字节
    System.out.println(toBinaryString(buff.get(k)));
```

从字节缓存中解压缩一个 VInt 编码的整数：

```java
buff.rewind(); //重置读取缓存的位置

byte b = buff.get(); // 读入一个字节
int i = b & 0x7F; // 取低 7 位的值

// 每个高位的字节多乘个 2^7，也就是 128
for (int shift = 7; (b & 0x80) != 0; shift += 7) { //如果最高位字节是 1，则继续
    if (buff.hasRemaining()) { //有更多的字节待读入
        b = buff.get(); //读入一个字节
        i |= (b & 0x7F) << shift; // 当前字节表示的位乘 2 的 shift 次方
    }
}
System.out.println(i); //解压缩后得到的数
```

org.apache.lucene.store.DataOutput.writeVInt(int i)方法包含了正整数编码成 VInt 的实现。
org.apache.lucene.store.DataInput.readVInt()方法包含了 VInt 解码方法的实现。

3.10.2 Gamma

Gamma 编码是基于位的变长编码。在介绍 Gamma 编码之前首先要了解一元编码和二进制编码。

一元编码:对于数字 N,一元编码的编码长度为 N,使用 N-1 个二进制 1 和末尾一个 0 表示,比如 3 表示成 110。对于大整数来说,一元编码的方式是非常不划算的。

至于二进制编码,就是它在计算机中实际储存的格式。对我们常见的一个十进制的整数用二进制的方法表示,如:3 的二进制表示为 11。

具体到 Gamma 的编码规则,即利用分解函数将待压缩的数字分解为两个因子,之后分别用一元编码和二进制编码来表示这两个因子,具体 3 个步骤如下:

(1) 对于数字 x 分解成 $x=2^N+M$;
(2) 对于 $N+1$ 使用一元编码;
(3) 对于 M 使用比特宽度为 N 的二进制编码。

其中,x 为待压缩的数字;N 和 M 分别为其因子。在分解出两个因子后,分别使用一元编码和二元编码进行压缩。比如:数字 9 经过分解因子为 $9=2^3+1$,对于 2 的指数 3,采取 3+1=4,然后对 4 进行一元编码,编码后为 1110。对于第二个因子 1,用比特宽度为 3 的二进制编码表示为 001,最后将两者结合,十进制 9 的 Gamma 编码结果就是 1110:001。

测试 Gamma 编码的代码示例如下:

```java
public class EliasGammaEncoding {
    public static String encode(int x) { //编码成字符串,测试用
        StringBuilder sb = new StringBuilder();
        while (x > 0) {
            sb.append(x % 2);
            x >>>= 1;
        }
        int N = sb.length();
        for (int i = 1; i < N; i++) sb.append(0);
        return sb.reverse().toString();
    }

    public static int decode(String s) {
        int x = 0, c = 0, i = 0, N = s.length();
        for (; i < N && s.charAt(i) == '0'; i++) c++;
        if (N != 2 * c + 1) throw new IllegalArgumentException("Incorrect encoding: " + s);
        for (; i < N; i++) x = (x << 1) + s.charAt(i) - '0';
        return x;
    }
}
```

当提到 Gamma 编码时,不得不提在此基础上的延伸出来的另一种编码方式——Delta 编码。实际上,Delta 编码就是两次 Gamma 编码的结果。在完成第一个步骤分解之后,我们对 $N+1$ 做一元编码之前,先进行一次 Gamma 编码。比如上文提到的十进制数字 9,第一次 Gamma 算法得到了 $9=2^3+1$,此时 3+1 需要再次 Gamma:$3+1=2^2+0$,因此,采用 Delta 算法得到的十进制数字 9 的最终编码出来就是 110:00:001。

3.10.3 PForDelta

VInt 对每个整数都选择不同的位数来编码,因为有判断分支,导致解码速度慢。和 VInt 按单个整数的压缩算法不同,PForDelta 压缩算法按批压缩整数。每批可以选取连续的 128 个整数。把 128 个整数中的 90%较小的数用统一长度的短编码表示。把剩下的 10%的整数看成是异常的大数,异常数用普通的 4 个字节的 int 类型来表示。一批数的空间有 128 个 b 位槽空间。使用没有用到的 b 位槽空间构造链接表,一个异常数的 b 位槽空间存储碰到下一个异常数的偏移量。异常数的原始值追加存储在后面。VInt 算法中每个整数的编码长度都不一定相同,而 PForDelta 压缩算法对于一批整数只有两个编码长度,分别是 b 和 4。

例如,可以做直方图来统计用多少位就可以表示这批数据中的 90%,如果统计出来是 b 位。也就是说,有 90%的数小于 2 的 b 次方。

假设在 128 个整数中,其中 90%的数是小于 32 的,另外 10%是大于 32 的,那么就把这 128 个整数以 32 为界,把小于 32 的作为一种处理,大于 32 的作为异常情况进行处理。32 是 2 的 5 次方,因此 b=5。整个压缩空间由两部分组成:分配 128×5 个位给 90%的正常数的空间,加上分配给剩下的 10%的异常数的空间。例如:b=5 整数序列是 23, 41, 8, 12, 30, 68, 18, 45, 21, 9, …压缩存储成如图 3-37 所示的形式。

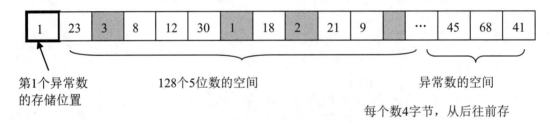

图 3-37 PForDelta 压缩空间示例

为了处理方便,如果有连续 32 个相邻的数是小于 32 的,那么把第 32 个数强制作为异常情况处理。

开源项目 Kamikaze(http://sna-projects.com/kamikaze/)中提供了 PforDelta 的 Lucene 实现版本。Kamikaze 来源于 LinkedIn 网站,是一个 Java 工具包,封装了文档 id 号集合的实现,它实现了排好序的整数段的 PForDelta 压缩算法。Kamikaze 也为 Lucene 提供了倒排索引列表的压缩。PForDelta 压缩算法实现在 com.kamikaze.docidset.compression.P4DsetNoBase 的 compressAlt()方法:

```
public long[] compressAlt(int[] input){
  //缺省情况下,b=32,编码长度是 32
  int BATCH_MAX = 1 << (_b - 1);

  //分配数据空间大小,这里 BASE_MASK=32
  long[] compressedSet = new long[(((( _batchSize) * _b + HEADER_MASK +
_exceptionCount * (BASE_MASK)))>>>6)+1];
```

```
   //把 b 加入数据空间
   copyBits(compressedSet, _b, 0, BYTE_MASK);

   // Offset 是下一个存储值的位置偏移量,在这里 HEADER_MASK 的值是 8
   int offset = HEADER_MASK;
   int exceptionOffset = _exceptionOffset;
   int exceptionIndex = 0;

   //遍历需要压缩的数组
   for (int i = 0; i < _batchSize; i++) {
      // 如果是正常数则放在前面的空间,否则复制到压缩空间最后的位置
      if (input[i] < BATCH_MAX) {
         copyBits(compressedSet, input[i] << 1, offset, _b);
      } else {
         // 记录异常点的编号到 b 位槽空间,并且增加一个位标志
         copyBits(compressedSet, ((exceptionIndex << 1) | 0x1), offset, _b);

         // 把异常数值放到数据空间的最后的位置
         copyBits(compressedSet, input[i], exceptionOffset, BASE_MASK);

         // 增加数据空间最后位置的长度
         exceptionOffset += BASE_MASK;
         exceptionIndex++;
      }

      offset += _b;
   }
   return compressedSet;
}
```

PforDelta 的解码过程:

```
public int[] decompress(long[] compressedSet) {
   int[] op = new int[_batchSize];
   // reuse o/p
   op[0] = _base;

   //异常列表的偏移量
   int exceptionOffset = HEADER_MASK + _b * _batchSize;

   //扩展和修补数据
   for (int i = 1; i < _batchSize; i++) {
      int val = getBitSlice(compressedSet, i * _b + HEADER_MASK, _b);

      if ((val & 0x1) != 0) {
         //碰到一个异常
         op[i] = getBitSlice(compressedSet, exceptionOffset, BASE_MASK);
         exceptionOffset += BASE_MASK;
```

```
    } else {
      op[i] = val >>> 1;
    }
    op[i] += op[i - 1];
  }
  return op;
}
```

PForDelta 的算法在实际的使用中，虽然压缩速度稍慢，但是有非常好的解压速度。因为压缩过程在索引阶段完成，解压过程在搜索阶段完成，所以在全文索引库实现中流行采用 PForDelta 压缩算法。解压速率快的原因是它符合现代计算机的流水线工艺——解压过程中没有判断语句，不会打断 CPU 的流水线；同时，每次 128 个数据可以缓存在 CPU 内。对比于 VInt 算法，由于 VInt 中的 128 是一个较小的数，无论怎样安排判断语句的顺序，都会造成 CPU 流水线的预测失误，从而造成较大的性能损失。

3.10.4 VSEncoding

前面已经总结过很多压缩算法了，总的来说分两类，一类是单个元素压缩，这个主要是考虑每个值出现的频率，很常用的一类编码是无前缀编码，简单来说，就是 N 个值的编码，都不会是另一个值的编码的前缀，这样确保解码的时候无二义性。根据不同的分布，有不同的算法。HUFFMAN 则根据统计信息，算出全局最优的。

另一类是一段的压缩，这一类主要是认为相邻的元素会比较相似。根据不同的情况，也会有不同的算法。那是否会有一个类似 HUFFMAN 的算法，根据统计信息算出最优的编码？

VSEncoding 就是这么一类算法，当然它不一定是最优的。

这个算法主要是把长度为 N 的拉链分成 M 部分，每部分有 3 个部分：一是 B，表示这部分的数字都用 B 个字节表示；一个是 L，表示这部分有多长；第 3 部分就是具体的数字了，共 B*L 个 bit。显然分段不能太长，这样 B 可能会很大，因为 B 要取整段里最大的数字的位数；分段也不能太短，这样 B 跟 L 的开销会比较大。B 与 K 的编码方式待定。

然后用动态规划去求解最优的分割。为了使得问题无后效性，B 与 K 的编码方式必须是基于单个元素的压缩，而且不能是 HUFFMAN 这种需要统计全局信息的。

如果续写一个序列，3 部分的开销都可以明确算出来，这个问题用动态规划就比较好解了，也就是复杂度高低的问题。

这个方法的优劣从理论上不好评价，明显可以看到的是，如果有很多很相近的大数字，这种算法是不如 PForDelta 的。但是这种思路很好，PFroDelta 之类分段压缩的算法，也可以改一下用动态规划的方法确定分段的方法，而不是预定义的方式。但是这样会带来压缩开销的大幅增加，不清楚是否可以接受。

任何整数列如下：考虑 L：

L = 1，3，2，3，5，9，1，3，2

任意分割成一个整数列，每套代表固定的连续化位设置的最低能力。

例如，在 L 上划分如下。

L'= [1，3，2] [3，5，9]，[1，3，2]

最左边的一对 2 位，4 位中的一对，3bit 可设置在所有整数的集合权分别代表，欧莱雅(2 位×3 + 4 位×3 + 3 位×3)的总 27bit 可序列化。如果你改变了分裂，但这种变化的总规模。欧莱雅'for 例如：如果总的大小是 24 位。

3.10.5 前缀压缩

因为索引词是排序后写入索引的，所以前后两个索引词词形差别往往不大。前缀压缩算法省略存储相邻两个单词的共同前缀。每个词的存储格式是：

<相同前缀的字符长度，不同的字符长度，不同的字符>

例如，顺序存储如下 3 个词：term、termagancy、termagant。

不用压缩算法的存储方式是：<4,term> <10,termagancy> <9,termagant>

如果应用前缀压缩算法，实际存储的内容如下：

<4,term> <4,6, agancy> <8,1,t>

Lucene 对于全文索引中的索引词也进行了压缩存储。索引词压缩算法采用了前面描述的前端编码(Front Encoding)。在 TermInfosWriter 类中实现的前缀压缩代码如下：

```
int start = 0;
final int limit = termBytesLength < lastTermBytesLength ? termBytesLength : lastTermBytesLength;
while(start < limit) {
  if (termBytes[start] != lastTermBytes[start])
    break;
  start++;
}

final int length = termBytesLength - start;
output.writeVInt(start);                          //写入相同前缀的长度
output.writeVInt(length);                         //写入差量长度
output.writeBytes(termBytes, start, length);      //写入差量字节
if (lastTermBytes.length < termBytesLength) {
  byte[] newArray = new byte[(int) (termBytesLength*1.5)];
  System.arraycopy(lastTermBytes, 0, newArray, 0, start);
  lastTermBytes = newArray;
}
System.arraycopy(termBytes, start, lastTermBytes, start, length);
lastTermBytesLength = termBytesLength;
```

在 TermVectorsReader 类中实现的解压缩代码如下：

```
start = tvf.readVInt();    //读入相同前缀的长度
deltaLength = tvf.readVInt();//读入差量长度
totalLength = start + deltaLength; //总长度
```

```
final String term; //要解压缩的索引词

if (byteBuffer.length < totalLength) {
  byte[] newByteBuffer = new byte[(int) (1.5*totalLength)];
  System.arraycopy(byteBuffer, 0, newByteBuffer, 0, start);
  byteBuffer = newByteBuffer;
}
tvf.readBytes(byteBuffer, start, deltaLength);
term = new String(byteBuffer, 0, totalLength, "UTF-8");
```

此外，字符串压缩算法还有 BurrowsWheeler 转换等。

3.10.6 差分编码

变长压缩算法对于较小的数字有较好的压缩比。差分编码(Differential Encoding)可以把数组中较大的数值用较小的数来表示，所以可以和变长压缩算法联合使用来实现压缩。差分编码中存储的是排好序的数组序列中前后两个数之间的差异。差分编码压缩过程示例如下：

$$X_1, X_2, ..., X_n \rightarrow X_1, X_2 - X_1, ..., X_n - X_{n-1}$$

差分编码解压缩的过程示例如下：

$$Y_1, Y_2, ..., Y_n \rightarrow Y_1, Y_2 + Y_1, ..., Y_n + Y_{n-1}$$

例如，对于排好序的 DocId 序列：

编码前是：345, 777, 11437, …

编码后是：345, 432, 10660, …

在压缩过程中，先取出一个已经从小到大排好序的数组的第 1 个数字，通过 longToBytes 方法把此数值转换成一个 byte 数组，用 BufferedOutputStream 实例把它写入文件，然后通过一个循环，用一个变量记录数组中后一个数减前一个数的差，如第 2 个数等于数组的第 2 个数减第 1 个数，再通过 writeVLong()方法将这个变量写入文件中。

在解压缩过程中，先用 DataInputStream 实例的 readLong()方法读出文件的第 1 个数字，写入数组，再通过 readVLong()方法把该文件的剩下的数字读取出来，同时把此数组剩下的数字都变成后一个数和前一个数字的和，还原成原来的数据。

```
//把一个 long 型的数据变成二进制
public static byte[] longToBytes(long n)   {
  byte[] buf=new byte[8];//新建一个 byte 数组
  for(int i=buf.length-1;i>=0;i--){
    buf[i]=(byte)(n&0x00000000000000ff);//取低 8 位的值
    n>>>=8;//右移 8 位
  }
  return buf;
}
//把一个 long 型的数据进行压缩
```

```java
public static void writeVLong(long i,BufferedOutputStream dos) throws IOException{
    while ((i & ~0x7F) != 0) {
        dos.write((byte)((i & 0x7f) | 0x80)); //写入低位字节
        i >>>= 7; //右移7位
    }
    dos.write((byte)i);
}
//把一个压缩后的long型的数据读取出来
private static long readVLong(DataInputStream dis) throws IOException {
    byte b = dis.readByte(); //读入一个字节
    int i = b & 0x7F; //取低7位的值
    //每个高位的字节多乘个2^7，也就是128
    for (int shift = 7; (b & 0x80) != 0; shift += 7) {
        if(dis.available()!=0){
            b = dis.readByte();
            i |= (b & 0x7F) << shift; //当前字节表示的位乘2的shift次方
        }
    }
    return i;//返回最终结果i
}
//把long型数组simHashSet写入fileName指定的文件中去
private static int write(long[] simHashSet,String fileName) {
    BufferedOutputStream dos =
            new BufferedOutputStream(new FileOutputStream(fileName));
    byte[] b = longToBytes(simHashSet[0]);// 数组的第一个数字转换成二进制
    dos.write(b);//把它写到文件中
    for (int i = 1; i < simHashSet.length; i++) {
        long deta=simHashSet[i]-simHashSet[i-1];//数组中后一个数减前一个数的差
        writeVLong(deta, dos);// 把这个差值写入文件
    }
    dos.close();
    return simHashSet.length;
}
//从fileName指定的文件中把long型数组读出来
private static long[] read(int len,String fileName) {
    DataInputStream dis = new DataInputStream(new BufferedInputStream(
            new FileInputStream(fileName)));
    long[] simHashSet = new long[len];
    simHashSet[0] = dis.readLong();//从文件读取第一个long型数字放入数组
    for (int i = 1; i < len; i++) {
        simHashSet[i] = readVLong(dis);//读取文件剩下的元素
//将元素都变成数组后一个数和前一个数字的和
        simHashSet[i] = simHashSet[i] + simHashSet[i - 1];
    }
    dis.close();
    return simHashSet;
}
```

3.10.7 静态索引裁剪

为了减少索引的大小，增加搜索性能，有时候使用有损压缩。为了不降低用户搜索的体验，保留前 k 个搜索结果。

基于词的裁剪保证每个词至少有一些项目留在索引中。

静态指事前裁剪，而不是在查询执行阶段。

对每个常见的查询，在完整的索引里执行。记录前 k 个匹配的文档。记录有贡献的，删除没有贡献的文档列表和词。

可以用静态索引裁剪的方法制作小到能够完全放入内存的第一层索引。

3.11 搜索中文

全文索引是按词组织的，词是怎么来的？对于中文文档来说，是中文分词分出来的。Lucene 主要在英文环境下开发和测试，如果用于处理中文，查全率一般，查准率比较差。所以中文信息检索值得特别关注。一般的考虑是，用中文分词来改进查准率。

为了能够保证查全率，往往采用全切分，返回所有可能的词，每个词增加一个概率，相当于返回文本是否包含词所代表的意思的概率分布。也许应该直接给词增加一个 WeightAttribute，但当前只能放在 PayLoad 中。

```
public final class CnTokenizer extends Tokenizer {
    //设置term跨基本词的范围，用于短语检索
    private final PositionLengthAttribute posLenAtt =
            addAttribute(PositionLengthAttribute.class);

    @Override
    public boolean incrementToken() throws IOException {
        //...
        posLenAtt.setPositionLength(word.getLength());
    }
}
```

如果查询用 AND 的方式包括全切分出来的词，则文档中的子串一般都可以找出来。例如：文档中包括"大学生活动中心"，切分成"大学生/活动/中心"。再加上字索引 "大/学/生/活/动/中/心"。用户输入"大学"，查询变成"大 AND 学 AND 大学"，将无法找出这个文档。如果使用全切分，则可以从文档中分出"大""学""大学"这 3 个词。实际使用"(大 AND 学)OR 大学"这样的查询。

搜索"瓜"时，最好不要出现"瓜子"，搜索"花"时，最好不要出现"花生"。把原子词放在原子词表 atomwords.txt 中。原子词表中的词不用再次切分，而复合词可以再次切分。保护原子词的 AtomFilter：

```
public final class AtomFilter extends TokenFilter {
    private CharTermAttribute termAtt;
```

```java
private OffsetAttribute offsetAtt;
private PositionIncrementAttribute posIncr;
private char[] curTermBuffer;
private int curPos;
private int curTermLength;
private int i = 0;

private List<char[]> atomBuffer;

public AtomFilter(TokenStream in) {
    super(in);
    this.termAtt = addAttribute(CharTermAttribute.class);
    this.offsetAtt = addAttribute(OffsetAttribute.class);
    this.posIncr = addAttribute(PositionIncrementAttribute.class);
}

public List<String> automSplit(String word) {
    AtomTernarySearchTrie atst = AtomTernarySearchTrie.getInstance();
    List<String> words = new ArrayList<String>();

    for (int i = 0; i < word.length();) {
        String w = atst.getMatch(word, i);
        if (!"".equals(w)) {
            words.add(w);
            i += w.length();
        } else {
            words.add(word.substring(i, i + 1));
            i++;
        }
    }

    return words;
}

public List<char[]> automSplit(char[] word, int len) {
    AtomTernarySearchTrie atst = AtomTernarySearchTrie.getInstance();
    ArrayList<char[]> words = new ArrayList<char[]>();

    for (int i = 0; i < len;) {
        char[] w = atst.getMatch(word, i, len);
        if (w != null) {
            words.add(w);
            i += w.length;
        } else {
            words.add(new char[] { word[i] });
            i++;
        }
    }
}
```

```java
        if (words == null || words.size() == 1)
            return null;

        return words;
    }

    @Override
    public boolean incrementToken() throws IOException {
        // 把词再次分成原子词或单字
        if (atomBuffer != null && (curPos < atomBuffer.size())) {
            offsetAtt.setOffset(i, i + atomBuffer.get(curPos).length);
            termAtt.copyBuffer(atomBuffer.get(curPos),
                        0, atomBuffer.get(curPos).length);
            posIncr.setPositionIncrement(0);
            i += atomBuffer.get(curPos).length;
            curPos++;
            return true;
        }

        if (input.incrementToken()) {
            curTermBuffer = termAtt.buffer();
            curTermLength = termAtt.length();
            i = offsetAtt.startOffset();
            curPos = 0;
            atomBuffer = automSplit(curTermBuffer, curTermLength);
            return true;
        }
        return false;
    }
}
```

在 NgramAnalyzer 里面调用 AtomFilter。

```java
public class NgramAnalyzer extends Analyzer {
    public TokenStream tokenStream(String fieldName, Reader reader) {
        BigramDictioanry dict = BigramDictioanry.getInstance("./dic/");
        TokenStream result = new NgramTokenizer(reader,dict);
        result = new AtomFilter(result);
        return result;
    }
}
```

《现代汉语语法信息词典》中说明了一个词是单纯词(原子词)还是合成词。
可以在索引阶段或查询阶段应用中文分词:
- 词索引 + 词查询：无法保证查全。
- 字索引 + 字词混合查询。
- 字词混合索引 + 字词混合查询。

- 不仅考虑词，而且还用词性来改进。

三级回退机制，词匹配上的文档最优先，然后是原子词匹配上的文档，最后是字匹配上的文档。

如"相机""照相机"这样的搜索；搜索"女"，要把所有带"女"字的都搜索出来，"女装"也要搜索出来。

3.11.1 Lucene 切分原理

通常不需要直接调用分词的处理类 Analyzer，而是由 Lucene 内部来调用，其中：

- 在做索引阶段调用 addDocument(doc)时，Lucene 内部使用 Analyzer 来处理每个需要索引的列，如图 3-38 所示。

```
IndexWriter index = new IndexWriter(indexDirectory,
                    new CnAnalyzer(),  //用支持分词的分析器
                    !incremental,
                    IndexWriter.MaxFieldLength.UNLIMITED);
```

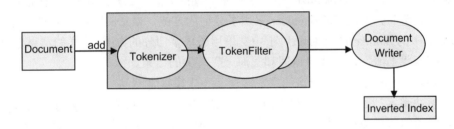

图 3-38 Lucene 对索引文本的处理

- 在搜索阶段调用 QueryParser.parse(queryText)来解析查询串时，QueryParser 会调用 Analyzer 来拆分查询字符串，但是对于通配符等查询不会调用 Analyzer。

```
Analyzer analyzer = new CnAnalyzer();        //支持中文的分词
QueryParser parser = new QueryParser("title", analyzer);
```

因为在索引和搜索阶段都调用了分词过程，索引和搜索的切分处理要尽量一致。所以分词效果改变后需要重建索引。另外，可以采用速度快的版本，用来在搜索阶段切分用户的查询词，另外用一个准确切分的慢速版本用在索引阶段的分词。

为了测试 Lucene 的切分效果，下面是直接调用 Analyzer 的例子：

```
Analyzer analyzer = new CnAnalyzer(); //创建一个中文分析器
//取得 Token 流
TokenStream ts = analyzer.tokenStream("myfield",new StringReader("待切分文本"));
while (ts.incrementToken()) {//取得下一个词
    System.out.println("token: "+ts));
}
```

测试 Analyzer 所调用的 CnTokenizer：

```
String t = "ATM接口测试";
CnTokenizer tokenizer = new CnTokenizer(new StringReader(t));

while(tokenizer.incrementToken()){
    CharTermAttribute termAtt = 
tokenizer.getAttribute(CharTermAttribute.class);
    System.out.println(termAtt); //打印词
}
```

3.11.2 Lucene 中的 Analyzer

Lucene 中处理中文的常用方法有 3 种。以"咬死猎人的狗"这句话的输出结果为例：
(1) 单字方式：[咬] [死] [猎] [人] [的] [狗]；
(2) 二元覆盖的方式：[咬死] [死猎] [猎人] [人的] [的狗]；
(3) 分词的方式：[咬] [死] [猎人] [的] [狗]。

例如：《新华字典》或者《现代汉语词典》是按字索引的。

Lucene 中的 StandardTokenizer 采用了单字分词的方式。CJKTokenizer 采用了二元覆盖的实现方式。笔者开发的 CnTokenizer 采用了分词的方式，本章将介绍部分实现方法。

Lucene 中的 StandardAnalyzer 对于中文采用了单字切分的方式。这样的结果是单字匹配的效果。搜索"上海"，可能会返回和"海上"有关的结果。

CJKAnalyzer 采用了二元覆盖的方式实现。小规模搜索网站可以采用二元覆盖的方法，这样可以解决单字搜索"上海"和"海上"混淆的问题。采用中文分词的方法适用于中大规模的搜索引擎。猎兔搜索提供了一个基于 Lucene 接口的 Java 版中文分词系统。

可以对不同的索引列使用不同的 Analyzer 来切分。例如可以对公司名采用 CompanyAnalyzer 来分析，对地址列采用 AddressAnalyzer 来分析。这样可以通过更细分化的切分方式来实现更准确合适的切分效果。

例如把"唐山聚源食品有限公司"拆分成以下所示的结果，见表 3-12。

表 3-12 公司名拆分结果表

词	开始位置	结束位置	标注类型
唐山	0	2	City
聚源	2	4	KeyWord
食品	4	6	Feature
有限公司	6	10	Function

这里的开始位置和结束位置是指词在文本中的位置信息，也叫作偏移量。例如"唐山"这个词在"唐山聚源食品有限公司"中的位置是 0~2。OffsetAttribute 属性保存了词出现在文档中的位置信息，也就是开始位置和结束位置。TypeAttribute 属性保存了词的类型信息。

切分公司名的流程如图 3-39 所示。

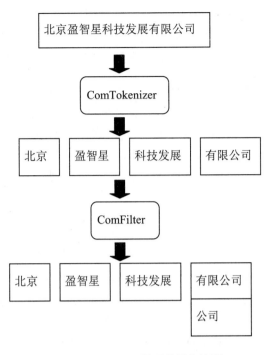

图 3-39 Lucene 处理公司名流程

专门用来处理公司名的 CompanyAnalyzer 实现如下：

```
public class CompanyAnalyzer extends Analyzer {
    @Override
    protected TokenStreamComponents createComponents(String f, Reader reader) {
        //调用 ComTokenizer 切分公司名
        final Tokenizer source =
      new ComTokenizer(AttributeFactory.DEFAULT_ATTRIBUTE_FACTORY, reader);
        return new TokenStreamComponents(source);
    }
}
```

URL 地址：http://news.bbc.co.uk/sport1/hi/football/internationals/8196322.stm。
可以分成如下的形式：

[http] [news.bbc.co.uk] [sport1] [hi] [football] [internationals] [8196322] [stm]

专门用来处理 URL 地址的 URLAnalyzer 实现如下：

```
public class URLAnalyzer extends Analyzer {
    @Override
    protected TokenStreamComponents createComponents(String fieldName,
            Reader reader) {
        // 以/切分 Token
        final Tokenizer source = new UrlTokenizer(
            AttributeFactory.DEFAULT_ATTRIBUTE_FACTORY, reader);
```

```
        // 小写化
        TokenStream result = new LowerCaseFilter(Version.LUCENE_43, source);
        return new TokenStreamComponents(source, result);
    }
}
```

对不同的数据写了不同用途的分析器,需要在同一个索引中针对不同的索引列使用。一般情况下只使用一个分析器。为了对不同的索引列使用不同的分析器,可以使用 PerFieldAnalyzerWrapper。在 PerFieldAnalyzerWrapper 中可以指定一个缺省的分析器,也可以通过 addAnalyzer()方法对不同的列使用不同的分析器。例如:

```
PerFieldAnalyzerWrapper aWrapper =
    new PerFieldAnalyzerWrapper(new CnAnalyzer());
  aWrapper.addAnalyzer("address", new AddAnalyzer());
  aWrapper.addAnalyzer("companyName", new CompanyAnalyzer());
```

在这个例子中,将对所有的列使用 CnAnalyzer,除了地址列使用 AddAnalyzer,公司名称列使用 CompanyAnalyzer,URL 地址使用 URLAnalyzer。像其他分析器一样,PerFieldAnalyzerWrapper 可以在索引或者查询解析阶段使用。PerFieldAnalyzerWrapper 是 AnalyzerWrapper 的子类。AnalyzerWrapper 使用 PerFieldReuseStrategy 存储每列对应的 TokenStreamComponents。

3.11.3 自己写 Analyzer

开发分词,除了要在分词项目中导入核心包 lucene-core-6.3.0.jar,还要导入 lucene-analyzers-common-6.3.0.jar。

分词做好以后,要套上 Tokenizer 的接口才能在 Lucene 中使用。最后用自定义的 Analyzer 调用支持中文的 Tokenizer。

一开始就初始化 CharTermAttribute 和 OffsetAttribute 这样的属性:

```
public final class CnTokenizer extends Tokenizer {
    private final CharTermAttribute termAtt = addAttribute(CharTermAttribute.class);//词
    private final OffsetAttribute offsetAtt = addAttribute(OffsetAttribute.class);//位置
}
```

CnTokenizer 的构造方法如下:

```
CnTokenizer(factory,input, dict);
```

下面是采用正向最大长度匹配实现的一个简单的 Tokenizer。

```
public class CnTokenizer extends Tokenizer {
    private static TernarySearchTrie dic = new TernarySearchTrie("SDIC.txt");//词典
    private CharTermAttribute termAtt;// 词属性
    private static final int IO_BUFFER_SIZE = 4096;
```

```java
private char[] ioBuffer = new char[IO_BUFFER_SIZE];

private boolean done;
private int i = 0;// i 是用来控制匹配的起始位置的变量
private int upto = 0;

public CnTokenizer(Reader reader) {
    super(reader);
    this.termAtt = addAttribute(CharTermAttribute.class);
    this.done = false;
}

public void resizeIOBuffer(int newSize) {
    if (ioBuffer.length < newSize) {
        // Not big enough; create a new array with slight
        // over allocation and preserve content
        final char[] newCharBuffer = new char[newSize];
        System.arraycopy(ioBuffer, 0, newCharBuffer, 0, ioBuffer.length);
        ioBuffer = newCharBuffer;
    }
}

@Override
public boolean incrementToken() throws IOException {
    if (!done) {
        clearAttributes();
        done = true;
        upto = 0;
        i = 0;
        while (true) {
            final int length = input.read(ioBuffer, upto, ioBuffer.length
                    - upto);
            if (length == -1)
                break;
            upto += length;
            if (upto == ioBuffer.length)
                resizeIOBuffer(upto * 2);
        }
    }

    if (i < upto) {
        char[] word = dic.matchLong(ioBuffer, i, upto);// 正向最大长度匹配
        if (word != null) {// 已经匹配上
            termAtt.setTermBuffer(word, 0, word.length);
            i += word.length;
        } else {
            termAtt.setTermBuffer(ioBuffer, i, 1);
```

```
            ++i;// 下次匹配点在这个字符之后
        }
        return true;
    }
    return false;
    }
}
```

为了能在 Lucene 中连续执行，CnTokenizer 需要重写 reset()方法。

```
@Override
public void reset() throws IOException {
  super.reset();
  this.i=0;
  this.done = false;
  this.upto = 0;
}

@Override
public void reset(Reader input) throws IOException {
  super.reset(input);
  reset();
}

@Override
public void end() throws IOException {
  // 设置最后的偏移量
  final int finalOffset = correctOffset(upto);
  offsetAtt.setOffset(finalOffset, finalOffset);
}
```

中文中的"的""了"等虚词在文档中出现频率高，但却不太可能搜索这些词，这样不被索引的词叫作停用词。Analyzer 可能会去掉一些停用词。

```
public class NgramAnalyzer extends Analyzer {
    @Override
    protected TokenStreamComponents createComponents(String fieldName,
                                                      Reader reader) {
        BigramDictioanry dict = BigramDictioanry.getInstance("./dic/");
        Tokenizer source = new NgramTokenizer(reader,dict);
        return new TokenStreamComponents(source);
    }
}
```

需要 Tokenizer 的工厂类 CnTokenizerFactory。

```
public class CnTokenizerFactory extends TokenizerFactory implements
        ResourceLoaderAware {
    public CnTokenizerFactory(Map<String, String> args) {
        super(args);//给超类构造函数传参
```

第 3 章　Lucene 的原理与应用

```
            dicPath = args.get("dicDir");
        }
        static BigramDictioanry dict; // 分词词典
        private String dicPath;

        @Override
        public Tokenizer create(AttributeFactory factory, Reader input) {
            if (dicPath != null && dict==null) {
                dict = BigramDictioanry.getInstance(dicPath);
            }
            return new NgramTokenizer(factory,input, dict);
        }
        @Override
        public void inform(ResourceLoader loader) {
            if (dicPath != null) {
                System.out.println("词典路径=" + dicPath);
                dict = BigramDictioanry.getInstance(dicPath);
            } else {
                System.out.println ("未设置词典路径");
            }
        }
}
```

3.11.4　Lietu 中文分词

Lietu 中文分词程序由 seg.jar 的程序包和一系列词典文件组成。通过系统参数 dic.dir 指定词典数据文件路径。我们可以写一个简单的分词测试代码：

```
String sentence = "有关刘晓庆偷税案";
//输入句子，返回单词组成的数组
String[] result = com.lietu.seg.result.Tagger.split(sentence);
for (int i=0; i<result.length;i++){
    System.out.println(result[i]);
}
```

使用分词的时候为了高亮显示关键字，必须保留词的位置信息。如果分词处理的网页有很多无意义的乱码，这些乱码导致的无意义的位置信息存储甚至可能会导致 5 倍以上的膨胀率。

3.11.5　字词混合索引

由于未登录词的识别准确率相对较低，当使用分词索引时，直接搜索商标名称、企业字号等词的时候，可能会漏掉一些错误切分的文档。例如输入"润泓"，它可能代表一个企业名称，但概率分词的结果难以预料，如果索引内容中没有正确切分出"润泓"，则不会返回"润泓"相关的文档，即使内容中已经提到这个词。如果仅仅按字对内容建立索引，却不会出现这样的问题。所以，一种解决方法是同样的内容存两列，一列按字索引，另外一列按词索引。这个问题本身是多粒度的语义理解问题。除了以字为粒度的理解和以词为粒度的理解，还可以增加词组级别或者更粗粒度的理解。

另外一种解决方法是某列按词索引的同时，也按字索引。具体实现上就是对内容按字切分的同时也按词切分。为了更准确地匹配结果，还需要区分出索引中包含一个 Term 的文档是按字切分出来的还是按词切分出来的。所以可以增加一个 SourceAttribute。

对英文来说，不可能把文档看成是一个字符一个字符的。对中文来说，因为很多字都有独立的意义，所以可以把文档看成是一个字符一个字符的。

字词混合索引对于小规模文档的网站搜索往往是必要的。为了实现字词混合搜索，需要修改 Analysis，在切分的时候，正确地输出多粒度切分结果。适当地修改对用户查询词的匹配方式，不要返回过多的查询结果。另外可以考虑修改搜索结果排序的打分方式。如果匹配上深加工的分词或语义信息，相关度算比较高，如果只是字面匹配上，相似度则比较低。可以把字面匹配上算作一种对搜索结果的平滑，赋予较低的相关度分值。

为了在分词的基础上实现字词混合索引，首先增加一个分词后处理的 Filter。SingleFilter 把词再次拆分成一个个的字。这样返回的 Token 序列会出现位置重叠的情况。PositionIncrementAttribute 属性为 0 用来说明当前 Token 和上一个 Token 处在重叠的位置。这里，字生成出的 Token 的 PositionIncrementAttribute 属性是 0，代表不增加偏移量。

一个词有多种词形，例如英语中的现在进行时或者过去式。文本中包含任何一个词形都应该能够匹配上词干相同的查询词。例如：索引的文本中出现了 bound to 这个短语，在对 bound 处理的时候会提取出两个词干，一个是 bound 本身，一个是 bond。假设返回的 Token 序列是[bound] [bond] [to]。这个时候如果不进行特殊处理，保持 PositionIncrementAttribute 的缺省值 1 不变，则由于额外生成的 Token[bond]导致[bound]与[to]的位置间隔不为 1 了，而是 2。在进行短语查询"bound to"时，由于在建立好的索引中 bound 和 to 之间的相对距离为 2，所以将不会返回结果，这样就导致了一种可能错误的结果，即用户希望查询时任何一个词干匹配成功都返回结果，而此时却不能实现。但是如果在处理 bound 产生的两个词干时把 bond 这个 Token 的 PositionIncrementAttribute 值设置为 0，则就会防止这种现象出现，因为两个词干中的任何一个与"to"的距离都是 1。

因此可以借助 Token 的 PositionIncrementAttribute 来实现，把派生出来的 Token 的 PositionIncrementAttribute 设置为 0，以此表示多个词出现在文本的同一个位置。把词再次拆分成字的 SingleFilter 实现如下：

```java
public final class SingleFilter extends TokenFilter {
    private CharTermAttribute termAtt; //词属性
    private OffsetAttribute offsetAtt; //位置属性
    private PositionIncrementAttribute posIncr; //位置增量属性
    private char[] curTermBuffer; //要再次切分开的词缓存
    private int curPos; //在词中的位置
    private int curTermLength; //长度缓存
    private int i = 0; //单字所在总的偏移量

    public SingleFilter(TokenStream in) {
        super(in);
```

```java
        this.termAtt = addAttribute(CharTermAttribute.class);
        this.offsetAtt = addAttribute(OffsetAttribute.class);
        this.posIncr = addAttribute(PositionIncrementAttribute.class);
    }

    @Override
    public boolean incrementToken() throws IOException {
        // 把词分成单个字
        if (curPos < curTermLength && curTermLength > 1) {
            offsetAtt.setOffset(i, i + 1);
            termAtt.copyBuffer(curTermBuffer, curPos , 1);
            posIncr.setPositionIncrement(0);
            curPos++;
            i++;
            return true;
        }
        if (input.incrementToken()) {
            curTermBuffer = termAtt.buffer();
            curTermLength = termAtt.length();
            curPos = 0;
            i = offsetAtt.startOffset();
            return true;
        }
        return false;
    }
}
```

对于数字串和英文字符,不需要再分拆分成单字。可以把 SingleFilter 和 CnTokenizer 搭配在一起使用:

```java
public class CnAnalyzer extends Analyzer {
    public TokenStream tokenStream(String fieldName, Reader reader) {
        TokenStream stream = new CnTokenizer(reader);//先分词
        stream = new SingleFilter(stream);//然后再次切分出单字
        return stream;
    }
}
```

或者可以把分词的方法和 n 元覆盖的方法联合在一起使用:

```java
public class CnAnalyzer extends Analyzer {
    public TokenStream tokenStream(String fieldName, Reader reader) {
        TokenStream stream = new CnTokenizer(reader);//先分词
        stream = new NGramTokenFilter(stream,1,10);//然后用n元覆盖的方法再次切分
        return stream;
    }
}
```

按字搜索采用有序的 SpanNearQuery 实现。SpanNearQuery 叫作跨度查询。其构造方

法是：

```
SpanNearQuery(SpanQuery[] clauses, int slop, boolean inOrder)
```

最后一个参数表示是否有序。如果是true，则在数组中的顺序和文档中的顺序必须一致。

```
//所有单字组成的SpanTermQuery
ArrayList<SpanTermQuery> s = new ArrayList<SpanTermQuery>(v.size());
for (int i = 0; i < v.size(); i++)   {
    Token token = v.get(i);
    if(token.termText().length()==1)   {
        SpanTermQuery singleTermQuery =
            new SpanTermQuery (new Term(field, token.termText()));
        s.add(singleTermQuery);
    }
}
//按字搜索的SpanNearQuery
SpanNearQuery nearQuery =
    new SpanNearQuery(s.toArray(new SpanQuery[s.size()]),s.size(),true);
```

SpanNearQuery 调用了 NearSpansOrdered 类中的方法 docSpansOrdered()来判断同一个文档中的两个 Span 是否是有序的。为了得到更加宽松的匹配环境，可以修改 docSpansOrdered() 方法：

```
return (start1 == start2)?(spans1.end() <= spans2.end()):(start1 < start2);
```

在 spans1.end() <= spans2.end()中间多加了一个等于，这样当两个 Span 都在同一位置时也算这两个 Span 是有序的。

最后在 Lucene 中增加 SingleQueryParser，按词搜索的同时也按字搜索。SingleQueryParser 中的 getFieldQuery()方法实现代码如下：

```
protected Query getFieldQuery(String field, String queryText)  throws
ParseException {
    TokenStream source = analyzer.tokenStream(field, new StringReader(queryText));
    ArrayList<Token> v = new ArrayList<Token>(10);
    Token t;
    while (true)  {
        try  {
            t = source.next();
        }
        catch (IOException e) {
            t = null;
        }
        if (t == null)
            break;
        v.add(t);
    }
    try  {
        source.close();
```

```
        }
        catch (IOException e)    {
            // ignore
        }

        if (v.size() == 0)
            return null;
        else if (v.size() == 1)
            return new TermQuery(new Term(field, ((Token)v.get(0)).termText()));
        else {
            PhraseQuery q = new PhraseQuery();
            q.setBoost(2048.0f);
            ArrayList<SpanQuery> s = new ArrayList<SpanQuery>(v.size());
            for (int i = 0; i < v.size(); i++)    {
                Token token = v.get(i);
                if(token.getPositionIncrement()>0)    {
                    q.add(new Term(field, token.termText()));
                }
                if(token.termText().length()==1)    {
                    SpanTermQuery tmp = new SpanTermQuery (new Term(field,
token.termText()));
                    s.add(tmp);
                }
            }
            BooleanQuery bQuery = new BooleanQuery();
            //用 OR 条件合并两个查询
            bQuery.add(q,BooleanClause.Occur.SHOULD);    //按词搜索
            SpanNearQuery nearQuery = new SpanNearQuery(s.toArray(new
SpanQuery[s.size()]),s.size(),true);
            nearQuery.setBoost(0.001f);
            bQuery.add(nearQuery,BooleanClause.Occur.SHOULD);    //按字搜索
            return bQuery;
        }
}
```

Token 的 positionIncrement 属性值除了会影响到短语查询 phraseQuery，也会影响到高亮显示。TokenSources.StoredTokenStream 没有分配 positionIncrement 信息。这意味着在 Token 流中的所有的 Token 都被认为是相邻的。当使用不连续的 Token 序列时，对 QueryScorer 中的短语高亮有影响。

例如：考虑一个 Token 流，创建 Token 为词干化和非词干化的版本：the fox (jump|jumped)。当使用 TokenSources.getTokenStream(tpv,false)，从索引中取得的 Token 流是：the fox jump jumped。

现在搜索"fox jumped"，然后高亮这个短语查询。搜索会正确地发现文档，而 highlighter 则不能高亮这个短语。因为它认为在"fox"和"jumped"之间有一个额外的单词。只有使用原来的 Analyzer 分析出来的 Token 流 highlighter 才能正常工作。

再考虑相反的情况：the fox did not jump。

"not"是一个停用词，有一个选择是通过增加位置来特别对待停用词，这时 Token 流是：(the,0) (fox,1) (did,2) (jump,4)。当使用 TokenSources.getTokenStream(tpv,false)从索引中重建时，Token 流是：(the,0) (fox,1) (did,2) (jump,3)。因此，短语查询"did jump" 会导致在文本"did not jump"中的"did"和"jump"高亮。

搜索部分需要对 highlighter 包修改，应用补丁 https://issues.apache.org/jira/browse/LUCENE-2035 来让 highlighter 包支持 SingleFilter。

可以对重要的网页建混合索引，对大量不重要的网页采用粗粒度切分。

3.12 搜索英文

为了准确地搜索英文文章，也需要预处理英文文章。对于英文文章，首先是断句，然后逐句分词。为了得到切分出来词的词性，大部分词采用词典匹配的方式返回，这样就能够根据词典中的词性来判断一个词在文本中所用的词性。

3.12.1 英文分词

句子切分并不是一个简单的问题。标点符号"?"和"!"的含义比较单一。但是"."有很多种不同的用法，并不一定在句子的结尾。例如："Mr. Vinken is chairman of Elsevier N.V., the Dutch publishing group."需要排除掉一部分情况。如果"."是某个短语中间的一部分，则它不是句子的结尾。这里的"Mr. Vinken"是一个人名短语。如果这个人名正好不在词典中，则可以根据上下文识别规则识别出这个短语。

```
String text= "Mr. Vinken is chairman of Elsevier N.V., the Dutch publishing
group.";
EnText enText = new EnText(text);
for(Sentence sent:enText){
    System.out.println(sent); //因为输入是一个句子,所以这里只会打印出一个句子
}
```

Java 中的 BreakIterator 类已经包含了切分句子的功能。用它实现一个英文句子迭代器：

```
private final static class SentBreakIterator implements Iterator<Sentence> {
    String text;
    int start;
    int end;
    // 根据英文标点符号切分
    static final BreakIterator boundary = BreakIterator
            .getSentenceInstance(Locale.ENGLISH);

    public SentBreakIterator(String t) {
        text = t;
        // 设置要处理的文本
        boundary.setText(text);
```

```
        start = boundary.first(); // 开始位置
        end = boundary.next();
    } // 用于迭代的类

    @Override
    public boolean hasNext() {
        return (end != BreakIterator.DONE);
    }

    @Override
    public Sentence next() {
        String sent = text.substring(start, end);

        Sentence sentence = new Sentence(sent, start, end);
        start = end;
        end = boundary.next();
        return sentence;
    }
}
```

BreakIterator 分得不太准确。所以我们自己写一个句子切分器。输入当前切分点，找下一个切分点的代码如下：

```
public static int nextPoint(String text, int lastEOS) {
    int i = lastEOS;
    while (i < text.length()) {
        // 跳过短语
        i = skipPhrase(text, i);

        // 然后再找标点符号
        String toFind = eosDic.matchLong(text, i); //匹配标点符号词典
        if (toFind != null) {
            //判断是否为有效的可切分点。例如，在括号中的标点符号不是有效的可切分点
            boolean isEndPoint = isSplitPoint(text, lastEOS, i);
            if (isEndPoint) {
                return i + toFind.length();
            }
            i = i + toFind.length();
        } else { //没找到
            i++;
        }
    }
    return text.length(); //返回最大长度
}
```

SentIterator 是一个用于迭代英文文本返回句子的内部类，实现代码如下：

```
private final static class SentIterator implements Iterator<Sentence> {
    String text;
```

```
    int lastEOS = 0;

    public SentIterator(String t) {
        text = t;
    }

    @Override
    public boolean hasNext() {
        return (lastEOS < text.length());
    }

    @Override
    public Sentence next() {
        int nextEOS = EnSentenceSpliter.nextPoint(text, lastEOS);
        String sent = text.substring(lastEOS, nextEOS);
        Sentence sentence = new Sentence(sent, lastEOS, nextEOS);
        lastEOS = nextEOS;
        return sentence;
    }
}
```

在英文分词中需要约定哪些匹配点可以作为匹配的结束边界。例如单词 apple 中的 a 后面不能作为匹配的结束边界。在切分方案中定义开始边界和结束边界：

```
public class SegScheme {
    public BitSet startPoints;// 可开始点
    public BitSet endPoints; // 可结束点
}
```

根据 FST 返回的 Token 集合设置开始边界和结束边界：

```
public static void addPoints(Collection<Token> tokens,BitSet startPoints,
                BitSet endPoints) {
    for(Token t:tokens){
        endPoints.set(t.end);
        startPoints.set(t.start);
    }
}
```

在匹配词典时使用结束边界去掉一些不可能的匹配：

```
public WordEntry matchWord(String key, int offset, BitSet endPoints) {
    WordEntry wordEntry = null; //词类型

    TSTNode currentNode = rootNode;
    int charIndex = offset;
    while (true) {
        if (currentNode == null) {
            return wordEntry;
        }
```

```
            int charComp = key.charAt(charIndex) - currentNode.spliter;
        if (charComp == 0) {
            if (currentNode.data != null) {
                if(endPoints.get(charIndex)){   //可结束点约束条件
                    wordEntry = currentNode.data; // 候选最长匹配词
                }
                //ret.end = charIndex;
            }
            charIndex++;
            if (charIndex == key.length()) {
                return wordEntry; // 已经匹配完
            }
            currentNode = currentNode.mid;
        } else if (charComp < 0) {
            currentNode = currentNode.left;
        } else {
            currentNode = currentNode.right;
        }
    }
}
```

3.12.2　词性标注

一段英文：Cats never fail to fascinate human beings. They can be friendly and affectionate towards humans, but they lead mysterious lives of their own as well.

标注词性后的结果是：

Cats(n.) never fail(v.) to(prep) fascinate(v.) human(n.) beings(n.). They(pron.) can(aux.) be(v.) friendly(adj.) and(conj.) affectionate(adj.) towards(prep) humans(n.)　but(conj.) they(n.) lead(v.) mysterious(adj.) lives(n.) of(prep) their(n.) own(n.) as well(adv.).

这里用编码来表示词性。括号中的输出是词性编码。汉语中的量词是英语中没有的。例如：件，个，艘。英语中也有一些独有的词性，例如冠词：a、an、the。英文词性编码表见表3-13。

表3-13　英文词性编码表

词　性	名　称
n	名词
adj	形容词
adv	副词
art	冠词
pos	所有格
pron	代词

续表

词 性	名 称
aux	情态助动词
conj	连接词
v	动词
num	数词
prep	介词
punct	标点符号
int	感叹词

词性标注的流程图如图 3-40 所示。

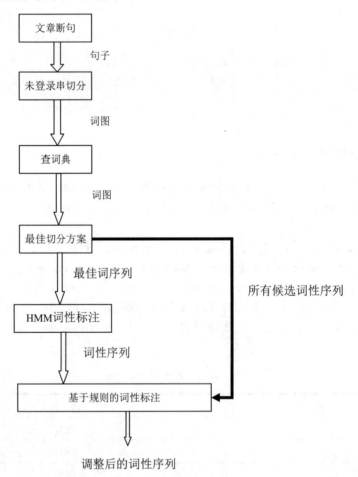

图 3-40　英文分词流程图

关于隐马尔可夫模型(HMM)做词性标注在中文词性标注实现中已经介绍过了。英文词性标注语料库和中文词性标注语料库不一样。可以从 GitHub 网站找到一些免费的。

标注规则：例如 I like it 对应的词性序列[r v r]。

```
key = new ArrayList<PartOfSpeech>();
key.add(PartOfSpeech.pron); //I
key.add(PartOfSpeech.v); //like
key.add(PartOfSpeech.pron); //it
posTrie.addProduct(key);
```

实现代码：

```
public static ArrayList<WordToken> getWords(Sentence sent){
    ArrayList<WordTokenInf> words = Segmenter.seg(sent); //先分词
    WordType[] tags = g.tag(words); //然后标注词性
    
    //再把词性和词本身结合起来，返回完整的词性标注结果
    int i=0;
    ArrayList<WordToken> tokens = new ArrayList<WordToken>();
    
    for(WordTokenInf w:words){
        WordToken t = new
WordToken(w.baseForm,w.termText,w.start,w.end,tags[i]);
        ++i;
        tokens.add(t);
    }
    
    return tokens;
}
```

3.12.3 原型化

同时查询原词和词干化后的词。

```
//两列存储同样的文本内容：一列存词干化后的词，另外一列存未被词干化的原词
private static Document createDocument(String id, String content) {
    Document doc = new Document();
    doc.add(new Field("id", id, StringField.TYPE_STORED));
    doc.add(new Field("stemmedText", content, TextField.TYPE_STORED));
//词干化
    doc.add(new Field("unstemmedText", content, TextField.TYPE_STORED));
//原词
    return doc;
}
```

对不同的列使用不同的分析器，也就是建立一个 PerFieldAnalyzerWrapper 分析器。

```
Map<String, Analyzer> analyzerList = new HashMap<String, Analyzer>();
analyzerList.put("stemmedText",
    new EnglishAnalyzer()); // 查询词干化后的词
analyzerList.put("unstemmedText", new StandardAnalyzer()); // 查询未被词干化
```

```
//的原文
PerFieldAnalyzerWrapper analyzer = new PerFieldAnalyzerWrapper(
        new StandardAnalyzer(), analyzerList);
```

在内存中建立索引，测试这个分析器。

```
Directory directory = new RAMDirectory();
IndexWriterConfig iwc = new IndexWriterConfig(analyzer);
IndexWriter writer = new IndexWriter(directory, iwc);

// 索引一些文档
writer.addDocument(createDocument("1", "foo stocks baz"));
writer.addDocument(createDocument("2", "red green blue"));
writer.addDocument(createDocument("3", "test foo stock test"));
writer.close();

// 查找文档
String queryString = "foo stock";
IndexReader reader = DirectoryReader.open(directory);
// 根据 IndexReader 创建 IndexSearcher
IndexSearcher searcher = new IndexSearcher(reader);
String defaultField = "stemmedText";
QueryParser parser = new QueryParser(defaultField, analyzer); // 用于解析查询语法
// 从查询字符串得到查询对象
Query query = parser.parse(queryString);

// 显示搜索结果
TopDocs topDocs = searcher.search(query, 10);
for (ScoreDoc scoreDoc : topDocs.scoreDocs) {
    Document doc = searcher.doc(scoreDoc.doc);
    System.out.println(doc);
}
```

3.13 索引数据库中的文本

如果要对很大的表建立索引，使用微软的 SQL Server 的 JDBC 驱动一次读取全部的数据到内存中，会导致内存溢出，如果内存足够大，增加 JVM 的使用内存就可以了。当碰到更大的表需要索引的时候，需要使用比较新的 SQL Server JDBC 驱动，设置 Connection 连接参数 responseBuffering=adaptive。

MySQL 数据库通过 JDBC 对大表进行查询时抛出 java.lang.OutOfMemoryError: Java heap space 异常。这是因为默认情况下，MySQL 的 JDBC 驱动会一下子把所有 row 都读取下来，这在一般情况下是最优的，因为可以减少 Client-Server 的通信开销。但是这样做也有一个问题，当数据库查询结果很大时，特别当不能全部放进内存时，就会产生性能问题。

本来，JDBC API 里在 Connection、Statement 和 ResultSet 上都有设置 fetchSize()的方法，

但是 MySQL 的 JDBC 驱动至少要 5.0 以上版本才支持。Statement 一定是 TYPE_FORWARD_ONLY 的，并发级别是 CONCUR_READ_ONLY(即创建 Statement 的默认参数)。以下两句语句选一即可：

(1) .statement.setFetchSize(Integer.MIN_VALUE);

(2) .((com.mysql.jdbc.Statement)stat).enableStreamingResults()。

这样会一行一行地从 Server 读取数据，因此通信开销很大，但可以解决内存问题。

为了同步索引，在需要索引的数据表上写触发器。

触发器监控原则：

无效数据到有效数据的变化；有效数据到无效数据的变化；有效数据到有效数据的变化。更新任务表：

```
CREATE TABLE dbo.SC_PRO_UPDATE (
      ID NUMERIC(12 , 0)  IDENTITY ,  // 自增长 ID,最多 12 位数字
      PRO_ID INT  NULL,  // 商品 ID 号
      OPTION_TYPE INT  NULL,  // 数据操作类型
      FLAG INT  NULL   // 数据表标志
  )
```

触发器：

```
CREATE TRIGGER dbo.TRIGGER_PD_PRODUCT
ON dbo.PD_PRODUCT
FOR  INSERT, UPDATE, DELETE
AS
IF NOT EXISTS (select 1 FROM deleted)
// if it is an insert option
BEGIN
    INSERT INTO SC_PRO_UPDATE(PRO_ID,OPTION_TYPE,FLAG) SELECT ID,1,1 FROM
inserted WHERE SALE_STATUS='0002' AND USE_FLAG='1'
END  // put the valid datas from inserted table into SC_PRO_UPDATE

IF EXISTS(SELECT 1 FROM inserted) AND EXISTS(SELECT 1 FROM deleted) AND
(UPDATE(PD_NAME_CN) OR UPDATE(BRAND_ID) OR UPDATE(VENDOR_ID) OR
UPDATE(SALE_STATUS) OR UPDATE(USE_FLAG) OR UPDATE(PICTURE_65) OR
UPDATE(PD_NAME_EN) OR UPDATE(CREATE_TIME) OR UPDATE(KEY1) OR UPDATE(KEY2)
OR UPDATE(KEY3) OR UPDATE(KEY4) OR UPDATE(KEY5) OR UPDATE(TRANSLATE_INTO))
// if update designated fields
BEGIN
  IF EXISTS(SELECT 1 FROM inserted WHERE SALE_STATUS='0002' AND USE_FLAG='1')
// if insert valid data
    INSERT INTO SC_PRO_UPDATE(PRO_ID,OPTION_TYPE,FLAG) SELECT ID,2,1 FROM
inserted where SALE_STATUS='0002' AND USE_FLAG='1'
  ELSE  // if delete valid data
    INSERT INTO SC_PRO_UPDATE(PRO_ID,OPTION_TYPE,FLAG) SELECT ID,3,1 FROM
deleted where SALE_STATUS='0002' AND USE_FLAG='1'
END   // it is a delete option
```

```
IF NOT EXISTS (SELECT 1 FROM inserted)
// if it is a delete option
BEGIN
    INSERT INTO SC_PRO_UPDATE(PRO_ID,OPTION_TYPE,FLAG) SELECT ID,3,1 FROM
deleted WHERE SALE_STATUS='0002' AND USE_FLAG='1'
END  // put the valid datas from delete table into SC_PRO_UPDATE
```

数据库商品主表，里面存放着商品的基本信息。

当一条数据的 SALE_STATUS(销售状态)= 0002 AND USE_FLAG (使用状态) = 1 时，这件商品才需要被加入到索引。

MySQL5 增加了对触发器的支持。

3.14 优化使用 Lucene

Lucene 优化包括性能的优化和搜索结果优化等方面。性能优化包括在索引阶段的优化和查询阶段的优化。搜索结果优化包括修改匹配结果的打分公式，以影响搜索结果排序，或修改 Tokenizer 等来影响搜索结果数量。

3.14.1 系统优化

Lucene 打开文件数量很多的时候，可能会达到操作系统允许打开文件数量的上限，导致写索引时出现"Too Many Open Files"的错误。Linux 允许打开文件数量的上限默认值是 1024，索引文件数量很容易达到这个限制值。可以通过 ulimit -a 命令来查看当前参数。

```
# ulimit -a
core file size          (blocks, -c) 0
data seg size           (kbytes, -d) unlimited
max nice                        (-e) 0
file size               (blocks, -f) unlimited
pending signals                 (-i) 81920
max locked memory       (kbytes, -l) 32
max memory size         (kbytes, -m) unlimited
open files                      (-n) 1024
pipe size            (512 bytes, -p) 8
POSIX message queues     (bytes, -q) 819200
max rt priority                 (-r) 0
stack size              (kbytes, -s) 10240
cpu time               (seconds, -t) unlimited
max user processes              (-u) 81920
virtual memory          (kbytes, -v) unlimited
file locks                      (-x) unlimited
```

可以使用命令 ulimit -n 65535 增大最大打开文件数量到 65535。为了保存配置参数，需要修改操作系统配置文件/etc/security/limits.conf，增加如下行：

```
*                    -        nofile               65535
```

3.14.2 查询优化

在英文中，用户的输入中包含空格，中文下用户的输入一般不包含空格。英文空格本身既作为单词的边界，在 QueryParser 解释查询语法的时候，又当作了 OR 来处理。例如，在 Lucene 搜索中输入"Olympic Torch" 和 "Olympic OR Torch"返回的结果一样。

因为中文用户的输入往往不包含空格，Lucene 自带的 QueryParser 是为解析英文查询设计的，用来解析中文查询时，返回的 Query 显得太严格了。比如搜索"日本小仓离合器"，只有连续出现"日本小仓离合器"的记录才能匹配上，将无法匹配到下面这段文字：

> 日本：
>   MUTOH 编码器、MUTOH 控制器
> OGURA(小仓)离合器

这时候，需要把查询约束条件改成"日本""小仓""离合器"三个词都出现在同一条记录即可匹配。下面的 BlankAndQueryParser 重写了 QueryParser 类的 getFieldQuery()方法。

```java
public class BlankAndQueryParser extends QueryParser{
    public BlankAndQueryParser(String field, Analyzer analyzer) {
       super(field, analyzer);
    }
    protected Query getFieldQuery(String field, String queryText, int slop)
throws ParseException {
 TokenStream source = analyzer.tokenStream(field, new StringReader(queryText));
        ArrayList<Token> v = new ArrayList<Token>(10);
        Token t;
        while (true) {
           try {
              t = source.next();
           }
           catch (IOException e)
           {
              t = null;
           }
           if (t == null)
              break;
           v.add(t);
        }
        try {
           source.close();
        }
        catch (IOException e)    {
        }
        if (v.size() == 0)
           return null;
        else if (v.size() == 1)
```

```
                return new TermQuery(new Term(field, ((Token)v.get(0)).termText()));
            else    {
                PhraseQuery q = new PhraseQuery();
                BooleanQuery b = new BooleanQuery();
                q.setBoost(2048.0f);
                b.setBoost(0.001f);
                for (int i = 0; i < v.size(); i++)  {
                    Token token = v.get(i);
                    q.add(new Term(field, token.termText()));
                    TermQuery tmp = new TermQuery(new Term(field, token.termText()));
                    tmp.setBoost(0.01f);
                    b.add(tmp, BooleanClause.Occur.SHOULD);
                }
                BooleanQuery bQuery = new BooleanQuery();
                // 用 OR 条件合并两个查询
                bQuery.add(q,BooleanClause.Occur.SHOULD);
                bQuery.add(b,BooleanClause.Occur.SHOULD);
                return bQuery;
            }
        }
        protected Query getFieldQuery(String field, String queryText) throws ParseException {
            return getFieldQuery(field, queryText, 0);
        }
}
```

当用户在搜索框中输入整句话"机床设备的最新退税率是多少？"，直接用上面的 BlankAndQueryParser 搜索可能不会返回任何结果。相对来讲，把"机床"和"退税率"当作搜索关键词是更好的选择。可以把"设备""的""最新""是""多少"这些意义不大的词从必选词中去掉。这些意义不大的词可以放在一个大的停用词表 stopSet 中。这样 getFieldQuery()方法中的循环条件改为：

```
for (int i = 0; i < v.size(); i++){
    Token token = v.get(i);
    q.add(new Term(field, token.termText()));
    //如果在停用词中，则不把这个词加入 AND 条件
    if( stopSet.contains(token.termText()) ) {
           continue;
    }
    TermQuery tmp = new TermQuery(new Term(field, token.termText()));
    tmp.setBoost(0.01f);
    b.add(tmp, BooleanClause.Occur.MUST);
}
```

如果用户输入较短的查询串，上面的布尔查询 b 容易匹配到一些看起来不太相关的长的文本，因为长文本中可能出现各种各样的 Term 组合。这时候，可能希望匹配的 Term 是分布在文本中较集中的区域，为了计算匹配的位置，可以利用 Lucene 中的 Span 对象，它

第 3 章 Lucene 的原理与应用

记录了匹配词在文档中所处的位置。SpanNearQuery 像一个 PhraseQuery 和 BooleanQuery 的组合，匹配文档中的查询词必须出现在一定间隔范围内。利用 SpanNearQuery，BlankAndQueryParser 可以进一步修改为：

```
PhraseQuery q = new PhraseQuery();
q.setBoost(2048.0f);
ArrayList<SpanQuery> s = new ArrayList<SpanQuery>(v.size());
for (int i = 0; i < v.size(); i++){
   Token token = v.get(i);
   q.add(new Term(field, token.termText()));
    if( stopSet.contains(token.termText()) )   {
          continue;
     }
   SpanTermQuery tmp = new SpanTermQuery (new Term(field,
token.termText()));
   s.add(tmp);
}
BooleanQuery bQuery = new BooleanQuery();
// 用 OR 条件合并 PhraseQuery 和 SpanNearQuery
bQuery.add(q,BooleanClause.Occur.SHOULD);
SpanNearQuery nearQuery =
 new SpanNearQuery(s.toArray(new SpanQuery[s.size()]),s.size(),false);
nearQuery.setBoost(0.001f);
bQuery.add(nearQuery,BooleanClause.Occur.SHOULD);
```

这样搜索"美丽人生"时，不会匹配"父母对子女的态度会影响他们日后的性格、感情，乃至整个人生……个性好的人更美丽"。

搜索"iphone4"，结果把"iphone4 贴膜"排在前面。而"苹果(APPLE)iPhone 4 16G 版 3G 手机(白色)WCDMA/GSM"位于后面了。解决方法是：搜索词优先匹配手机类别，而把手机附件类别的匹配结果放在后面。或者把 iphone 扩展出 APPLE 这样的查询词。

3.14.3 实现时间加权排序

情境搜索是对用户搜索的各类数据和信息进行深入理解后，提供给用户最贴切的搜索服务。根据用户搜索行为的时间、地点、输入、需求、习惯、背景等因素，由情境计算得到最适合的搜索结果。在索引列中存储文档的日期、时间和类别等属性。根据用户搜索的时间把索引库中时间最近的文档放在前面。也就是根据非全文性的值来影响排序结果。

有时候希望当两个搜索结果的相关度分值 score 差不多的时候，新的搜索结果显示在前面。这个特性无法用查询语言来实现。FieldScoreQuery 是对 Lucene 的一个最新的增加类。Lucene 2.3 以后才有这个版本，基本上，FieldScoreQuery 的功能是把索引中的一列解释成浮点数并且获得一个分值(score)。然后使用 CustomScoreQuery 把这个分值与最初查询的分值合并起来。索引中增加一个浮点数的列"timestampscore"，这个列的格式是"0." +时间戳，其中时间戳格式为"yyyyMMddhhmm 字符串"(Lucene 索引中只能存储字符串列)。实现代码如下：

```
String query="foo"
QueryParser parser =new QueryParser("name", new StandardAnalyzer());
Query q = parser.parse(query);
Sort updatedSort = new Sort();
FieldScoreQuery dateBooster = new FieldScoreQuery("timestampscore",
FieldScoreQuery.Type.FLOAT);
CustomScoreQuery customQuery = new CustomScoreQuery(q, dateBooster);
Hits results = getSearcher().search(customQuery, updatedSort);
```

结果，最新时间戳有一个稍微高的分数。或者还可以通过权重来进一步调整查询。

尽管这个模型可以运行，但是并不完美。不能同样对待一周前的文档和两周前的文档之间的差距与一年前的文档和一年零一周前的文档之间的差距。可以通过对文档年份取对数后导出的分值来提高这个模型。为了达到这个目标，先深入了解一下 function 包。FieldScoreQuery 基于类 ValueSourceQuery，这个类用一个 ValueSource 得到文档的值。如下的例子创建一个定制的 ValueSource，叫作 AgeFieldSource。这个 ValueSource 和上面的 FieldScoreQuery 类似。然而，AgeFieldSource 并不是仅仅返回索引列中的值，它返回一个基于文档年份取对数后的倒数的权重。

```
public class AgeFieldSource extends FieldCacheSource {
 private int now;

 public AgeFieldSource(String field) {
   super(field);
   now = (int)(System.currentTimeMillis() / 1000);
 }

 @Override
 public boolean cachedFieldSourceEquals(FieldCacheSource other) {
   return other.getClass() == MyFieldSource.class;
 }

 @Override
 public int cachedFieldSourceHashCode() {
   return Integer.class.hashCode();
 }

 @Override
 public DocValues getCachedFieldValues(FieldCache cache,
String field,
IndexReader reader) throws IOException {
   int[] times   = cache.getInts(reader, field);
   float[] weights = new float[times.length];
   for (int i=0; i<times.length; i++) {
```

```
    //权重是文档年份取对数后的倒数
    weights[i] = new Double(1/Math.log((now - times[i]) / 3600)).floatValue();
  }
  final float[] arr = weights;
  return new DocValues() {
    public float floatVal(int doc) {
      return (float) arr[doc];
    }

    public int intVal(int doc) {
      return (int) arr[doc];
    }

    public String toString(int doc) {
      return description() + '=' + intVal(doc);
    }
  };
 }
}
```

索引表示时间的列：

```
//从数据库中取得的值转换成 1970 年 1 月 1 日以来的秒数
int createdTime=(int)(rs.getTimestamp("create_time").getTime()/1000);
doc.add(new NumericField("created").setInt(createdTime));
```

下面的例子显示如何使用这个类：

```
TermQuery termQuery = new TermQuery(new Term("title", "NBA"));
ValueSourceQuery valueQuery = new ValueSourceQuery(new
AgeFieldSource("created"));
CustomScoreQuery query = new CustomScoreQuery(termQuery, valueQuery);
```

类似地，可以实现 pageRankFieldScoreQuery。在索引中建立一个"pagerank"列，用来存储每个文档的 PageRank 值。然后通过 pageRankFieldScoreQuery 来改变排序结果。

使用 FieldScoreQuery，采用列值作为文档的分值。

```
new FieldScoreQuery(PAGE_RANK_FIELD_NAME, FieldScoreQuery.Type.FLOAT);
```

可以使用一个 CustomScoreQuery 把它和其他的查询类型合并到一起，例如：

```
new CustomScoreQuery(otherQuery, pageRankFieldScoreQuery);
```

在 Solr 中可以用 FunctionQuery 来实现，例如：

```
queryWord+= " AND
_val_:\"linear(recip(rord(timestamp),1,10000,10000),10000000,0)\"";
```

Solr 中的函数在 Lucene 中有对应的实现类，见表 3-14。

表 3-14 FieldScoreQuery 实现表

Lucene 类	Solr 函数	描　述
OrdFieldSource	ord	返回根据 Unicode 排序的值。例如，如果索引列仅有 3 个值："apple"、"banana"和"pear"。则 ord("apple")=1，ord("banana")=2，ord("pear")=3
ReverseOrdFieldSource	rord	返回根据 Unicode 排序的值。例如，如果索引列仅有 3 个值："apple"、"banana"和"pear"。则 rord("apple")=3，rord("banana")=2，rord("pear")=1
ReciprocalFloatFunction	recip	recip(x,m,a,b)实现 a/(m*x+b)
LinearFloatFunction	linear	linear(x,m,c)实现 m*x+c。这里，m 和 c 是常量，而 x 是一个任意的函数

Solr 中实现时间加权的 FunctionQuery：

val:\"linear(recip(rord(timestamp),1,10000,10000),10000000,0)\"

在 Lucene 中对应的实现：

```
//主查询对象
TermQuery termQuery = new TermQuery(new Term("title", "NBA"));
//时间加权
ValueSource timeBoostQuery = new LinearFloatFunction (
            new ReciprocalFloatFunction(
            new ReverseOrdFieldSource("timestamp")
            ,1,1000,1000)
             ,10000000,0);

CustomScoreQuery query =
        new CustomScoreQuery(termQuery,
                        new ValueSourceQuery(timeBoostQuery));
```

可以在索引库中记录每条记录的点击次数，然后把点击次数多一些的文档排在前面。

如果有两部分得分 score1 和 score2。score1 很大，比如 128.0，score2 很小，比如 0.21。如果相对这两部分得分进行融合，使用什么办法比较好？

score1 越小越好，score2 越大越好。

```
w1*1/score1 + w2*score2
```

这样试过，不过 score2 比重过大。log 或者对两个 score 做个归一化。log 是熵的形式，不是 *或者 + 能代替的。

3.14.4 词性标注

定义属性：

```
/**
```

```
 * Part of Speech Attribute.
 */
public interface PosAttribute extends Attribute {
  public void setPos(Pos pos);
  public Pos getPos();
}
```

定义实现：

```
public class PosAttributeImpl extends AttributeImpl implements PosAttribute {
  private static final long serialVersionUID = -8416041956010464591L;
  private Pos pos = Pos.OTHER;

  public Pos getPos() {
    return pos;
  }

  public void setPos(Pos pos) {
    this.pos = pos;
  }

  @Override
  public void clear() {
    this.pos = Pos.OTHER;
  }

  @Override
  public void copyTo(AttributeImpl target) {
    ((PosAttributeImpl) target).setPos(pos);
  }

  @Override
  public boolean equals(Object other) {
    if (other == this) {
      return true;
    }
    if (other instanceof PosAttributeImpl) {
      return pos == ((PosAttributeImpl) other).getPos();
    }
    return false;
  }

  @Override
  public int hashCode() {
    return pos.ordinal();
  }
}
```

缺省情况下，Lucene 并不会把词性保存到索引库，需要自己保存词性信息。除此外，在搜索时也需要使用 Payload。org.apache.lucene.search.payloads 提供了发现和使用查询机制。

```java
public class SenPayloadTokenizer extends SenTokenizer {
   public SenPayloadTokenizer(Reader in, String configFile) throws IOException {

    super( in, configFile );

   }

   public Token next() throws IOException {

    Token token = input.next();

    if( token == null ) return null;

    String pos = token.type().trim();

    token.setPayload( new Payload( pos.getBytes() ) );

    return token;

   }
}
```

把词性保存到索引：

```java
public class TypeSavingAnalyzer extends Analyzer {
    @Override
    public TokenStream tokenStream(String fieldName, Reader reader) {
        TokenStream result = new SenTokenizer(reader);
        return new TypeAsPayloadTokenFilter(result);
    }
}
```

TermPositions 提供了一个方法用来遍历一个词的<文档，频率，<位置>* >元组。用 TermEnum 遍历索引库中所有的词，然后再用 TermPositions 遍历词的每个位置：

```java
Directory dir = FSDirectory.open(new File(indexDir));
IndexReader reader = IndexReader.open( dir );

TermEnum te = reader.terms();

String fName = "content";   //只输出内容列的词信息
while( te.next() ){
    Term t = te.term();
    String field = t.field();
```

```
    if( !field.equals( fName ) ) continue;
    TermPositions tps = reader.termPositions( t );
    tps.next(); //下一个文档
    tps.nextPosition(); //下一个位置
    if( tps.isPayloadAvailable() ){
        int len = tps.getPayloadLength();
        byte[] payload = new byte[len];
        tps.getPayload( payload, 0 );
        int freq = tps.freq();
        String pos = new String( payload ); //词性
        System.out.println( t.text() + "freq:" + freq + " pos:" + pos );
    }
    tps.close();
}
te.close();
reader.close();
```

3.14.5 个性化搜索

个性化搜索对搜索缓存造成压力，对用户聚类，按类别缓存。

3.15 实时搜索

实时搜索问题的来源是：搜索界面中的 IndexReader 一旦打开，就相当于对当前的索引建立了一个快照，仅能看到此时间之前提交过的文档，除非此 IndexReader 重新打开，否则即便 IndexWriter 添加了文档，并且提交了，也不能够被搜到。

为了提高性能，索引首先缓存在内存中，如果缓存达到了预定的内存数量，就会写入硬盘。然而，即使 IndexWriter 从缓存中写出这些文档的索引，在没有提交之前 IndexReader 也不能看到这些新加入的文档。但是如果频繁地调用 IndexWriter.commit 就会降低索引的通量。所以不要过于频繁地提交索引。可以通过测试来决定具体加入多少篇文档后再提交索引。

通常情况下，IndexWriter 添加文档后，如果不提交，则不能被新打开的 IndexReader 看到从上次提交至今新增的文档。为了支持实时索引搜索，Lucene 3.0 提供了 IndexWriter.getReader()方法，使得还没有提交的文档也能够被看到。准实时搜索(Near Realtime Search)通过 IndexWriter.getReader()取得索引。

IndexWriter.getReader()返回一个只读的 IndexReader 对象，覆盖了所有索引中提交了和没提交的改变。这个方法在功能上等价于调用 commit，然后使用 IndexReader.open()方法打开一个新索引。但这个方法使用的总时间更少，因为避免了可能很耗时的 commit(long)方法。

```
IndexReader reader = indexWriter.getReader();
...
IndexReader newReader = indexWriter.reopen();
```

```
if (reader != newReader) {//如果新对象和旧的不一样,则关闭旧的对象
    reader.close();
    reader = newReader;
}
```

IndexWriter.getReader()返回的 IndexReader 对象支持重新打开,只要打开了的这个 IndexReader 的 IndexWriter 没有提交。一旦 IndexWriter 已经提交,重新打开它产生的 IndexReader 会报 AlreadyClosedExecption 错误。

```
try {
    if (! reader.isCurrent()){//判断索引是否有更新
        IndexReader newReader = reader.reopen();
        if (newReader != reader) {
            //已经重新打开 reader
            searcher.close();
            reader.close();
        }
        reader = newReader;
        searcher = new IndexSearcher(reader);
        if (searcher == null) throw new ServletException("searcher is null (init a)" );
    }
} catch (CorruptIndexException e){
    throw new ServletException( "Could not refresh reader?"+e.getMessage() );
} catch (IOException e){
    throw new ServletException( "Could not refresh reader?"+e.getMessage() );
} catch (Exception e){
    throw new ServletException( "Could not re-open reader?"+e.getMessage() );
}
```

打开大的索引是个很费时的过程,openIfChanged 会比 open 省时些,因为它只加载改变过的段。

```
IndexReader r = IndexReader.open(dir);
IndexReader nr = IndexReader.openIfChanged(r);
```

通常的检索系统中,建索引和查询是分开的,即建索引是离线的,新的索引会以一定频率(比如每隔5分钟)供查询端使用。对于一些站内检索来说,这种延迟性使得不需要建索引的速度足够快,只要能跟得上文档提交频率就行,查询的效果不必完全精确。

而要取得实时检索效果,典型的思路是:建索引和查询是在一个进程内,这样每一次往索引添加的文档都会被下一次的查询用到。

解决即时搜索麻烦的根源在于:对 Lucene 来说,建索引和查询经常不在同一个进程内。如果建索引和查询由同一个对象管理,解决即时搜索这个问题也许会简单。

可以使用 Zoie(http://sna-projects.com/zoie/)实现实时搜索。Zoie 来源于 SNS 网站

LinkedIn.com 开发的实时检索系统。

Zoie 基于 Lucene 的解决方法是：把索引分成两种，内存索引和硬盘索引。建索引的过程是：首先建立内存索引，因为是内存操作，这个过程通常较快，建完后会重新打开 IndexReader，使查询端能看到最新的索引；当内存中的索引文档数达到阈值(例如 10000)或者间隔时间达到阈值后，一个后台线程就将内存索引合并到硬盘索引里去，完成后清空已经无用的内存索引，并重新打开硬盘索引的 IndexReader 供查询使用。对新打开的索引有个预热 IndexReader 的过程。预热 IndexReader 也就是先用一个常用的查询词搜一次索引。

因为索引合并操作可能会很耗时，所以 Zoie 的内存索引有两个，这使得当一个内存索引在和硬盘索引做合并操作时，另一个内存索引仍能提供建索引的操作。对于查询，使用的索引就包括两个内存索引和一个硬盘索引，所以只要索引在内存里建好，就能查询到最新的数据。

3.16 语义搜索

搜索"海信(Hisense) TLM32V68A"，怎样把"海信 32 英寸液晶电视 TLM32V68A"这个结果搜索出来？增加同义词 "Hisense → 海信"。将鲍鱼归类到海鲜中，如果有人找海鲜类菜谱，包含鲍鱼的菜谱就会被搜索出来。

一个词可能有好几个意思。例如"地道"有两个意思：作为名词时表示在地面下掘成的交通坑道；作为形容词时表示纯粹的，真正的。另外有些词可以表示相同的意思，如"西红柿"和"番茄"是同义词；"招商行"是"招商银行"的简称，也算同义词；"照相机"和"相机"是同义词；"照相机械"和"照相机机械"是同义词；"刮胡刀"和"剃须刀"是同义词。

同义词扩展是必不可少的东西，搜"北大"的时候，把"北京大学"扩展出来，同义词的扩展，有的时候有上下文无关的同义词，有的时候是上下文相关的同义词，那么"小猪"什么情况下扩展到"罗志祥"？"小猪演唱会"是二元关键词，语料也出现过，如果搜"小猪演唱会"，就会把"小猪"扩展到"罗志祥"。

需要把文本中的词义变成可计算的。用 Protégé 这样的专门软件来对词义建模，这样显得太严格了。可以从词典中的解释来提取语义关系。OWL 提取语义。

3.16.1 发现同义词

可以从网络挖掘出同义词词库。例如有下面一些链接到同一个网站的标签。

```
<a href="http://www.cmbchina.com/" target="_blank">中国招商银行</a>
<a href="http://www.cmbchina.com/" target="_blank">招行</a>
<a href="http://www.cmbchina.com/" target="_blank">招商银行</a>
```

从上面的链接中可以提取出同义词：
中国招商银行 招行 招商银行

还可以从语料库挖掘同义词。先找这个词与哪些词有比较强的关联，得到一个向量，所有的词通过向量比较就能知道同义程度。

另外可以通过句子模板发现同义词。从 Google 搜索"大豆 又称",可以发现"大豆(又称黄豆)"。这样可以找到"大豆"的同义词"黄豆"。

英文的同义词词库最著名的是 WordNet,它的下载地址为 http://wordnet.princeton.edu。可以通过 JWNL 读取 WordNet 词典。JWNL 提供了 API 寻找同义词集和下义词等类似的事情。WordNet 的 Prolog 同义词数据库 wn_s.pl 格式:

```
s(synset_id,w_num,'word',ss_type,sense_number,tag_count).
```

第 1 列是语义编码,同一个语义有唯一的语义编码列;第 2 列是单词编号;第 3 列是单词本身;第 4 列是词性;第 5 列代表该词第几个语义;第 6 列是该词在语料库中出现的频率。例如:

```
s(100006026,1,'person',n,1,7229).
s(100006026,2,'individual',n,1,51).
s(100006026,3,'someone',n,1,17).
s(100006026,4,'somebody',n,1,0).
s(100006026,5,'mortal',n,1,2).
s(100006026,6,'human',n,1,7).
s(100006026,7,'soul',n,2,6).
```

中文方面,"同义词词林"和 HowNet(知网)都有现成的同义词库。以"同义词词林"为例,它用树形结构表示了词的同义和上下位关系。

下面是同义词的例子:

Bp31B 表
Bp31B01= 表 手表
Bp31B02= 马表 跑表 停表
Bp31B03= 怀表 挂表
Bp31B04= 防水表 游泳表
Bp31B05= 表针 指针
Bp31B06= 表盘 表面
Bp31B07# 夜光表 秒表 电子表 日历表 自动表
Bp31B08= LONGINES 浪琴
Bp31B09= Rema 瑞尔玛
Bp31B10= ROSSINI 罗西尼
Bp31B11= SEIKO 精工
Bp31B12= SUUNTO 松拓
Bp31B13= Tissot 天梭
Bp31B14= CASIO 卡西欧

整理出这样的词表后,我们通过下面的程序转换成 WordNet 的 Prolog 同义词数据库 wn_s.pl 格式。同义词词林按照该格式整理如下:

```
s(Dk02A12,1,'中学',n,1,0).
s(Dk02A12,2,'国学',n,1,0).
s(Dk02A12,3,'旧学',n,1,0).
```

这个辞典格式还没有词性信息，而词典中的词有词性却没有词义信息。可以借助词典中的词性猜测同义词词林中的义项的词性。

需要给词典文件中的每个词条标注语义。词典文件的格式是：

词:词性:词频:拼音:语义 1,语义 2

例如：

一举一动:26880:2:yijuyidong:Di20B01

实现过程：

(1) 给同义词词林的每个义项增加词性。为了更好地消除歧义，可以给每个义项增加词性。例如："行"作为名词用的时候可能是"行业"的意思，语义编码是 Di18B01，"行"作为量词用的时候可能是"队列"的意思，语义编码是 Dd07B01。可以参考基本词典中对应的每个词的词性，然后统计最有可能的词性。

(2) 给基本词典中词频大于 1 的常用词加上义项到同义词词林。只需要把每个常用词分类到二级子类即可。例如归类到"Be"这类就可以。

"Be 地 貌 地 形"

全部内容如下：

Be01 陆地 原野 沙漠 陆空

Be01A01= 陆地 大陆 陆 洲 地 陆上 次大陆 新大陆 大洲 沂

...

Be03 滩 岸

Be04A50= 恒山 北岳

...

Be05 海洋 江河 溪涧 濒海

Be05A01= 海洋 大海 大洋 五洋 沧海 海洋 溟 瀛 瀛海 沧溟 重溟 大壑 水宗 海域 浅海 汪洋大海 深海

Be05A02= 远洋 重洋

Be05A03= 洋流 海流

Be05A04= 大陆架 陆棚 大陆棚 陆架 大陆坡

Be05A05= 内海 内陆海 陆海

Be05A06= 海岭 海脊

Be05A07# 海湾 海峡 海沟 海弯 海床

Be05A08= 国际海域 公海

最后可以采用字面搜索分类的方法给未标注语义的词标注上语义。

利用 Lucene 筛选最相关词的方法：先对同义词词林中所有的词按字建索引，索引库分为两列，第一列是词本身，第二列是每个词所属的类别；然后把要归类的词作为搜索关键词查找同义词索引库。

例如"海勒斯台高勒河"不在已有的同义词表中。看起来既有"海"又有"河"。所以 Be 类的词匹配的可能比较多。搜索后统计匹配上的词的类别。匹配上结果最多的类作为"海勒斯台高勒河"的类别。

同义词词林格式到 WordNet 格式实现转换的代码如下：

```java
public static void CL2WordNet(String sSourceFile,String save_filename,String dic_file){
      DicCore dic = new DicCore(dic_file);
      BufferedReader fpSource = null;
      BufferedWriter output=null;
      String sLine;
      StringTokenizer st = null;
      String number= null;

      fpSource= new BufferedReader(new FileReader(sSourceFile));
      output = new BufferedWriter(new FileWriter(save_filename));
      while( (sLine = fpSource.readLine()) != null ) {
          if(sLine.length()>9 )  {
             int pos = sLine.indexOf('=');
             if (pos<=0)  {
                continue;
             }
             number = sLine.substring(0,pos);
             st = new StringTokenizer(sLine.substring(pos+1).trim()," \t\n\r");
             String word;
             int count = 0;
             String wordPOS = "n";

             while(st.hasMoreElements()){
                   count++;
                   word = st.nextToken();
                   POSQueue posQ = dic.get(word);
                   if(posQ==null)        {
                      System.out.print(word+":"+wordPOS+"\n");
                   }
                   else if (posQ.size()==1)   {
                      wordPOS = DicCore.gePOSName(posQ.getHead().item.nPOS);
                   }
                   output.write("s("+number+","+count+",'"+word+"',"+wordPOS+",1,0)." +"\n");
             }
          }
      }
      output.close();
      fpSource.close();
   }
```

3.16.2 垂直领域同义词

某些词在一个领域中算同义词，在另外一个领域可能不是。把自定义的同义词放在 synonyms.txt。SolrSynonymParser 解析这个文件格式。

这个文件格式是：空行和以"#"起头的行是注释行。有两类行：一类是显式映射行，还有一类不是映射行。

显式映射行的"=>"的左边的是 LHS，右边的是 RHS。例如：

  i-pod, i pod => ipod

又如，下面的映射行把"西红柿"和"番茄"都转换成"西红柿"。

  西红柿,番茄 => 西红柿

LHS 定义了源词。RHS 定义了所有的可选目标词。匹配所有的标注序列，然后把 LHS 中的词替换成 RHS 中的词。

等价的同义词用逗号分隔开，并且不做显式的映射。在这个情况下，映射行为能够被从创建器的扩展参数中取出。例如：

  ipod, i-pod, i pod

会合并同一个词的多个同义词映射条目。例如：

  foo => foo bar
  foo => baz

等价于：

  foo => foo bar, baz

可以定义局部模板。例如：

  [什么时间|多久]派送

这样同时匹配"包裹什么时间派送"和"包裹多久派送"这两个不同的问句。

3.16.3 同义词扩展

这里通过同义词搜索来尝试语义扩展的搜索，也就是通过查询扩展的方式实现语义搜索。当用户输入"计算机"进行搜索的同时，程序通过查找同义词库，按照"计算机""电脑""微机"等多个同义词查找。当用户输入"轿车"的同时，也能按照"奥迪""奔驰"等进行下位词扩展。

维基百科分类中包含了一些层次关系的词。例如：https://zh.wikipedia.org/wiki/Category:%E6%B9%AF。

对于有很多不同义项的词，一般不要用同义词扩展搜索。例如"人民大学"和"人大"是同义词，"人大"和"人民代表大会"也是同义词。"人民大学"和"人民代表大会"意思就完全不一样。最好是用没有歧义的词替换有歧义的词，这样能消除歧义。对于义项单一的词，也可以削减词形。比如把出现"番茄"的地方都用"西红柿"替换。

简单的做法是把同义词保存在一个散列表中。如果同义词词表很大，可以把同义词保存在一个索引库中。

SynonymEngine 是一个很简单的接口，输入一个词，返回它的同义词。

```
public interface SynonymEngine {
  String[] getSynonyms(String s) throws IOException;
}
```

DexSynonymEngine 使用一个简单的存储在散列表中的同义词实现。

```
public class DexSynonymEngine implements SynonymEngine {
    //词和对应的同义词数组
    private static Map<String, String[]> map = new HashMap<String, String[]>();

    static {
        //数字同义
        map.put("1", new String[] { "一" });
        map.put("2", new String[] { "二" });
        map.put("3", new String[] { "三" });
        map.put("4", new String[] { "四" });
        map.put("5", new String[] { "五" });
        map.put("6", new String[] { "六"});
        map.put("7", new String[] { "七" });
        map.put("8", new String[] { "八" });
        map.put("9", new String[] { "九" });
        map.put("10", new String[] { "十" });

        //日期同义
        map.put("非周末", new String[] { "周一","周二","周三","周四","周五" });
        map.put("周末", new String[] { "周六","周日" });

        //词同义
        map.put("西红柿", new String[] { "番茄" });
        map.put("黄豆", new String[] { "大豆" });
    }

    public String[] getSynonyms(String word) throws IOException {
        return map.get(word);
    }
}
```

为了节省内存，可以把同义词放在 FST。org.apache.lucene.analysis.synonym.SynonymMap 类的使用方法如下：

```
SynonymMap.Builder builder = new SynonymMap.Builder(true);
builder.add(new CharsRef("fast"), new CharsRef("quick"), true);
builder.add(new CharsRef("jumps"), new CharsRef("hops"), true);
SynonymMap map = builder.build();
```

构建出 SynonymMap 后，在 SynonymAnalyzer 中使用这个 SynonymMap。

```java
public class SynonymAnalyzer extends Analyzer {
    private SynonymMap engine;// 保存了一个词的同义词

    public SynonymAnalyzer(SynonymMap engine) {
        this.engine = engine;
    }

    @Override
    protected TokenStreamComponents createComponents(String f,
                                                     Reader reader) {
        final Tokenizer source = new StandardTokenizer(reader);// 先分词

        SynonymFilter result =
            new SynonymFilter(source, engine, false);// 在TokenStream 中增加同义词
        return new TokenStreamComponents(source, result);
    }
}
```

使用 SynonymFilter 扩展同义词。应该把 SynonymFilter 放在停用词之前处理。

```java
public class SynonymAnalyzer extends Analyzer {
    private SynonymMap engine;

    public SynonymAnalyzer(SynonymMap engine) {
        this.engine = engine;
    }

    @Override
    protected TokenStreamComponents createComponents(String f, Reader in) {
        Tokenizer source = new StandardTokenizer(in);
        TokenFilter result = new SynonymFilter(source, engine, false);
        result = new StandardFilter(result);
        result = new LowerCaseFilter(result);
        result = new StopFilter(result,
                StopAnalyzer.ENGLISH_STOP_WORDS_SET);
        return new TokenStreamComponents(source, result);
    }
}
```

在索引和搜索过程中使用 SynonymAnalyzer：

```java
//构建同义词表
SynonymMap.Builder builder = new SynonymMap.Builder(true);
builder.add(new CharsRef("fast"), new CharsRef("quick"), false);
builder.add(new CharsRef("jumps"), new CharsRef("hops"), false);
SynonymMap map = builder.build();
```

```
SynonymAnalyzer analyzer = new SynonymAnalyzer(map);

RAMDirectory directory = new RAMDirectory();
IndexWriterConfig config = new IndexWriterConfig(analyzer);
IndexWriter writer = new IndexWriter(directory, config);
Document doc = new Document();
doc.add(new TextField("content",
    "The quick brown fox jumps over the lazy dog", Store.YES));
writer.addDocument(doc);
writer.close();
IndexReader reader = DirectoryReader.open(directory);

IndexSearcher searcher = new IndexSearcher(reader);
// 查询匹配"hops"的文档
TermQuery termQuery = new TermQuery(new Term("content", "hops"));
System.out.println(searcher.search(termQuery, 10).totalHits);
// 查询匹配短语"fox hops"的文档
PhraseQuery phraseQuery = new PhraseQuery();
phraseQuery.add(new Term("content", "fox"));
phraseQuery.add(new Term("content", "hops"));
System.out.println(searcher.search(phraseQuery, 10).totalHits);

reader.close();
```

在索引过程中使用 StandardAnalyzer，同时在搜索过程中定义 SynonymAnalyzer：

```
RAMDirectory directory = new RAMDirectory();

Analyzer analyzer = new StandardAnalyzer();

IndexWriterConfig config = new IndexWriterConfig(analyzer);
IndexWriter writer = new IndexWriter(directory, config);
Document doc1 = new Document();

Field f1 = new TextField("content", "quick fast", Store.YES);
doc1.add(f1);
writer.addDocument(doc1);
Document doc2 = new Document();
Field f2 = new TextField("content", "quick hops", Store.YES);
doc2.add(f2);
writer.addDocument(doc2);
writer.close();
IndexReader reader = DirectoryReader.open(directory);

IndexSearcher searcher = new IndexSearcher(reader);

SynonymMap.Builder builder = new SynonymMap.Builder(true);
builder.add(new CharsRef("fast"), new CharsRef("quick"), true);
builder.add(new CharsRef("jumps"), new CharsRef("hops"), true);
```

```
SynonymMap map = builder.build();

analyzer = new SynonymAnalyzer(map);
QueryParser parser = new QueryParser("content", analyzer);
Query query = parser.parse("fast jumps");
TopDocs docs = searcher.search(query, 2);
for (ScoreDoc doc : docs.scoreDocs) {
    System.out.print("docid : " + doc.doc + " score : " + doc.score + " content:");
    Document hitDoc = searcher.doc(doc.doc);
    System.out.println(hitDoc.get("content"));  //输出文档
}
```

两个文档的分值不同,说明 SynonymAnalyzer 可以正确处理协调因子 coord。

查看生成的查询:

```
QueryParser parser = new QueryParser("content", analyzer);
Query query = parser.parse("fast jumps");
System.out.println(query.toString("content"));   //输出(fast quick) (jumps hops)
System.out.println(query.getClass().getName());
//输出 org.apache.lucene.search.BooleanQuery
```

Searcher.explain(Query query, int doc)方法可以显示协调因子的值。

```
TopDocs docs = searcher.search(query, 2);
for (ScoreDoc doc : docs.scoreDocs) {
    System.out.println("----------");
    System.out.print("docid : " + doc.doc + " score : " + doc.score + " content:");
    Document hitDoc = searcher.doc(doc.doc);
    System.out.println(hitDoc.get("content"));  //输出文档
    Explanation explanation = searcher.explain(query, doc.doc);
    System.out.println("explain: "+explanation.toString());
}
```

第 1 个文档的协调因子是 1,第 2 个文档的协调因子是 1/2。

索引时使用 SynonymAnalyzer 把同义词扩展放到索引库,在搜索时使用下面这个 CnAnalyzer,不做同义词扩展。

```
public class CnAnalyzer extends Analyzer {
    public TokenStream tokenStream(String fieldName, Reader reader) {
        TokenStream stream = new CnTokenizer(reader);
        stream = new SingleFilter(stream);
        return stream;
    }
}
```

实现这个功能的基本步骤如下:

(1) 准备语义词库;

(2) 把语义词库转换成同义词索引库;

(3) 在 SynonymAnalyzer 中使用同义词索引库。

然后通过程序 Syns2Index，把 wn_s_cn.pl 转换成 Lucene 同义词索引库。通过 Luke 索引工具可以察看到索引结构是一个 word 值对应的多个 syn 同义词。

最后我们通过 SynonymAnalyzer 来调用这个同义词索引库实现同义词扩展查找。但是 SynonymAnalyzer 只是简单地通过词本身来扩展同义词，这样并不一定准确，尤其是对词义有很多的词来说。在 Lucene4.0 的灵活索引出来以前，不允许在索引中任意存储信息。但是 Lucene 的 2.2 版本以后支持把对词的额外描述存储在 Payload 属性中。所以可以把一个词的语义编码存储在词的 Payload 属性中。

可以思考动态同义词的实现。搜索"星期日"的时候同时返回下一个最近的日期，例如离 2010 年 11 月 5 日最近的一个星期日是 2010 年 11 月 7 日。

扩展查询词以后，会出现一些不太相关的词。需要调整相关度计算方法。

根据每个扩展查询词的来源词设置 Payload 值。不同来源词的扩展查询词的 Payload 值不同。n 个查询词都包括的文档才不扣分，否则都扣分。少一个查询词扣 1 分，少 2 个查询词扣 2 分，依次类推。

测试 SynonymAnalyzer：

```
String text = "The quick brown fox jumps over the lazy dog";

SynonymMap.Builder builder = new SynonymMap.Builder(true);
builder.add(new CharsRef("fast"), new CharsRef("quick"), true);
builder.add(new CharsRef("jumps"), new CharsRef("hops"), true);
SynonymMap map = builder.build();

SynonymAnalyzer analyzer = new SynonymAnalyzer(map);

TokenStream stream = analyzer.tokenStream("title", new StringReader(
        text));
stream.reset();    //重置 TokenStream
// 增加 Token 表示的字符串属性
CharTermAttribute term = stream.addAttribute(CharTermAttribute.class);
TypeAttribute type = stream.addAttribute(TypeAttribute.class);

while (stream.incrementToken()) {
    System.out.println(term + " "+ type.type()); // 逐个单词输出
}
analyzer.close();
```

输出结果：

```
The <ALPHANUM>
quick <ALPHANUM>
brown <ALPHANUM>
fox <ALPHANUM>
jumps <ALPHANUM>
hops SYNONYM
over <ALPHANUM>
the <ALPHANUM>
```

```
lazy <ALPHANUM>
dog <ALPHANUM>
```

判断 Token 类型是不是 Synonym。如果是 Synonym，则设置 PayLoad 值。调用 PayloadAttribute.setPayload()方法设置值，参数是一个 BytesRef 类型的对象。

```
public class SynonymPayloadFilter extends TokenFilter {
    private final PayloadAttribute attr = addAttribute(PayloadAttribute.class);

    protected SynonymPayloadFilter(TokenStream input) {
        super(input);
    }

    @Override
    public boolean incrementToken() throws IOException {
        BytesRef p = new BytesRef(PayloadHelper.encodeInt(1));
        attr.setPayload(p);
        return false;
    }
}
```

3.16.4 语义标注

对文本中一些有歧义的词，根据语义库标注语义编码。"我们伟大祖国在新的一年"用同义词词林标注语义的结果是：我们/r/Aa02B01 伟大/a/Ed20A01 祖国/n/Di02A18 在/p/Jd01A01 新/a/Eb28A01 的/u/Kd01A01 一/m/Dn04A02 年/q/Ca18A01。

3.17 本章小结

本章介绍了 Lucene 全文索引库的基本使用方法和常用的定制修改方法。介绍了索引文件的格式，以及通过分发索引文件到其他服务器来实现分布式搜索。为了实现更好的搜索准确性，可以改进检索模型。

Lucene 的 Java Doc 说明文档在 http://lucene.apache.org/core/documentation.html。

Cutting 在 1999 年写 Lucene 以前，也只是一个用 C++开发搜索引擎的普通青年。Lucene 是他写的第一个 Java 软件。在后来的 10 多年里，Lucene 越来越流行，成为开源组织 Apache 基金会的项目，并在维基百科网站等项目中得到广泛使用。Cutting 后来开发 MapReduce 的 Java 版本 Hadoop 也同样成功。Cutting 因此进入 Apache 基金董事会，并在 2010 年成为董事会主席。

有时候少就是多，用户输入一个查询词，返回的结果可能很多，搜索引擎需要挑选出用户最感兴趣的少数几个文档。

字符串数组排序后使用前缀压缩，整数数组排序后使用差分编码压缩。和压缩相关的类有：用于压缩单个整数的 VInt 类和压缩排好序的整数数组的 SortedVIntList 类。为了实现更好的压缩，在有些地方可以使用 PForDelta 来代替 VInt。

多词查询假设如果一篇文档包含一些这样的句子：至少两个查询词在同一句话中出现，则这篇文档更相关。也就是说多个查询词之间要有更短的距离，例如"Pisa Tower"。如果词出现的位置之间的距离增加，则底层意思可能已经变了。

BM25 源自 20 世纪八九十年代伦敦城市大学第一个实现该函数的系统——Okapi 信息检索系统。它是基于 20 世纪七八十年代由 Stephen E. Robertson, Karen Spärck Jones 等人开发的概率检索框架。从 20 世纪 80 年代末开始，概率模型(特别是以 Okapi 系统为代表的 BM25 系列算法)出现并逐渐分享了经典模型在信息检索模型领域的地位，成为新兴的、功能强大、表现越来越出色的模型。

BM25 和 BM25F 两种模型在 TREC 文本检索评测会议中都较其他模型有优越的表现并且被公认是目前 IR 范围内最为先进的检索模型。BM25 适用于没有结构的全文检索，而 BM25F 适用于结构化的文档检索，也就是用于有好几个全文搜索列的情况。

在 RankNet 模型的基础上改进，形成了 LambdaRank 模型。在 LambdaRank 模型上又发展出了 LambdaMART 机器学习评分检索模型。后续章节介绍机器学习检索模型在 Solr 和 ElasticSearch 搜索软件上的应用。

除了用于检索文档，机器学习评分算法还可以用于推荐引擎。例如，购物网站给访问网站的用户推荐商品。

LIRE(Lucene Image REtrieval)是一个开源的图像搜索项目。它建立在 Lucene 基础上。是一个基于内容的图像检索库。可以从 https://github.com/dermotte/lire 下载源代码。

Cassandra 是一套开源分布式 NoSQL 数据库系统。把存储和搜索功能相结合。Stratio 公司已经开发了一个基于 Lucene 的自定义二级索引实现作为核心大数据平台的一部分。它是开源的，地址为 https://github.com/Stratio/cassandra-lucene-index。

在下一章，我们将介绍索引库在用户界面中的调用方法。

3.18 术 语 表

Learn To Rank LTR：根据用户行为学习文档和查询词的相关性。

随机性差异模型(divergence from randomness，DFR)：一个优秀的文档集合，见证了富含信息的词。文档内的词频和文档内的词频会有差异，这个差异越大，则在文档 d 中的词 t 蕴含着更多的信息。

泊松分布(Poisson distribution)：描述单位时间内随机事件发生的次数。假设一个人每段间隔内读一篇文档，读到某个词的次数组成泊松分布。

伯努利试验(Bernoulli trial)：只有两种可能结果的单次随机试验。

二项分布(Binomial distribution)。

玻色-爱因斯坦分布(Bose–Einstein distribution)。

二项独立模型(Binary Independence Model)。

相关性反馈(Relevance feedback)。

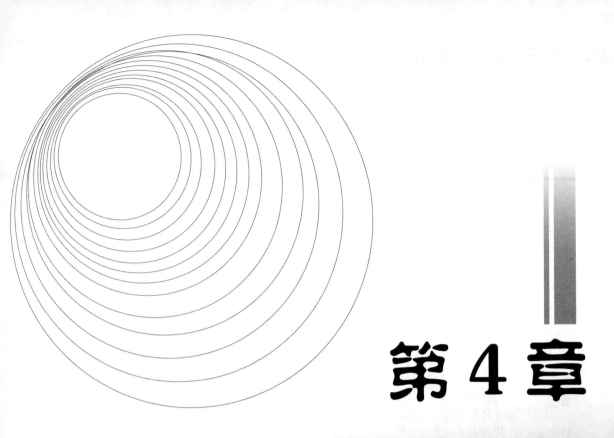

第4章

搜索引擎用户界面

在界面设计上，要想办法节约用户的时间。例如，手机上的锁状态，需要用户额外输入才能解锁，用触感指纹或者手的设计代替。

新闻搜索结果页中显示小的缩略图片。因为 HTML5 中的 src 属性可以直接存表示图片的二进制序列，所以把这个二进制序列直接放入索引中。这样使用<img>标签：

```
<img src="data:image/gif;base64,×××××">
```

这里，×××××部分是 GIF 图片数据的 base64 编码。

看搜索日志可能会想，搜索访问量较卖快餐的量差太多了。用户难得搜索一次，所以要把返回页面的价值最大化。所以返回结果的信息可能是各种各样的，但一定都是用户可能想看到的信息，尽量不要列出任何无关的信息。现在各大搜索引擎的第一页基本都是综合页，同时返回新闻、图片或视频的搜索结果，即都是聚合搜索(aggregated search)。还有类似于 Naver(http://www.naver.com)那样的综合页面。

在搜索结果页显示个性化的背景图。背景图根据什么规则来替换？比如说喜欢足球的，就用足球背景小图标；喜欢户外活动的，用风景图小图标；喜欢书法的，用毛笔涂小图标；女性用鲜花……类似博客背景那样的。

如何判断他喜欢足球？根据不同 profile 推出不同查询结果页背景，类似于根据用户特征数据做个性化推荐。

对于互联网搜索来说，搜索结果界面往往采用 JSP、ASP.NET、PHP、Python 等技术来实现。搜索联想词的页面效果可以用 Ajax 来实现。为了实现更好的封装，可以结合 Spring 或 Struts 框架封装搜索请求，展现搜索结果。

这里介绍采用 JSP 实现搜索界面。需要区分哪些类是静态的，是全局唯一的；哪些对象需要在页面内即时创建和使用；哪些对象在整个用户会话期间内有效。

4.1 实现 Lucene 搜索

有些热词直接跳转。例如搜索手机直接跳转到相关手机的界面，而不执行相关性查询。有个散列表对应查询词和跳转的页面。

首先在控制台测试搜索功能，然后再考虑在 Web 环境中加载索引库。

4.1.1 测试搜索功能

在开发搜索 Web 界面之前，首先写一个控制台方式运行的搜索程序测试一下索引库：

```
Searcher searcher = new IndexSearcher(indexPath);
IndexReader reader = IndexReader.open(indexPath);
System.out.println("索引中的文档数量: " + reader.numDocs());
Query bodyQuery =null,titleQuery=null,query=null;
//对内容列的查询
QueryParser parser = new QueryParser("body", analyzer);
bodyQuery =  parser.parse(queryString);
//对标题列的查询
```

第 4 章 搜索引擎用户界面

```
parser = new QueryParser("title", analyzer);
titleQuery = parser.parse(queryString);//解析查询词
System.out.println("Searching for: " + bodyQuery.toString("body"));
BooleanQuery bodyOrTitle = new BooleanQuery();
bodyOrTitle.add(bodyQuery, BooleanClause.Occur.SHOULD);
bodyOrTitle.add(titleQuery, BooleanClause.Occur.SHOULD);
query=bodyOrTitle.rewrite(reader);
//需要这行语句来扩展搜索词
//设置排序方式,比如说高优先级的先显示
SortField classSortField = new SortField("class",SortField.INT,true);
Sort classSort = new Sort(new SortField[] {classSortField});
TopDocs hits = searcher.search(query,null,1000, classSort);
System.out.println("返回结果总条数:"+ hits.totalHits);
Highlighter highlighter =new Highlighter(new SimpleFormatter(),new
QueryScorer(query));
String text;
TokenStream tokenStream;
for (int i = 0; i< hits.scoreDocs.length; i++){
    Document hitDoc = searcher.doc(hits.scoreDocs[i].doc);
    text = hitDoc.get("body");
    //内容列的高亮显示
    TermPositionVector tpv = (TermPositionVector)
            reader.getTermFreqVector(hits.scoreDocs[i].doc,"body");
    tokenStream=TokenSources.getTokenStream(tpv,false);
    String result = highlighter.getBestFragment(tokenStream,text);
    System.out.println("body:"+result);
    //标题列的高亮显示
    text = TextHtml.text2html(hitDoc.get("title"));
    tokenStream=analyzer.tokenStream("title",new StringReader(text));
    result = highlighter.getBestFragment(tokenStream,text);
    if (result == null)
        System.out.println(hitDoc.get("title"));
    else
        System.out.println("title:"+result);
}
searcher.close();
```

查询对象 Query 的 rewrite()方法把复杂的查询条件重写成简单的查询条件。例如：MultiTermQuery 会扩展成很多的 TermQuery。MultiTermQuery 需要根据索引库中的值来扩展，所以 rewrite()方法需要传入 IndexReader 参数。

4.1.2 加载索引

索引路径必须是可以配置的。所以考虑把配置信息放在一个叫作 conf.properties 的配置文件中。

```
indexDir = D:/priceIndex/indexdir
```

但是在程序中如何找到 conf.properties 配置文件呢？假设 Web 应用部署在根路径，则考虑把配置文件放在 webapps\ROOT\WEB-INF\classes 路径。在一个类的静态块中可以通过如下行访问到配置文件。

```
ClassName.class.getClassLoader().getResourceAsStream("conf.properties");
```

SearchByQuery 类中加载索引相关的代码如下：

```
public class SearchByQuery {
    private static Directory directory = null;
    private static IndexSearcher isearcher = null;
    private static String indexDir = "";

    static{
        InputStream inputFile=null;
        Properties propertie = new Properties();
        try {
            inputFile =
    SearchByQuery.class.getClassLoader().getResourceAsStream("conf.properties");
            propertie.load(inputFile);
            indexDir = propertie.getProperty("indexDir");
        } catch (FileNotFoundException ex) {
            System.out.println("读取属性文件失败！原因：文件路径错误或者文件不存在");
            ex.printStackTrace();
        } catch (IOException ex) {
            System.out.println("装载文件失败!");
            ex.printStackTrace();
        } finally {
            try {
                if(inputFile!=null)
                    inputFile.close();
            } catch (IOException e) {
                e.printStackTrace();
            }
        }

        if (isearcher != null) {
            try {
                isearcher.close();
            } catch (IOException e) {
                e.printStackTrace();
            }
        }
        if (directory != null) {
            try {
                directory.close();
            } catch (IOException e) {
                e.printStackTrace();
```

```
            }
        }
        try {
            directory = FSDirectory.open(new File(indexDir));
            isearcher = new IndexSearcher(directory, true);
        } catch (IOException e) {
            e.printStackTrace();
        }
    }
}
```

4.2 搜索页面设计

JSP 只是输出一个字符串,然后交给 Web 服务器,最后由用户的浏览器显示网页。

解决 Tomcat 乱码问题:一般使用 UTF-8 编码。为了正确地接收 HTTP 参数,在 JSP 页面中设置如下:

```
request.setCharacterEncoding("utf-8");
```

org.apache.commons.lang3.StringUtils 中的 defaultIfEmpty 方法返回空值的替代值。

```
String query = StringUtils.defaultIfEmpty(request.getParameter("query"),"");
```

执行搜索的 Java Bean。

```
public class SearchDoc {
    private Client client;   //在 Web 容器内全局唯一

    //只调用一次
    public void init(String host) throws Exception {
        Settings settings = ImmutableSettings.settingsBuilder()
            .put("client.transport.sniff", true).build();
        client = new TransportClient(settings)
            .addTransportAddress(new InetSocketTransportAddress(
                host, 9300));   //"localhost"
    }
}
```

<jsp:useBean>标签创建一个 Bean 实例并指定它的名字和作用范围。

```
<jsp:useBean id="searchInf" class="com.lietu.search.SearchDoc"
scope="application">
<!-- 声明索引存放的机器 -->
<% searchInf.init("localhost"); %>
</jsp:useBean>
```

传中文字符作为参数时,需要用对应字符集编码这个字符串,为 URL 编码专门写一个自定义标签 iteratePropURLEncodeTag。使用 iterateURLEncodeProp:

```
<a href="folder.jsp?folder=<list:iterateURLEncodeProp
property="folder"/>&docType=<list:iterateProp property="docType"/>"
```

用 JSTL 把 Java 代码用标签来实现。下载 JSTL 的 jar 包：http://tomcat.apache.org/download-taglibs.cgi#Standard-1.2.1。

为了方便开发和部署到 Tomcat 等 Web 应用服务器，建议在 MyEclipse 中开发搜索页面。MyEclipse 8.0 版本开始支持 Struts 2 的开发。

为了防止报 "Bad version number in .class file" 错误，需要 MyEclipse 和 Tomcat 使用同一个 JDK。当安装 All in One 版本的 MyEclipse 时，就会自带了它的 JDK。建议将 MyEclipse 中的 JDK 路径修改为自己的 JDK。具体修改方法如下：

(1) 单击 "Window" → "Preferences" → "Java" → "Installed JREs"；

(2) 在右边单击 "Add..."，在弹出的对话框中单击 "Browse..." → "选择你的 JDK 路径"（如 C:\Program Files\Java\jdk1.6.0_10），选择 "OK"；

(3) 在刚配置的 JDK 路径上打钩，单击 "OK" 设置为默认 JDK 路径。

搜索相关页面主要包括首页和搜索结果页。如果用户输入搜索词是空，则可以显示一个对信息分类导航的页面。

首页最主要包含搜索条区域。此外可以包括一些推荐信息，以及当前热门信息。

为了对其他的搜索引擎友好(SEO)，搜索关键词通过 GET 方式取得参数，而不要通过 POST 方式，翻页参数也最好通过 GET 方式传递。同样为了 SEO，搜索词会出现在搜索结果页的标题中。查询关键词的参数一般用 q 或 query 表示。

为了防止 JavaScript 脚本注入，需要使用 StringEscapeUtils.escapeHtml 对用户查询转码。

如果把搜索功能作为单独的一个域名，例如 http://so.1798hw.com，不要忘记添加网站头像：favicon.ico 文件。在首页的 <head></head> 之间添加一行代码：

```
<link rel="Shortcut Icon" href="./img/favicon.ico"/>
```

4.2.1 Struts2 实现的搜索界面

Struts2 中的 jar 包很多，但这里只需要两个。

- struts2-core.jar：核心包；
- ognl.jar：用于 ognl 表达式计算。

Struts2 把页面显示和控制逻辑分离。由 SearchAction 执行搜索并把搜索结果传递给负责显示的 JSP 页面。为了简化实现，先只定义必须实现的搜索结果显示页面 result.jsp 和错误处理页面 index.jsp。

配置文件 struts.xml 的内容如下：

```
<struts>
    <package name="default" namespace="/" extends="struts-default">
        <action name="SearchAction" class="com.lietu.action.SearchAction">
            <result name="success">/result.jsp</result>
            <result name="error">/index.jsp</result>
```

```
         </action>
      </package>
</struts>
```

Struts2 则基于接口编程，自己开发的 Action 类可以实现一个 Action 接口，也可以实现其他接口。Struts2 提供了一个 ActionSupport 基类去实现这些常用的接口。在 Struts2 中 Action 接口不是必需的，任何有 execute()方法的 POJO 对象，都可以作为 Action 类来使用。Struts2 的 Action 对象为每个请求产生一个实例，因此没有线程安全的问题。

表示控制的 SearchAction 实现如下：

```
public class SearchAction extends ActionSupport {
    private String query;   //通过参数绑定赋值

    public String execute() {
        if(query == null || query.equals("")) {
            return "error";
        }
        SearchByQuery ts = new SearchByQuery();
        ArrayList<Goods> result = ts.getResults(query); //返回搜索结果

        HttpServletRequest request = ServletActionContext.getRequest();
        request.setAttribute("result" , searchResult);
        request.setAttribute("query" , message);
        return "success";
    }
}
```

Struts2 会直接给 Action 中的变量赋值。如 http://www.lietu.com/SearchAction.action?query=NBA，会将 action 中的 query 属性赋值。

一般来说，搜索结果页标题中要包括搜索关键词。可以用 Struts2 的 Tag 中的 OGNL 表达式在 result.jsp 页面中显示搜索关键词。OGNL 表达式将查找 action 中的变量，例如查找用户输入的查询词%{query}。

搜索结果页返回一个可能和用户输入内容相关的信息列表(常常会是很长一个列表，例如包含 1 万个条目)。这个列表中的每一条目代表一篇网页，至少有 3 个元素：

(1) 标题：以某种方式得到的网页内容的标题。最简单的方式就是从网页的<title></title>标签中提取内容。已经介绍了通过信息提取的方法形成"标题"的方法。

(2) URL：该网页对应的"访问地址"。有经验的 Web 用户常常可以通过这个元素对网页内容的权威性进行判断，例如 http://www.amazon.com 上面的内容通常就比 http://notresponsible.net(某个假想的个人网站)上的要更权威些。

(3) 摘要：以某种方式得到的网页内容的摘要。最简单的一种方式就是将网页内容的头若干字节(例如 512)截取下来作为摘要。已经介绍了根据搜索关键词形成动态摘要的方法。为了在页面中正常显示特殊符号，需要用 StringEscapeUtils.escapeHtml 对摘要结果转码。

Struts2 中 action 的名称一般以.action 结尾。

```
<form method="get" action="./SearchAction.action">
         <input name="query" type="text" />
         <input id="searchsubmit" value="Find It!" type="submit">
</form>
```

在 Web 应用的 WEB-INF/web.xml 文件中增加如下配置：

```
<filter>
   <filter-name>struts2</filter-name>

<filter-class>org.apache.struts2.dispatcher.FilterDispatcher</filter-class>
</filter>
<filter-mapping>
   <filter-name>struts2</filter-name>
   <url-pattern>/*</url-pattern>
</filter-mapping>
```

地址栏显示的是 SearchAction.action，而页面显示的是 result.jsp，Struts 处理请求的方式是请求转发，而不是重定向。

4.2.2 用于显示搜索结果的 Taglib

搜索结果页是一个表格型的数据。Listlib 实现了对数据的封装和抽象，可以通过它来控制显示的结果数量，比如可以指定每页显示 20 条记录或 10 条记录。实际执行 Lucene 搜索的类继承 ListCreator 接口，并把搜索结果通过 ListContainer 类的实例返回即可。

`init`

Listlib 的起始 Tag，创建一个 ListCreator 对象，并且运行该对象的 execute() 方法，把它存储在 HttpServletRequest 属性中。这是一个容器 Tag，所以在 JSP 页面中使用时，其他的 Tag 都必须嵌套在这个 Tag 中。

它的主要属性有：通过 name 指定一个名字，因为需要通过这个名字来把 ListCreator 对象存储在 HttpServletRequest 属性；通过 listCreator 来指定创建 ListCreator 的对象；通过 max 来声明每页必须显示的记录条数。

`hasResults`

如果 list 有结果，就会执行这个 Tag，否则会跳过。

`hasNoResults`

如果 list 没有结果就会执行这个 Tag，否则会跳过。

`prop`

返回 list 中的属性值。和搜索结果总体相关的信息可以通过它来显示，例如搜索提示词、搜索结果分类统计等等。

`hasPrev`

如果还可以继续往回遍历，就会显示这个 Tag 中的内容，否则就跳过。

`hasNext`

如果还可以继续向下遍历，就会显示这个 Tag 中的内容，否则就跳过。

`iterate`

遍历 ListContainer 中的元素。

`iterateProp`

从 Iterator 的当前对象返回属性，比如返回标题通过<list:iterateProp property="title"/>。实现搜索的主要代码如下：

```jsp
<!--创建一个 Lucene 搜索对象 - 它实现了 ListCreator 的 execute 方法 -->
<jsp:useBean id="searchInf" class="com.bitmechanic.spindle.SearchInfo" scope="application">
    <!--指定存储 Lucene 索引的路径-->
    <% searchInf.init("d:/tomcat5.5/webapps/search/lietu/index/info/"); %>
</jsp:useBean>
    <!--把查询从 http 的 get 参数设置到搜索对象中去-->
    <jsp:setProperty name="searchInf" property="query" value="<%=query%>"/>
    <!--执行搜索并把返回结果封装到 ListContainer -->
<%long start = System.currentTimeMillis();%>
<list:init name="information" listCreator="searchInf" max="20">
<% long end = System.currentTimeMillis();%>
<!--记录搜索执行的时间 -->
```

4.2.3 实现翻页

翻页链接需要指定相对路径。<base>标签为所有链接指定相对路径。首先通过 Java 代码取得相对路径。

```jsp
<%
String path = request.getContextPath();
String basePath = request.getScheme()
        +"://"+request.getServerName()
        +":"+request.getServerPort()+path+"/";
%>
```

然后在<base>标签中指定相对路径是 basePath：

```html
<head>
  <meta http-equiv="Content-Type" content="text/html; charset=utf-8">
  <base href="<%=basePath%>">
  <title>旅游活动搜索</title>
</head>
```

需要在分页器的构造方法中告诉分页器符合查询条件的结果总共有多少条。分页器则告诉查询对象,从第几条结果开始返回。另外一个预先固定设置好的值是每页最多显示的结果数。

推荐使用翻页组件 pager-taglib (http://jsptags.com/tags/navigation/pager)。Pager-taglib 是一个 JSP 标签库,支持多种风格的分页显示。为在 JSP 上显示分页信息而设计的一套标签,通过这些标签的不同的组合,会形成多种不一样的分页页面,风格各异,它自带的 DEMO 就有 7 种左右的分页风格,包括 Google 的分页风格。而需要订制自己的风格的分页页面也非常简单。使用 pager-taglib 的流程如下:

(1) 复制 pager-taglib.jar 包到 lib 目录下,不需要修改 web.xml。
(2) 在 JSP 页面中使用 taglib 指令导入 pager-taglib 标签库。
(3) 使用 pager-taglib 标签库进行分页处理。

通过 maxPageItems 参数设定每页最多显示的结果数。在 JSP 页面中使用翻页标签库的例子如下:

```
<pg:pager   url="SearchAction.action"
        items="<%=Integer.parseInt(listSize)%>"
        maxPageItems="20"
        maxIndexPages="10"
        export="currentPageNumber=pageNumber"
        scope="request">
 <pg:param name="query" value="<%=query%>"/>
…
</pg:pager>
```

其中在<pg:pager>标签中定义了 action 的 url 地址,<pg:param>标签中定义了查询参数 query。

假设按 10 条记录分页,显示第一页的实现如下:

```
QueryParser parser = new QueryParser(defaultField,
                                    analyzer);

Query query = parser.parse(queryString);
//最多返回 10 个文档,也就是返回和查询词最匹配的前 10 个文档
TopDocs hits = searcher.search(query, 10);
System.out.println("返回结果总条数:"+hits.totalHits);
for (int i = 0; i < hits.scoreDocs.length; i++) {
    Document hitDoc = searcher.doc(hits.scoreDocs[i].doc);
    System.out.println("第"+i+"条:"+hitDoc.get("title"));
}
if (hits.totalHits > hits.scoreDocs.length){
    System.out.println("还有更多结果在后面排队");
}else{
    System.out.println("结果显示完毕");
}
```

返回从 offset 开始的 rows 行的实现如下：

```
searcher.search(query,offset+rows);   //最多返回查询的前(offset+rows)条
System.out.println("返回结果总条数："+hits.totalHits);
for (int i = offset; i < hits.scoreDocs.length; i++) {
    Document hitDoc = searcher.doc(hits.scoreDocs[i].doc);
    System.out.println("第"+i+"条："+hitDoc.get("title"));
}
```

pager-taglib 在输出的页面中生成链接 search.jsp?query=%E7%9A%84&pager.offset=10，其中包含了开始位置的参数。在 InitTag 类中得到开始返回结果的位置。

```
public static final String OFFSET_KEY = "pager.offset";

public int doStartTag() throws JspException {
    String offsetStr = pageContext.getRequest().getParameter(OFFSET_KEY);
    int offset = Integer.parseInt(offsetStr);
}
```

Web 服务器在加载 search.jsp 页面时会执行 doStartTag()方法。

IterateTag 是 BodyTagSupport 的子类。其中的方法执行顺序是 doStartTag()→setBodyContent()→doInitBody()→doAfterBody()→doEndTag()。

不建议把搜索结果放在 session 中，这样让服务器长时间记住每个用户的搜索结果，会消耗 Web 服务器的内存。由服务器端对象 request 传递搜索结果。客户端每次提交请求 Servlet 都会生成一个新的 request 对象，服务器会回收不用的 request 对象。把搜索结果存放在 request 对象中。Struts2 的 action 代码如下：

```
@Override
public String execute(){
    //..
    servletRequest.setAttribute("searchResultPagerSize" , searchResultPagerSize);
    servletRequest.setAttribute("searchResult" , searchResult);
    servletRequest.setAttribute("numberOfItemsPerPage" , numberOfItemsPerPage);
    servletRequest.setAttribute("maxNumberOfPagesToShow" , maxNumberOfPagesToShow);

    return SUCCESS;
}
```

在 JSP 页面取得搜索结果：

```
ArrayList<Goods> searchResult = (ArrayList)
request.getAttribute("searchResult");
int numberOfItemsPerPage = (Integer)
request.getAttribute("numberOfItemsPerPage");
```

```
int maxNumberOfPagesToShow =
 (Integer) request.getAttribute("maxNumberOfPagesToShow");
int searchResultPagerSize = (Integer)
request.getAttribute("searchResultPagerSize");
```

假设搜索结果返回的是一个商品列表。Goods 类是一个 POJO 类，包含取得商品名的方法 getGoodName()。

使用 Struts2 的迭代标签显示搜索结果：

```
<%@ taglib prefix="s" uri="/struts-tags"%>
...
<s:iterator value="searchResult">
<s:property value="GoodName" escape="false" />
</s:iterator>
```

注意这里的 escape="false"，缺省把所有的网页相关的字符转义，但是为了高亮显示，设置 escape="false"。

从 Tomcat 7 开始，<taglib>需要位于<jsp-config>里面。可以把 tld 文件放到 jar 文件中的 META-INF 目录。

4.3 实现搜索接口

本节介绍从基本的布尔逻辑查询开始，到指定范围的查询以及搜索结果排序等的实现方法。

4.3.1 编码识别

搜索引擎的查询关键词是很重要的一个参数，这个参数是一个查询字符串的 URL 编码。一个非 ASCII 字符的 URL 编码由一个 "%" 符号后面跟着两个十六进制的数字组成。中文搜索需要判断传入的这个字符串的 URL 编码是 GBK 还是 UTF-8 格式。

符合 J2EE 标准的 Web 服务器(例如 Tomcat)通过调用 request.getQueryString() 方法可以得到原始提交的参数。比如发送：http://localhost/search.do?query=%B0%A1，getQueryString() 方法得到的字符串是：query=%B0%A1，然后调用编码识别方法，用正确的编码来解码。

```
String input = "%E6%B5%B7%E6%8A%A5%E7%BD%91";
String codingName=getEncoding(input);//判断编码
System.out.println(URLDecoder.decode(input, codingName)); //用正确的编码来解码
```

主要的开发工作是根据输入字符串判断编码。

GB2312 的字符编码范围为%B0%A1～%F7%FE，见表 4-1。

汉字 Unicode 编码范围为\u4e00～\u9fa5。UTF-8 下汉字 URL 编码后的取值范围为：

%E4%B8%80 - %E4%BF%BF

%E5%B8%80 - %E5%BF%BF

%E6%B8%80 - %E6%BF%BF

%E7%80%80 - %E7%BF%BF
%E8%80%80 - %E8%BF%BF
%E9%80%80 - %E9%BE%A5

表 4-1 汉字编码对照表

字 符	编 码
啊	%B0%A1
阿	%B0%A2
鞍	%B0%B0
齄	%F7%FE

像左括号和右括号这样的 ASCII 编码,小于 128 的字符编码都小于%80,例如:左括号字符编码是%28,右括号字符编码是%29。而所有的汉字编码,无论是 UTF-8 还是 GBK,每个字节的编码都大于或等于%80。

```
//判断是否可能为UTF-8编码的汉字
public static boolean isUtf8(String code1,String code2,String code3) {
    if (code1.compareTo("E4") >= 0 && code1.compareTo("E9") <= 0 &&
        code2.compareTo("80") >= 0 && code2.compareTo("BF") <= 0 &&
        code3.compareTo("80")>=0 &&code3.compareTo("BF")<=0) {
      return true;
    }
    return false;
}
//判断是否可能为GB2312编码的汉字
public static boolean isGb2312(String code1,String code2) {
    if (code1.compareTo("B0") >= 0 && code1.compareTo("F7") <= 0 &&
        code2.compareTo("A0")>=0 &&code2.compareTo("FF")<=0) {
      return true;
    }
    return false;
}
//根据字符列表猜测字符编码
public static String getEncodeByList(List<String> code) {
    if(code.size() >= 2 && code.size()%2 == 1 && code.size()%3 == 0) {
      return "utf8";
    }
    else if(code.size() >= 2 && code.size()%2 == 0 && code.size()%3 != 0) {
      return "gbk";
    }
    else if(code.size()%6 == 0) {
      for(int m=0;m<code.size();m = m+6) {
        if( ! isUtf8(code.get(m), code.get(m+1), code.get(m+2)) &&
          isGbk(code.get(m), code.get(m+1)) &&
          isGbk(code.get(m+2), code.get(m+3)) ) {
```

```
            return "gbk";
        } else if(isUtf8(code.get(m), code.get(m+1), code.get(m+2)) &&
            ! isGbk(code.get(m), code.get(m+1)) ) {
            return "utf8";
        }
        if(! isUtf8(code.get(m+3), code.get(m+4), code.get(m+5)) &&
          isGbk(code.get(m+2), code.get(m+3)) &&
          isGbk(code.get(m+4), code.get(m+5)) ) {
            return "gbk";
        } else if(isUtf8(code.get(m+3), code.get(m+4), code.get(m+5)) &&
            !isGbk(code.get(m+2), code.get(m+3))) {
            return "utf8";
        }
      }
    }
    return "utf8";
}
```

根据有限状态机的思想把字符串切分成数组。首先定义状态类。

```
public enum CharType {
    Enter,//碰到%
    Code1,//碰到%后的第1个字符
    Code2,//碰到%后的第2个字符
}
```

然后根据上一个状态以及当前的字符决定下一个状态。进入下一个状态时,有可能执行判断字符编码的动作。

```
public static String getURLEncoding(String url) {
    List<String> codes = new ArrayList<String>();
    CharType currentSate = null;
    char c1='\0';
    char c2='\0';
    for(int i=0; i<url.length(); ++i) {
        char currentChar = url.charAt(i);
        if(currentChar == '%') {
            if(currentSate == CharType.Code2 ) {
                char[] s1 = {c1,c2};
                codes.add(new String(s1));
            }
            currentSate = CharType.Enter;
        }else if(currentSate==CharType.Enter) {
            c1 = currentChar;
            currentSate = CharType.Code1;
        }else if(currentSate==CharType.Code1) {
            c2 = currentChar;
            currentSate = CharType.Code2;
        }else if(currentSate==CharType.Code2) {
```

```
            char[] s1 = {c1,c2};
            codes.add(new String(s1));
            currentSate = null;
            return getEncodeByList(codes);
        }
    }
    if(currentSate==CharType.Code2) {
        char[] s1 = {c1,c2};
        codes.add(new String(s1));
    }
    return getEncodeByList(codes);
}
```

4.3.2 布尔搜索

用布尔查询来合并多个查询条件,最常见的例子是搜索标题或正文。

```
BooleanQuery bodyOrTitle = new BooleanQuery();
bodyOrTitle.add(bodyQuery, BooleanClause.Occur.SHOULD);
bodyOrTitle.add(titleQuery, BooleanClause.Occur.SHOULD);
```

这里 BooleanClause.Occur.SHOULD 代表这个查询条件是可选的,如果一个查询条件是必须满足的,就用 BooleanClause.Occur.MUST。

也可以使用 MultiFieldQueryParser 来合并对多个列的搜索,比如下面实现对"body"和"title"两列的查找。

```
Query query = MultiFieldQueryParser.Parse(queryWord,
                        new string[]{"body","title" },
                        analyzer);
```

4.3.3 指定范围搜索

在商品搜索中,经常需要指定按时间条件或价格等数值条件查找,如图 4-1 所示。

图 4-1 指定范围搜索实例

可以通过 RangeQuery 来实现这样的时间条件区间条件查找:

```
java.util.Calendar upper = GregorianCalendar.getInstance();
upper.add(java.util.Calendar.YEAR, +100);
```

```
String t2 = formatter.format(upper.getTime());

if ("1".equals(dateRange)){
    //一周内
    now.add(java.util.Calendar.DATE, -7);
    String t1 = formatter.format(now.getTime());
ConstantScoreRangeQuery dateQuery =
 new ConstantScoreRangeQuery("time", t1, t2, true, true);
}
else if ("2".equals(dateRange)){
    //一月内
    now.add(java.util.Calendar.MONTH, -1);
    String t1 = formatter.format(now.getTime());
ConstantScoreRangeQuery dateQuery =
 new ConstantScoreRangeQuery("time", t1, t2, true, true);
}
else if ("3".equals(dateRange)){
    //三月内
    now.add(java.util.Calendar.MONTH, -3);
    String t1 = formatter.format(now.getTime());
ConstantScoreRangeQuery dateQuery =
 new ConstantScoreRangeQuery("time", t1, t2, true, true);
}
else if ("4".equals(dateRange)){
    //六月内
    now.add(java.util.Calendar.MONTH, -6);
    String t1 = formatter.format(now.getTime());
ConstantScoreRangeQuery dateQuery =
 new ConstantScoreRangeQuery("time", t1, t2, true, true);
}
```

如果是在 Solr 界面中，区间条件的查询语法例子如下：

```
+汽车 +expiretime:[2007-08-13T00:00:00Z TO 2008-08-13T00:00:00Z]
```

如果要搜索单个日期值，需要对":"转义，例如：

```
postdate:2007-08-13T00\:00\:00Z
```

4.3.4 搜索结果排序

可以按单列或者多列排序，但是需要保证排序列是不做切分处理的，也就是对该列做索引的时候设置 Field.Index.NOT_ANALYZED。例如"url"网址列没有做过切分，可以按该列排序，而标题列"title"做过切分，不能按该列排序。

经常需要按日期倒排序，为了支持对日期列排序，需要把日期转换成统一的字符串格式"yyyyMMddHHmmssSSS"。如果精度低，字符串长度相应变短。

索引日期的例子：

```
Date pubDate = rs.getDate("pubDate");
Field f = new Field("pubDate",
        DateTools.dateToString(pubDate, DateTools.Resolution.DAY),//精度到天
        Field.Store.YES,
        Field.Index.NOT_ANALYZED);
```

按日期倒排序的例子：

```
Sort sort= new Sort(new SortField("pubDate",SortField.STRING,true));
ScoreDoc[] hits = searcher.search(query,null,1000,sort).scoreDocs;
```

也可以对多个字段排序，比如先按地区"area"排序，然后按类别"type"排序：

```
Sort sort= new Sort(new SortField[]{new SortField("area"),new SortField("type")});
ScoreDoc[] hits = searcher.search(query,null,1000,sort).scoreDocs;
```

也可以通过 SortComparatorSource 自定义排序方法。

4.3.5 索引缓存与更新

索引一般是一个比较大的文件。一般从几百 M 到几个 G 不等。页面执行搜索的时候打开大的索引往往是一个非常耗时的过程。一般情况下需要缓存 IndexReader 和 Searcher，不要每次响应用户的搜索请求都重新打开索引后再执行搜索。

因为后台在更新索引，前台的缓存会导致无法及时搜到已经更新的内容，这时候就需要重新装载索引库。下面的代码实现过一段时间就检查索引的版本号，如果有更新则重新加载索引库。

```
private void refreshIndexReader() {
        //如果检查时间已经到了，就检查当前索引的版本号
        if ((LastCheckTime+interview) < System.currentTimeMillis()) {
           long newIndexVersion;
           newIndexVersion = IndexReader.getCurrentVersion(_dir);
            //如果索引已经是最新的，就重新设置检查时间
    if (newIndexVersion == currentIndexVersion) {
            LastCheckTime = System.currentTimeMillis();
            return;
        }

        synchronized (this) {//同步
            LastCheckTime = System.currentTimeMillis();
              searcher.close();
              reader.close(); //旧的 IndexReader 有可能还在被其他线程使用着
              reader = IndexReader.open(_dir);
              searcher = new IndexSearcher(reader);
              currentIndexVersion = newIndexVersion;
         }
        }
}
```

需要防止关闭还在被其他线程使用着的 IndexReader。IndexSearcher 不会增加对 IndexReader 的引用，也就是说不会调用 IndexReader 的 incRef() 方法。IndexReader 的 close() 方法只是调用 decRef() 方法，减少对 IndexReader 实例的引用。如果引用计数降低到 0，则会实际关闭 reader。设计一个 SearcherManager 来管理 IndexReader 的缓存。

```java
public class SearcherManager {
    private IndexSearcher currentSearcher; // 当前的 IndexSearcher
    private Directory dir;

    public SearcherManager(Directory dir) throws IOException {
        this.dir = dir;
        //创建初始的 IndexSearcher
        currentSearcher = new IndexSearcher(IndexReader.open(dir));
    }

    public void warm(IndexSearcher searcher) {
    } // 预热新的 IndexSearcher

    private boolean reopening;

    private synchronized void startReopen() throws InterruptedException {
        while (reopening) {
            wait();
        }
        reopening = true;
    }

    private synchronized void doneReopen() {
        reopening = false;
        notifyAll();
    }

    // 重新打开 IndexSearcher
    public void maybeReopen() throws InterruptedException, IOException {
        startReopen();
        try {
            final IndexSearcher searcher = get();
            try {
                long currentVersion =
currentSearcher.getIndexReader().getVersion();
                if (IndexReader.getCurrentVersion(dir) != currentVersion) {
                    IndexReader newReader = currentSearcher.getIndexReader()
                            .reopen();
                    assert newReader != currentSearcher.getIndexReader();
                    IndexSearcher newSearcher = new IndexSearcher(newReader);
                    warm(newSearcher);
```

```
                    swapSearcher(newSearcher);
                }
            } finally {
                release(searcher);
            }
        } finally {
            doneReopen();
        }
    }

    public synchronized IndexSearcher get() { // 返回当前的 IndexSearcher
        currentSearcher.getIndexReader().incRef();
        return currentSearcher;
    }

    public synchronized void release(IndexSearcher searcher) // 释放 IndexSearcher
            throws IOException {
        searcher.getIndexReader().decRef();
    }

    private synchronized void swapSearcher(IndexSearcher newSearcher)
            throws IOException {
        release(currentSearcher);
        currentSearcher = newSearcher;
    }
}
```

在 Web 服务器中使用 SearcherManager 类的方法如下面的代码所示：

```
//多个线程之间共享同一个 SearcherManager
static SearcherManager searcherManager = new SearcherManager(indexdir);

//在每次搜索时调用如下方法
searcherManager.maybeReopen();
IndexSearcher searcher = searcherManager.get();
try {
    // 执行搜索和结果输出
} finally {
    searcherManager.release(searcher);
}
```

完整的代码如下：

```
public class SearchByQuery {
    private static SearcherManager searcherManager;
    private Logger logger = LoggerFactory.getLogger(this.getClass());

    static {
        String indexDir = null;
```

```java
        InputStream inputFile = null;
        Properties propertie = new Properties();
        try {
            inputFile = SearchByQuery.class.getClassLoader()
                    .getResourceAsStream("conf.properties");
            propertie.load(inputFile);
            indexDir = propertie.getProperty("indexDir");
            System.out.println(indexDir);
        } catch (FileNotFoundException ex) {
            System.out.println(
                "读取属性文件失败! 原因：文件路径错误或者文件不存在");
            ex.printStackTrace();
        } catch (IOException ex) {
            System.out.println("装载文件--->失败!");
            ex.printStackTrace();
        } finally {
            try {
                if (inputFile != null)
                    inputFile.close();
            } catch (IOException e) {
                e.printStackTrace();
            }
        }

        try {
            if (indexDir != null) {
                Directory directory =
                        FSDirectory.open(new File(indexDir));
                searcherManager = new SearcherManager(directory);
            }
        } catch (IOException e) {
            e.printStackTrace();
        }
    }

    /**
     * 根据查找关键词返回结果集
     *
     * @param word
     * @return
     * @throws ParseException
     * @throws IOException
     * @throws InterruptedException
     * @throws Exception
     */
    public List<GoodInfo> getResults(String word) throws ParseException,
            InterruptedException, IOException {
```

```java
Analyzer analyzer = new StandardAnalyzer();
QueryParser parser = new QueryParser("title", analyzer);
Query titleQuery = parser.parse(word);

searcherManager.maybeReopen();
IndexSearcher searcher = searcherManager.get();
try {
        ScoreDoc[] hits = 
                    searcher.search(titleQuery, 10000).scoreDocs;
        List<GoodInfo> lst = new ArrayList<GoodInfo>();
        // 遍历结果
        for (int i = 0; i < hits.length; i++) {
                Document hitDoc = searcher.doc(hits[i].doc);
                String title = hitDoc.get("title");
                String url = hitDoc.get("url");
                String des = hitDoc.get("body");

                String date = hitDoc.get("date");
                Date d = DateTools.stringToDate(date);
                Date nowDay = new Date();
                SimpleHTMLFormatter simpleHTMLFormatter = 
   new SimpleHTMLFormatter( "<font color='red'><b>", "</b></font>");
                Highlighter highlighter =
                        new Highlighter(simpleHTMLFormatter,
                                  new QueryScorer(titleQuery));
                highlighter.setTextFragmenter(
                                  new SimpleFragmenter());
                TokenStream tokenStream =
                    new StandardAnalyzer().
                        tokenStream("title",
                            new StringReader(title));
                String highLightText =
                        highlighter.getBestFragment
                              (tokenStream, title);

                GoodInfo good = new GoodInfo();
                good.setGoodsName(highLightText);
                good.setGoodsNameURL(url);
                good.setGoodsDescription(des);
                lst.add(good);
        }
        return lst;
} catch (Exception e) {
        e.printStackTrace();
        logger.error("searchbyQuery",e);
} finally {
        searcherManager.release(searcher);
```

```
            }
            return null;
        }
}
```

为了防止关闭还在被其他线程使用着的 IndexReader。搜索界面可以先借出 IndexReader，然后再还回，这样告诉 IndexReader 对象缓存池，已经可以关闭这个对象了。类似于借书，先借出去一本书，然后把这本书还回来。

```
public class ReaderPool {
    private final Queue<IndexReader> objects;
    private IndexReader _reader;

    public ReaderPool(IndexReader r) {
        objects = new ConcurrentLinkedQueue<IndexReader>();
        _reader = r;
    }

    //借出对象
    public IndexReader borrow() throws Exception {
        IndexReader t;
        if ((t = objects.poll()) == null) {
            t = _reader.reopen();
        }
        return t;
    }

    //还回对象
    public void giveBack(IndexReader object) {
        this.objects.offer(object);
    }
}
```

搜索界面使用的例子：

```
//借出对象
IndexReader reader = readerPool.borrow();

// 执行搜索
IndexSearcher searcher = new IndexSearcher(reader);
//...

//还回对象
readerPool.returnIndexReaders(reader);
```

可以用一个专门的守护线程检查缓存池中的 IndexReader 是否是新的。

4.4 实现分类统计视图

一个职位搜索网站需要统计出某一关键词下的要求本科学历的有多少岗位，要求专科学历的有多少岗位，薪资范围在 4000～6000 元/月的有多少岗位，薪资范围在 6000～8000 元/月的有多少岗位。从术语上讲，就是要从各个角度(维)进行分类并统计搜索结果数在相关分类中的分布情况。这个功能叫作搜索结果分类统计搜索(Faceted search)。分类可以是多层次的，用户可以沿着某一类继续细化，这有点像数据仓库中的向下钻取，但它不是用数据库而是用 Lucene 完成的。这也是 Lucene 的一个很有特色的应用案例。

分类统计搜索的基本功能有：

- 根据刻面(facet)把搜索结果分组；
- 显示每个刻面值命中的总数；
- 可以通过刻面值细化搜索结果。

首先定义一个 XML 文件存储分类目录。可以使用 W3C 的 DOM 创建这个类别文件或者遍历文件。

```xml
<?xml version="1.0" encoding="UTF-8"?>
<categories>
    <category name="Video"/>
    <category name="Office"/>
    <category name="Music"/>
    <category name="Appliances"/>
    <category name="Furniture"/>
    <category name="Magazines"/>
    …
</categories>
```

这个项目需要和 xercesImpl.jar 用于 DOM 接口，serializer.jar 用于保存 XML 文件。DOM 提供了很多方便的类来创建 XML 文件。首先要使用 DocumentBuilder 创建一个 Document，定义所有的 XML 内容-节点、Element 类上的属性。

写 XML 文件的代码如下：

```java
Set<String> categoriesSet = readerCategory(); //读入类别到 categoriesSet
//新建工厂类
DocumentBuilderFactory docFactory = DocumentBuilderFactory.newInstance();
//新建文档构建器
DocumentBuilder builder = docFactory.newDocumentBuilder();
Document doc = builder.newDocument(); //创建根节点
doc.setXmlVersion("1.0");
Element catRoot = doc.createElement("categories"); //根节点下创建元素
doc.appendChild(catRoot);   //元素节点增加到根节点
appendChild(doc, catRoot, categoriesSet); //增加孩子节点
writeDom(doc); //写文件
```

增加孩子的方法实现如下：

```
private static void appendChild(Document doc, Element node, Set<String> categories) {
    for (String line : categorys) {
        Element element = doc.createElement("category"); //创建元素
        element.setAttribute("name", line); //设置元素节点的内容
        node.appendChild(element); //增加孩子节点
    }
}
```

使用 Transformer 输出所有的 XML 内容到一个文件。写 XML 文件的方法实现如下：

```
public static void writeDom(Document doc) throws ParserConfigurationException,
        IOException, TransformerException {
    TransformerFactory factory = TransformerFactory.newInstance();
    Transformer trans = factory.newTransformer();//获得一个转换器

    //使用一个 DOM 树作为来源对象
    DOMSource source = new DOMSource(doc);

    File file = new File(FILE_CATEGORY_XML_PATH);
    if (!file.exists()) {
        file.createNewFile();
    }

    FileOutputStream out = new FileOutputStream(file);
    StreamResult result = new StreamResult(out);
    //把来源对象转换到输出文件
    trans.transform(source, result);
}
```

按指定类别搜索，对搜索结果的分类统计，这两个功能是不一样的。按类别搜索类似于 SQL 语句的 where 条件，分类统计类似于 SQL 语句中的 group by 功能。

最基本的想法是：按类别搜索和按关键词搜索，然后把这两个搜索结果做与运算，最后计算结果数量，如图 4-2 所示。

图 4-2　计算中间的交集大小

把计算交集大小最终转换成计算二进制位中 1 的个数。用一个 bit 位来表示一个文档是否属于集合。一个这样的集合就是一个 BitSet。计算一个刻面值的流程如图 4-3 所示。

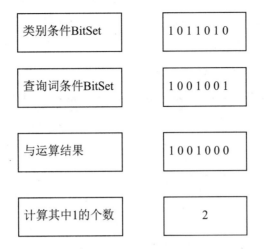

图 4-3　分类统计计算流程

可以利用 QueryFilter 来实现搜索结果分类统计。QueryFilter 有个 bits()方法返回一个 BitSet 集，这个 BitSet 的大小是所针对的 Lucene 库的大小(也就是 new BitSet(reader.maxDoc()))，凡符合 filter 条件的文档在集合中相应位置上置为 1(true)。这个 BitSet 集对特定 QueryFilter 对象来说是 cache 保存的，下次调用不会重新计算。然后利用这个 BitSet 集，将满足各个基本属性值的 BitSet 值计算出，根据特定用户需要进行相关的 BitSet 与(交集)操作，最后利用 BitSet 集的 cardinality()方法就可计算出满足该类的总数。

用 QueryFilter 实现搜索结果分类统计的参考代码如下：

```
String[] cats = {"001004003","001008003021","001004014" }; //类别数组
long[] catCounts = new long[cats.length];//分类统计结果

//原始查询
Filter all = new QueryWrapperFilter(q);

//用 AND 逻辑合并 Filter
ChainedFilter.DEFAULT = ChainedFilter.AND;
for (int i=0;i<cats.length;++i) {
    //分类统计查询条件
    Filter these = new QueryWrapperFilter(new TermQuery(new Term("cat",cats[i])));
    ChainedFilter chainedFilter = new ChainedFilter(
            new Filter[]{all,these}
        );
    //计算 Filter 中的 BitSet 的 1 的个数
    catCounts[i] = chainedFilter.getCardinality(reader);
}
return catCounts;
```

在 Apache 的另外一个企业搜索项目 Solr 中，通过优化后的 BitSet 实现了一个 DocSetHitCollector 来做分组求和。这个优化后的 BitSet 叫作 OpenBitSet。但是这个 OpenBitSet 只是在 64 位的机器上，当返回的结果数量很多的时候才比 Java 内部的 BitSet 类更快。

计算一个二进制的数组的 1 的个数(叫作 PopCount 或者 cardinality)在整个计算中对性能有比较重要的影响。这里把这个功能叫作计算数组的二进制势。通过查表可以快速地计算从 0 到 255 的二进制势，见表 4-2。

表 4-2　计算从 0~255 数值的二进制数据中 1 的出现次数

数　　值	0	1	2	3	4	…	255
二进制形式	00000000	00000001	00000010	00000011	00000100	…	11111111
1 的个数	0	1	1	2	1	…	8

下面是查表法实现的计算数组的二进制势的程序：

```java
private static int[] _bitsSetArray65536 = null;
static {
    _bitsSetArray65536 = new int[65536]; //16 位整数的二进制势表
    byte[] _bitsSetArray256 = { 0, 1, 1, 2, 1, 2, 2, 3, 1, 2, 2,
        3, 2, 3, 3, 4, 1, 2, 2, 3, 2, 3, 3, 4, 2, 3, 3, 4, 3, 4, 4, 5, 1,
        2, 2, 3, 2, 3, 3, 4, 2, 3, 3, 4, 3, 4, 4, 5, 2, 3, 3, 4, 3, 4, 4,
        5, 3, 4, 4, 5, 4, 5, 5, 6, 1, 2, 2, 3, 2, 3, 3, 4, 2, 3, 3, 4, 3,
        4, 4, 5, 2, 3, 3, 4, 3, 4, 4, 5, 3, 4, 4, 5, 4, 5, 5, 6, 2, 3, 3,
        4, 3, 4, 4, 5, 3, 4, 4, 5, 4, 5, 5, 6, 3, 4, 4, 5, 4, 5, 5, 6, 4,
        5, 5, 6, 5, 6, 6, 7, 1, 2, 2, 3, 2, 3, 3, 4, 2, 3, 3, 4, 3, 4, 4,
        5, 2, 3, 3, 4, 3, 4, 4, 5, 3, 4, 4, 5, 4, 5, 5, 6, 2, 3, 3, 4, 3,
        4, 4, 5, 3, 4, 4, 5, 4, 5, 5, 6, 3, 4, 4, 5, 4, 5, 5, 6, 4, 5, 5,
        6, 5, 6, 6, 7, 2, 3, 3, 4, 3, 4, 4, 5, 3, 4, 4, 5, 4, 5, 5, 6, 3,
        4, 4, 5, 4, 5, 5, 6, 4, 5, 5, 6, 5, 6, 6, 7, 3, 4, 4, 5, 4, 5, 5,
        6, 4, 5, 5, 6, 5, 6, 6, 7, 4, 5, 5, 6, 5, 6, 6, 7, 5, 6, 6, 7, 6,
        7, 7, 8 }; //8 位整数的二进制势表
    //根据 8 位整数的二进制势表生成 16 位整数的二进制势表
    for (int j = 0; j < 65536; j++) {
        _bitsSetArray65536[j] = _bitsSetArray256[j & 0xff]
                + _bitsSetArray256[(j>>> 8 )& 0xff];
    }
}

//计算给定数组 A 的二进制势表
public static long pop_array2(long A[],int wlen) {
    long _count = 0;
    for (int i = 0; i < wlen; i++) {
        _count += _bitsSetArray65536[(int) (A[i]& 0Xffff)]
                + _bitsSetArray65536[(int) ((A[i] >>> 16 )& 0xffff)]
                + _bitsSetArray65536[(int) ((A[i] >>> 32) & 0xffff)]
                + _bitsSetArray65536[(int) ((A[i] >>> 48) & 0xffff)];
    }
    return _count;
}
```

查表法和 Java 内部实现同样功能的 bitCount()方法测试比较性能：

```
long x = 100000000000000001;
for(int i=0;i<1000;i++){
    pop(x);//查表法实现的位计算
}
long end = System.nanoTime();
System.out.println(end-start);//输出计算时间

long start2 = System.nanoTime();
for(int i=0;i<1000;i++){
    Long.bitCount(x);//java内部实现的实现的位计算
}
long end2 = System.nanoTime();
System.out.println(end2-start2);//输出计算时间
```

下面的搜索结果分类统计功能实现通过 DocSet 的 intersectionSize()方法减少计算步骤，又比上面的实现至少快了百分之几。

```
String[] cats = new System.String[] {"体育", "商业", "艺术", "教育"}; //类别数组

DocSetHitCollector all = new DocSetHitCollector(reader.MaxDoc());
searcher.Search(q, all);
DocSet allDocSet = all.DocSet;

int[] catCounts = new int[cats.Length];

for (int i = 0; i < mfgs.Length; ++i){
    DocSetHitCollector these = new DocSetHitCollector(reader.MaxDoc());
    searcher.Search(new TermQuery(new Term("type", cats)), these);
    //集合求交运算和计算集合大小运算两个操作
//在QueryFilter()方法中是分开计算和独立优化的
//现在把这两个操作放入一个函数中整体优化来提高程序运算效率
    catCounts[i] = these.DocSet.intersectionSize(allDocSet);
}
```

上面这个实现比起最初的 QueryFilter 实现，在于合并了以下两个步骤：

```
these.and(all);
mfg_counts[i] = these.cardinality();
```

这样得到搜索结果在类别中的分布图，如图 4-4 所示。
封装成一个方法：

```
public static long[] bitSetCounter(IndexSearcher searcher, Query q) {
        //类别数组
        String cats[] = { "体育", "商业", "艺术", "教育" };
        IndexReader reader = searcher.getIndexReader();
        long[] catCounts = new long[cats.length];//分类统计结果

        DocSetCollector all = new DocSetCollector(100, reader.maxDoc());
```

```
            searcher.search(q, all);
            DocSet allDocSet = all.getDocSet();
            for (int i = 0; i < catCounts.length; i++) {
                DocSetCollector these = new DocSetCollector(100, reader.maxDoc());
                searcher.search(new TermQuery(new Term("type", cats[i])), these);
                catCounts[i] = these.getDocSet().intersectionSize(allDocSet);
                System.err.println((new StringBuilder
(String.valueOf(cats[i]))).append("=").append(catCounts[i]).toString());
            }
            return catCounts;
}
```

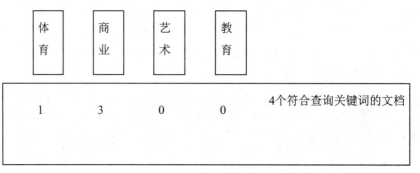

图 4-4 搜索结果在类别中的分布图

使用 Filter 来通过刻面值细化搜索结果。

```
//按类别条件过滤搜索结果
TermQuery categoryQuery = new TermQuery(new Term("category", "art"));
Filter categoryFilter = new QueryWrapperFilter(categoryQuery);
CachingWrapperFilter cachingWrapperFilter = new
CachingWrapperFilter(categoryFilter);
TopDocs hits = searcher.search(allBooks, cachingWrapperFilter, 20);
```

Lucene 3.4 版本支持层次分类统计。实现代码位于 contrib/facet，索引阶段：

```
IndexWriter writer = ...
TaxonomyWriter taxo =
new LuceneTaxonomyWriter(taxoDir, OpenMode.CREATE);
...
Document doc = new Document();
doc.add(new Field(
"title", titleText, Store.YES, Index.ANALYZED));
...
List<CategoryPath> categories = new ArrayList<CategoryPath>();
categories.add(new CategoryPath("author", "Mark Twain"));
categories.add(new CategoryPath("year", "2010"));
...
DocumentBuilder categoryDocBuilder =
new CategoryDocumentBuilder(taxo);
```

```
categoryDocBuilder.setCategoryPaths(categories);
categoryDocBuilder.build(doc);
writer.addDocument(doc);
```

搜索阶段：

```
String indexPath = "d:/index";
Directory indexDir= FSDirectory.open(new File(indexPath));
IndexReader indexReader = IndexReader.open(indexDir);
IndexSearcher searcher = new IndexSearcher(indexReader);
TaxonomyReader taxo = new LuceneTaxonomyReader(taxoDir);
...
Query q = new TermQuery(new Term(SimpleUtils.TEXT, "white"));
TopScoreDocCollector tdc = TopScoreDocCollector.create(10, true);
...
FacetSearchParams facetSearchParams = new FacetSearchParams();
facetSearchParams.addFacetRequest(
new CountFacetRequest(
new CategoryPath("author"), 10));
FacetsCollector facetsCollector =
new FacetsCollector(facetSearchParams, indexReader, taxo);
searcher.search(q,
MultiCollector.wrap(topDocsCollector, facetsCollector));
List<FacetResult> res = facetsCollector.getFacetResults();
```

4.4.1 单值列分类统计

如果列只有 1 个值，则可以把这个列中所有文档的值用一个数组表示，这就是 FieldCache。采用 FieldCache 实现单值列分类统计的算法如图 4-5 所示。

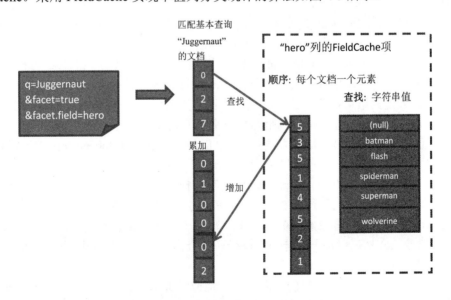

图 4-5　单值列分类统计算法

在图4-5中，基本查询找到3个文档{0,2,7}。FieldCache表示的值数组是{5,3,5,1,4,5,2,1}，取这个数组中的第0个、第2个以及第7个元素，得到命中文档集合对应的"hero"列的值{5,5,1}。累加得到数组：{0,1,0,0,0,2}。

4.4.2 侧钻

在典型的分面搜索界面，你能在搜索结果的左侧看到一列，可以用来过滤属性，如：价格、制造商等。这些属性有些是单值，有些是范围值，然后还展示出每个属性的数量。

这个数量就是告诉你，如果通过单击这个属性过滤了你的搜索结果后，还会有多少结果符合搜索条件。

如果你在Overstock.com上买LED Television的话，就会发现左侧像drill down(下钻)一样。

如果在Overstock.com单击其中一个属性，例如Samsung品牌，你会发现这个属性从下钻选项中消失，然后跳到页面上部，作为搜索标签的一部分。还可以通过单击关闭按钮去掉这个过滤来上钻。

如果在几个属性中下钻，但是想看这些属性交叉不同值的结果，怎么办？

如果不能真的做到这一点，用户体验就太差了——要过度地使用后退键，退回上一个页面来选择不同的选项。上钻和下钻只能在一个页面完成，这种用户界面限制性太强了。Overstock.com是一个纯上钻和纯下钻用户界面的例子。

其他网站通过提供侧钻(drill sideways)，提供单击可选的选项，或者提供下钻之前的选择作为附加选项，来实现这个功能。

例如：在亚马逊上搜LED Television，在商标选项右边的图片是多选的UI，能让你不止选择一个值。当你选中一个选项(勾选那个选项框，或者单击那个选项)，搜索结果就会被过滤掉，但是此时，这个选择项并没有消失，它们仍然在那里，让你可以横向选择(侧钻)其他选项。

LinkdIn的分面搜索在页面的左边。它还有这些特性：不仅仅是所有选项都能横向选择和多选，而且在底部还有文本框让你输入没有列出来的值。

简单来说，单个选项一次只能过滤一个条件；多选选项可以一次过滤多个条件。比如：向下钻取就是用一个过滤条件过滤搜索结果，减少符合条件结果；向上钻取就是去除一个过滤条件，扩大搜索结果；侧钻就是改变一个已有过滤条件——在单选条件下就是选择一个不同的过滤条件；在多选条件下就是增加一个过滤条件。

是否提供横向选择和是否提供多选是正交的。例如：Search-lucene.com上的日期选项，它是单选的，但是可以横向选择(但是技术好像还有Bug)。

怎么实现横向选择呢？当你对一个选项向下选择后，其他选项当然是没有被向下选择的，所以其他选项的计数也只能为0。所以，怎么才能计算这些数量呢？

一个直接的解决办法是去掉一个的方式：如果直接说搜索是"foo"，有3个选项可以向下选择(A:a,B:b,C:c)，然后对每一个选项都进行一次查询。每一次查询都有一部分过滤条件，然后计算出相应的向下查询数量。

例如：对于A选项向下选择的分面计数，是搜索foo，选中B:b和C:c(A:a被踢出)的结果计数；对于B选项向下选择的分面计数，是搜索foo，选中A:a和C:c(B:b被踢出)的结果计数。你也需要特别地用所有过滤条件(foo,A:a,B:b,C:c)来搜索出一个命中数。Solr 的 part of the facet request 是一个非常清晰的、体验去掉一个的例子。

即使这个方法能计算出正确的数量,但是它太费时了。因为要做 4 次搜索才行。可以用位集合来提速:把所有符合"foo"搜索的结果放在一个位集合中,然后对于符合 A,B,C 的都做一次向下选择,结果放在各自的位集合中,再用它们的交集计算分面结果的数量。Solr 用了一个类似的方法,缓存了请求的位集合。

当然还有其他方法,在 LUCENE-4748 上讨论过,就是执行一次搜索,这个搜索匹配所有条件和 near-misses 选项——文档只符合其中之一的过滤条件。它是用标准的 BooleanQuery 来执行的。原始查询是 MUST 条件,所有的过滤项是 SHOULD 条件,还用 minNumberShouldMatch 限制向下的过滤条件最小匹配数是 1。所以,所有符合条件的结果都做收集处理,如果过滤条件符合,就是搜索命中,相应地收录到这个条件,然后增加向下搜索计数。如果不符合,就增加排除此选项后横向选择的计数。

注意,当第一次对一个选项向下选择的时候,在向下选择之前,这个选择的横向选择计数和向下选择计数是一样的。如果用户界面可以缓存这个状态,就可以做一个优化:通过重用以前的向下选择的数量,来避免计算此时向下选择项的横向选择数量。

比如说在硬盘里面搜索"希捷",那"希捷"就是关键字,有下钻、上钻、侧钻 3 种操作。

硬盘有多个属性,比如说"容量大小"有 0~80G、80~200G、200~500G 和 500G 以上。这些属性一般是放在页面左侧,让用户选择。如果选了其中一个,比如 0~80G 就是下钻。其实就是限定了一个范围。结果就显示所有 0~80G 的结果。如果去掉这个限定,就是上钻。

侧钻就是要搜索出这几个属性的数量。每个属性有多少个匹配的结果。原始的方法就是每一个都搜索一次,而新的方法是,使用 BooleanQuery,搜索关键字是"must",其他属性用 should,还用 minNumberShouldMatch 来限制最少匹配是 1。再对结果进行处理,如果和这个属性匹配,这个属性的计数就+1。

4.5 实现相似文档搜索

有时候需要检索与给定文档(例如 BBS 讨论区内某一帖子)相似的文档。打开一个新闻网页,往往在下面有块区域,显示和这篇新闻相关的新闻。

在 Lucene 的外围资源中有个 MoreLikeThis 类,可以实现对索引内部的文档查询相似文档。顾名思义,这个类的作用就是找出更多类似于这个(This)文档的结果。

```
MoreLikeThis      mlt = new MoreLikeThis(reader);
mlt.setFieldNames(new String[] {"title", "content"});
mlt.setMaxQueryTerms(5);
if(queryString.startsWith("related:")){
        int docId = Integer.parseInt(queryString.substring(8));
        query = mlt.like(docId);
}
```

另外举个例子说明这个类的作用。比如对一个卖商品的网站来说，当顾客正在浏览一件商品时，如果能把和这件商品性能、作用很相近的商品也同时罗列在网页的左边，万一顾客想要的商品正好就在其中，那么这个网站的营业额肯定会有所提高。

MoreLikeThis 类有一个主要的方法：like(int docNum)，这个方法的参数还可以是 File、InputStream、Reader 或 URL，返回值是一个 Query 对象，MoreLikeThis 类的构造函数即 MoreLikeThis(IndexReader ir)，它需要传进一个 IndexReader。下面举例说明 like()方法的用法。就拿第一段的需求为例：

```
public static void main(String[] a) throws Throwable {
    String indexName = "indexpath";
    IndexReader r = IndexReader.open(indexName);
    PrintStream o = System.out;
    o.println("Open index " + indexName + " which has " + r.numDocs() + " docs");
    MoreLikeThis mlt = new MoreLikeThis(r);
    mlt.setMaxQueryTerms(5);
    o.println("Query generation parameters:");
    o.println(mlt.describeParams());
    o.println();
    String keygoodid = "";
    String similarGoodsid = "";
    String keygoodName = "";
    String similarGoodsName = "";
    if(!r.isDeleted(100)){
    Document keyDoc = r.document(j);
    keygoodid = keyDoc.get("id");
    keygoodName = keyDoc.get("name");
    Query query = mlt.like(100);
    IndexSearcher searcher = new IndexSearcher(indexName);
    Hits hits = searcher.search(query);
    int len = hits.length();
    for (int i = 0; i < Math.min(5, len); i++) {
        Document d = hits.doc(i);
        similarGoodsid += d.get("id") + ",";
        similarGoodsName += d.get("name") + ",";
    }
    o.println("keygoodid:" + keygoodid + "|" + "similarGoodsid:" + similarGoodsid);
    o.println("keygoodName:" + keygoodName + "|" + "similarGoodsName:" + similarGoodsName);
}
```

Indexpath 下面存放的是对所有商品的索引，构造一个 MoreLikeThis 的对象 mlt，然后调用 mlt.like(100)，这里的 100 是 Lucene 内部的 docNum。最后搜索一下，取前几个结果就是与此文档最为相似的。

like(int docNum)方法返回的 Query 是怎么产生的呢？它首先根据传入的 docNum 找出该文档里去除停用词后的高频词，然后用这些高频词生成 Queue，最后把 Queue 传进 search()

方法得到最后结果。主要思想就是认为这些高频词足以表示文档信息，然后通过搜索得到最后与此文档类似的结果。

缺省的 MoreLikeThis 没有定义停用词，也不支持中文分词，这都是需要进一步完善的。

为了把这个功能集成到界面，可以修改查询语法，当用户输入"related:doc_id"的时候返回索引库中的相关文档。

4.6 实现 AJAX 搜索联想词

搜索输入框中的下拉提示给用户一个有参考意义的搜索词表，有时候还提供给用户搜索该词预期的结果数量。这个功能有时候也叫作自动完成(AutoCompleter)，其一般是由浏览器端的 AJAX 代码完成的。

搜索词表可以从用户搜索日志中统计出来，搜索次数多的词排在前面。除了按一般的设计，为每个用户提供统一的词表，还可以对每个用户提供个性化的推荐词表。例如：用户输入"汽"时，"汽车"的搜索次数比"汽油发电机"多，所以排在提示词列表的前面。如果搜索日志比较少，无法挖掘出足够多的推荐搜索词，可以考虑从文本中挖掘一些关键词作为推荐搜索词。因为后台索引一般都在不断地变化，推荐搜索词右侧显示的"**结果"并不是实时搜索出的结果，只是一个估计值，只具有参考价值。为了实现搜索提示效果，需要用到词典的前缀匹配。

4.6.1 估计查询词的文档频率

为了对于用户输入任何词都可以显示一个估计的搜索结果数量，需要计算这个词的文档频率。例如用户输入"NBA 直播"，可以根据"NBA"和"直播"的文档频率估计"NBA 直播"的文档频率。假设"NBA"和"直播"之间的出现没有依赖关系，则可以简化计算如下：

$$P("NBA" \cap "直播") = P("NBA") \bullet P("直播")$$

其中，$P("NBA" \cap "直播")$ 是联合概率，可以认为是"NBA 直播"的出现概率，而 $P("NBA")$ 和 $P("直播")$ 是每个词出现的概率。假设索引中的总文档数量是 N，而 $P("NBA") = Freq("NBA")/N$，$P("直播") = Freq("直播")/N$。

因此"NBA 直播"的文档频率

$$Freq("NBA直播") = N \times \frac{Freq("NBA")}{N} \times \frac{Freq("直播")}{N}$$

而 Freq("NBA") 和 Freq("直播") 所代表的文档频率在 Lucene 中可以通过 org.apache.lucene.search.Searcher 的 docFreq(Term term) 方法得到。因为多个词之间并不一定满足独立出现的假设，因此这个估计值有可能偏低。

4.6.2 搜索联想词总体结构

当用户在浏览器的输入框输入查询词时，JavaScript 代码捕获用户即时输入的数据并向服务器发送请求。自动完成的功能的总体结构如图4-6所示。

图 4-6 自动完成功能总体结构

如果在使用,则不要让Struts2的FilterDispatcher把所有的URL都拦截了。

```
<!-- 定义Struts2的FilterDispathcer的Filter -->
   <filter>
      <filter-name>struts2</filter-name>
      <filter-class>org.apache.struts2.dispatcher.FilterDispatcher</filter-class>
   </filter>

<!-- FilterDispatcher用来初始化Struts2并且处理.action和.jsp的Web请求 -->
   <filter-mapping>
      <filter-name>struts2</filter-name>
      <url-pattern>*.action</url-pattern>
   </filter-mapping>
```

4.6.3 服务器端处理

当用户输入一个搜索字的同时由Web服务器中的一个Servelet(AutoCompleteServlet)从后台取数。可以直接在内存中管理查询,而不是访问数据库取数。先设计词典格式。它由3列组成,第一列是词,第二列是搜索返回结果数量,第三列是用户搜索次数,中间用%隔开,例如:

综合教程第一册%34%2

搜索词是"综合教程第一册",搜索返回结果数量是34,用户搜索了2次。这样把用户搜索次数多的关键词放在前面优先显示。我们构造一个在"词典查找算法"部分已经介绍过的"Trie"树词典来实现快速地前缀匹配查找:

```
/**
 * 返回以一个前缀开始的所有关键词的数组
 *
 *@param   prefix          前缀
 *@param   numReturnValues 返回数组的最大长度
 *@return                  返回数组结果
 */
public TSTItem[] matchPrefix(String prefix, int numReturnValues) {
    TSTNode startNode = getNode(prefix);
    if (startNode == null) {
        return null;
```

第 4 章 搜索引擎用户界面

```java
    }
    ArrayList<TSTItem> sortKeysResult = new ArrayList<TSTItem>();

    ArrayList<TSTItem> wordTable = sortKeysRecursion(
            startNode.EQKID,
            ((numReturnValues < 0) ? -1 : numReturnValues),
            sortKeysResult);
    int retNum = Math.min(numReturnValues,wordTable.size());

    Select.selectRandom(wordTable,wordTable.size(),retNum,0);
    TSTItem[] fullResults = new TSTItem[retNum];
    for(int i=0;i<retNum;++i)   {
        fullResults[i] = wordTable.get(i);
    }

    return fullResults;
}

/**
 * 按顺序返回关键词，包括当前节点和与当前节点相关的所有节点对应的关键词
 * 关键词将按顺序追加到结果的尾数
 *@param  currentNode         当前节点
 *@param  sortKeysNumReturnValues    最多返回结果数
 *@param  sortKeysResult2            到目前为止的结果
 *@return   一个列表
 */
private ArrayList<TSTItem> sortKeysRecursion(
    TSTNode currentNode,
    int sortKeysNumReturnValues,
    ArrayList<TSTItem> sortKeysResult2) {

    if (currentNode == null) {
        return sortKeysResult2;
    }

    ArrayList<TSTItem> sortKeysResult =
        sortKeysRecursion(
            currentNode.LOKID,
            sortKeysNumReturnValues,
            sortKeysResult2);

    if (currentNode.data != 0) {
        sortKeysResult.add(
            new TSTItem(getKey(currentNode),
                    currentNode.data,
                    currentNode.weight)
        );
```

```
        }
        sortKeysResult =
            sortKeysRecursion(
                currentNode.EQKID,
                sortKeysNumReturnValues,
                sortKeysResult);

        return sortKeysRecursion(
            currentNode.HIKID,
            sortKeysNumReturnValues,
            sortKeysResult);
    }
```

可以写一个简单的测试代码:

```
public static void main(String[] args) {
    SuggestDic sugDic = SuggestDic.getInstance();
    String prefix = "m";
    TSTItem[] ret = sugDic.matchPrefix(prefix, 10);
    for(TSTItem i:ret ) {
        System.out.println(i.key+":"+i.data+":"+i.weight);
    }
}
```

服务器传递给客户端的 JSON 数据格式是一个数组,例如:

```
["lietu","lucene"]
```

因为 JSON 格式是一个比较简单的数据传输格式,所以采用了 JSON.org 提供的一个简单的生成包。

```
JSONArray jsonarray = new JSONArray();
jsonarray.put("lietu");
jsonarray.put("lucene");
System.out.println(jsonarray.toString());
```

根据对象生成 JSON 串的 POJO 类:

```
public class SuggestItem {
    public String w; //词
    public int c; //结果数量

    public int getC() {
        return c;
    }
    public void setC(int c) {
        this.c = c;
    }
    public String getW() {
```

```
        return w;
    }
    public void setW(String w) {
        this.w = w;
    }
}
```

通过 Servlet 输出 JSON 时，需要设置正确的 MIME 类型(application/json)和字符编码。假定服务器使用 UTF-8 编码，则可以使用以下代码输出编码后的 JSON 文本：

```
response.setContentType("application/json;charset=UTF-8");
response.setCharacterEncoding("UTF-8");
```

这样自动完成的 Servlet 类可以写成下面这样：

```
String val = request.getParameter("q");

String message = null;
try {
    SuggestDicByWord sugDic = SuggestDicByWord.getInstance();
    SuggestItem[] items = sugDic.matchPrefix(val, 10);

    JSONValue lMyValue = JSONMapper.toJSON(items);
    message = Escape.toUnicodeEscapeString(lMyValue.render(false));
    message = lMyValue.render(false);
} catch (Exception e) {
    e.printStackTrace();
}

response.setContentType("application/json; charset=utf-8");
PrintWriter out = response.getWriter();
if (message != null) {
    out.println(message);
} else {
    out.println("");
}
```

把 AutoCompleteServlet 通过 web.xml 部署到 URL 地址"/autoComplete"。

```
<servlet>
  <servlet-name>AutoCompleteServlet</servlet-name>

<servlet-class>com.lietu.autocomplete.AutoCompleteServlet</servlet-class>
</servlet>
  <servlet-mapping>
  <servlet-name>AutoCompleteServlet</servlet-name>
  <url-pattern>/autoComplete</url-pattern>
</servlet-mapping>
```

服务器端修改成：

```java
public void doPost(HttpServletRequest request,
           HttpServletResponse response)
    throws IOException {
        String val = request.getParameter("q");
        String message = null;
        SuggestItem[] items = null;
        try {
            SuggestDic sugDic = SuggestDic.getInstance();

            items = sugDic.matchPrefix(val, 10);//查找Trie树

            JSONValue lMyValue = JSONMapper.toJSON(items);
            message = Escape.toUnicodeEscapeString( lMyValue.render(false) );

        } catch (Exception e) {
            e.printStackTrace();
        }

        response.setContentType("text/html; charset=utf-8");
        PrintWriter out = response.getWriter();
        if(message!=null)          {
            out.println(message);
        } else {
            out.println("");
        }
}
```

可以通过 EasyMock 测试这个 Servlet 的返回值：

```
String queryWord = "P";
//录制mock对象
HttpServletRequest request = createMock(HttpServletRequest.class);
HttpServletResponse response = createMock(HttpServletResponse.class);
ServletConfig servletConfig = createMock(ServletConfig.class);
ServletContext servletContext = createMock(ServletContext.class);

AutoCompleteServlet instance = new AutoCompleteServlet();

//初始化servlet,一般由容器承担，用servletConfig作为参数初始化，此处模拟容器行为
instance.init(servletConfig);
//在某些方法被调用时设置期望的返回值
//如下这样就不会去实际调用servletConfig的getServletContext()方法，而是直接返回
//servletContext，由于servletConfig是mock出来的，所以可以完全控制
expect(servletConfig.getServletContext()).andReturn(servletContext).anyTimes();

expect(request.getParameter("q")).andReturn(queryWord);

PrintWriter pw=new PrintWriter(System.out,true);
```

```
expect(response.getWriter()).andReturn(pw).anyTimes();
response.setContentType("text/html; charset=utf-8");

//重放 mock 对象
replay(request);
replay(response);
replay(servletConfig);
replay(servletContext);

instance.doPost(request, response);

pw.flush();

//检查预期和实际结果
verify(request);
verify(response);
verify(servletConfig);
verify(servletContext);
```

返回一个 JSON 数组格式的数据，如果要支持中文，还要对汉字编码。

4.6.4 浏览器端处理

剩下的就是在前台通过 AJAX 组件库 jQuery 中的 Autocomplete 插件 (http://jqueryui.com/demos/autocomplete/)来完成显示了。

先到官方网站下载 JQuery 的最新版本。然后需要 JQuery UI 的 3 个核心组件：Core、Widget、Position，还有 AutoComplete 插件。

将插件中的 JavaScript 文件和 CSS 文件分别置于 Web 项目中的 "js" 文件夹和 "css" 文件夹中。最后将这些文件导入到需要搜索联想词的页面，一般是搜索首页和搜索结果页面。也就是网页的头信息中包含这些文件。

```html
<head>
    <link rel="stylesheet" href="css/jquery.ui.all.css">
    <script language="javascript" src="js/jquery-1.5.1.js"></script>
    <script language="javascript" src="js/jquery.ui.core.js"></script>
    <script language="javascript" src="js/jquery.ui.widget.js"></script>
    <script language="javascript" src="js/jquery.ui.position.js"></script>
    <script language="javascript" src="js/jquery.ui.autocomplete.js"></script>
</head>
```

AutoComplete 插件与一个 HTML 的<input>标签相结合。

现在网页加一个输入标签：

```html
<input id="query" autocomplete="off" />
```

注意对输入标签来说需要增加 autocomplete="off"。如果不加，浏览器可能不会提交 HTTP 请求到后台。

当 DOM(文档对象模型)已经加载,并且页面(包括图像)已经完全呈现时,会发生 ready 事件。在 JQuery 中,使用$(function)定义 ready 事件的处理函数。

在 ready 事件中增加 autocomplete 控件:

```
<script>
$(function() {
   var availableTags = ["ActionScript", "AppleScript", "Asp"];
   $( "#query" ).autocomplete({
      source: availableTags
   });
});
</script>
```

通过 AJAX 方式取得数据,访问后台 URL 地址位于"./autoComplete"的数据源的代码:

```
<script>
$(function() {
   $( "#query" ).autocomplete({
      source: "./autoComplete",
      minLength: 2
   });
});
</script>
```

它自动传递一个叫作"term"的参数,这个参数中包括"query"输入框中的值。例如,当用户在搜索框输入"o"这个字母,浏览器就会发送下面这个请求给服务器:

```
HTTP GET autoComplete?term=o
```

服务器传递给客户端的 JSON 数据格式是一个搜索词组成的数组,例如:

```
["open office","online backup","opera","onkyo",...]
```

用户选择词后,一般直接跳转到这个词的搜索结果,而不是只在输入框显示这个搜索词,之后用户需要再次按搜索按钮才返回搜索结果。有时候需要支持用户选择提示词直接跳转到搜索结果页面。在 JavaScript 中,通过 location.href 实现跳转。用户选择提示词后直接搜索的实现如下:

```
$(function() {
   $("#query").autocomplete({
      source: "./autoComplete",
      minLength: 2,
      select: function( event, ui ) {
         window.location.href="./searchAction.action?query=" + ui.item.value;
      }
   });
});
```

在这里,通过 ui.item.value 得到用户选择的搜索词。

如果不能正常工作，首先看浏览器是否已经发送 HTTP 请求。然后再看 Servlet 返回的结果。为了方便跟踪错误，可以使用 FireFox 中的 Firebug 插件调试网页中的 JavaScript 代码。在 Firebug 中，可以为 JavaScript 设置断点，可以暂停执行 JavaScript 并且看到每个变量的当前值。如果代码速度慢，还可以通过 JavaScript 配置器查看性能，快速发现性能瓶颈。

4.6.5 拼音提示

为了支持汉语拼音感应，需要把所有的词生成出拼音列。Trie 树可以看成关键词和值的映射。拼音列和词本身都可以作为关键词，值这一列则存放词原型。例如对于"厦门"这个词，会存储两个关键词和值的映射。

xiamen → 厦门

厦门 → 厦门

这样当用户输入"厦"或"xia"都可能提示出"厦门"这个词。

对于基本的中文词提示来说，关键词和值都是一样。另外，注音程序把中文词转换成拼音，这部分数据支持汉语拼音感应功能。

因为存在多音字，按词注音会有好的结果。可以在 Trie 树的值域中存储一个词对应的拼音。

```java
public static String yin(String sentence){//传入一个字符串作为要处理的对象
    int senLen = sentence.length();//首先计算出传入的这句话的字符长度
    int i = 0;//用来控制匹配的起始位置的变量
    StringBuilder result = new StringBuilder(senLen);
    TernarySearchTrie.MatchRet matchRet = new TernarySearchTrie.MatchRet("",0);
    while (i < senLen){// 如果 i 小于此句话的长度就进入循环
        boolean match = dic.matchLong(sentence, i, matchRet);//正向最大长度匹配
        if (match){//已经匹配上，按词注音
            i = matchRet.end;
            result.append(matchRet.data);
        } else//如果没有找到匹配上的词，就按单字注音
        {
            result.append(ziYin.zi2Yin(sentence.charAt(i)));
            ++i;// 下次匹配点在这个字符之后
        }
    }
    return result.toString();
}
```

4.6.6 部署总结

提示词词典 suggestDic.txt 可以放在 WEB-INF/classes/dic/路径下。AutoCompleteServlet 可以放在 WEB-INF/lib/路径下，通过 web.xml 发布。界面用到的 JavaScript 脚本 jquery.js、jquery.ajaxQueue.js、jquery.autocomplete.css 和 jquery.autocomplete.js 可以放在 ROOT/js 路径。

改进方法可以包括：对于不同的用户，提示也不一样。同样输入"大"字，对于影迷，提示"大话西游"，对于美食爱好者提示"大福"(一种日式甜品)。

4.7 推荐搜索词

搜索引擎中往往有个可选搜索词的列表,当搜索结果太少的时候,可以帮助用户扩展搜索内容,或者当搜索结果过多的时候,可以帮助用户深入定向搜索。这叫作查询推荐(Query Suggestion)。

一种方法是从搜索日志中挖掘字面相似的词作为相关搜索词列表。从一个给定的词语挖掘多个相关搜索词,可以用编辑距离为主的方法查找一个词的字面相似词,如果候选的相关搜索词很多,就要筛选出最相关的 10 个词。还可以从查询日志中聚类出一些相关的查询。

有的搜索词是有歧义的。例如搜索"李娜",相关搜索词应该提示:歌手、网球运动员还是跳水运动员。

4.7.1 挖掘相关搜索词

下面是利用 Lucene 筛选给定词的最相关词的方法。

```
private static final String TEXT_FIELD = "text";
/**
 * @param words 候选相关词列表
 * @param word 要找相关搜索词的种子词
 * @return
 * @throws IOException
 * @throws ParseException
 */
static String[] filterRelated(HashSet<String> words, String word) {
    StringBuilder sb = new StringBuilder();
    for(int i=0;i<word.length();++i)    {
        sb.append(word.charAt(i));
        sb.append(" ");
    }
    RAMDirectory store = new RAMDirectory();
    //按字生成索引和查找,也可以按细粒度的词分开
    IndexWriter writer = new IndexWriter(store, new StandardAnalyzer(), true);
    for(String text:words) {
        Document document = new Document();
        Field textField = new Field(TEXT_FIELD, text,
Field.Store.YES, Field.Index.TOKENIZED);
        document.add(textField);
        writer.addDocument(document);
    }
    writer.close();
    IndexSearcher searcher = new IndexSearcher(store);
    QueryParser queryParser = new QueryParser(TEXT_FIELD,
```

```
new StandardAnalyzer());
      Query query = queryParser.parse(sb.toString());
      Hits hits = searcher.search(query);
      int maxRet = Math.min(10, hits.length());
      String[] relatedWords = new String[maxRet];
      for (int i = 0; i < maxRet ; i++) {
        Document document = hits.doc(i);
        String text = document.get(TEXT_FIELD);
        System.out.println(text);
        relatedWords[i]=text;
      }
      searcher.close();
      store.close();
      return relatedWords;
}
```

上述代码整理出这样的相关词表：第一列是关键词，后续是 10 个以内的相关搜索词：

集福轩婚礼%集福轩

手机定位跟踪系统%手机定位系统%手机定位%手机定位仪器

喷绘材料卖店电话%我要喷绘材料卖店电话

厦门房产%厦门租房%厦门新闻%厦门桑拿%房产%青岛房产%厦门%恒雄房产

送水果%送水%水果

三星传真机%三星手机

另外一种方法，可以把多个用户共同查询的词看成相关搜索词，需要有记录用户 IP 的搜索日志才能实现。

可以通过 RelatedEngine 类查找某个关键词的相关词。

```
RelatedEngine re =new RelatedEngine(
            new File("D:/lg/work/xiaoxishu/dic/relatedwords.txt"));
String word = "徐家汇";
String[] relatedWords = re.getRelated(word);
for(String w : relatedWords) {
    System.out.println(w);
}
```

输出如下：

```
上海徐家汇
徐汇
徐家汇价格是
上房徐家汇路附近有吗
```

当我们需要为新建立的搜索引擎开发相关搜索时，如果没有搜索日志而用户文本很多的时候，可以：

(1) 首先运行 IndexMaker，从待搜索的文档中提取关键词并生成索引；

(2) 然后运行 RelatedWords，从索引生成相关词表。

另外还考虑用日志记录用户对相关搜索词的选择。如果用户也选择了这个词，那搜索词肯定和这个选择词相关了。

隐含语义索引(Latent Semantic Indexing，LSI)的原理是在相同的上下文中的词有相似的含义。Java 版本的实现参考 http://code.google.com/p/airhead-research/。

将词表示成向量可参考 https://code.google.com/p/word2vec/，然后找出和这个词语义相近的词，并给出相似度。

4.7.2 使用多线程计算相关搜索词

如果要计算任意两个查询词之间的相关性，则发现相关搜索词的时间复杂度是 $O(n^2)$。例如要分析 10M 多的相关搜索词的用时，则单线程的计算量可能长达 24 小时。

我们使用 Java 中自带的轻量级线程池来实现数据分析。JDK1.5 以后的版本提供了一个轻量级线程池 ThreadPool。可以使用线程池执行一组任务，最简单的任务不返回值给主调线程。要返回值的任务可以实现 Callable<T>接口，线程池执行任务并通过 Future<T>的实例获取返回值。

实现 Callable()方法的任务类的主要实现：

```java
public class FindSimCall implements Callable<String[]> {
    private HashSet<String> words; // 总的搜索词集合
    private String s;               // 待发现相关词的词

    public FindSimCall(HashSet<String> w ,String source)   {
        words = w;
        s = source;
    }

    @Override
    public String[] call() throws Exception {
        System.out.println(s);
        // 形成 related words 列表
        // ...
        return relatedWords;
    }
}
```

主线程类的实现如下：

```java
int threads = 4;
ExecutorService es = Executors.newFixedThreadPool(threads);

Set<Future<String[]>> set = new HashSet<Future<String[]>>();

for (final String s : words) {
    FindSimCall task = new FindSimCall(words,s);
    Future<String[]> future = es.submit(task);
    set.add(future);
```

```
}
FileOutputStream fos = new FileOutputStream(relatedWordsFile);
OutputStreamWriter osw = new OutputStreamWriter(fos,"GBK");
BufferedWriter writer = new BufferedWriter(osw);

for (Future<String[]> future : set) {
    String[] ret = future.get();

    for(String word:ret) {
        writer.write("%"+word);
    }
    writer.write( "\r\n" );
}
writer.close();
```

采用线程池可以充分利用多核 CPU 的计算能力,并且简化了多线程的实现。

4.8 查询意图理解

首先对用户查询预处理,如果无结果,再调整用户查询词来返回用户可能想要的文档。

4.8.1 拼音搜索

当用户输入一个全拼或都拼音的简写,都可以搜索到数据。例如,用户输入"大衣","dayi","dy","day"都能搜索到。

可以通过查询扩展,把拼音转成中文搜索词。例如,"dayi"转换成搜索词"大衣"。

拼音串作为键建立 Trie 树,如图 4-7 所示。

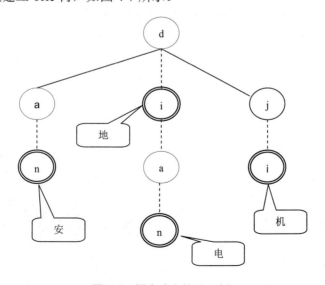

图 4-7 拼音建立的 Trie 树

也可以把拼音存到索引，做索引时加拼音。

需要有个 Analyzer，能在做分词的时候加注拼音。

4.8.2 无结果处理

用户输入"天梭的我表"，输入没有返回结果。系统提示根据"天梭表""我的"找到的结果。

拿着这几个字去索引里面查相关的词。根据"天\梭\表"三个字找到"天梭表"，根据"的"和"我"两个字，找到"我的"。

4.9 集成其他功能

搜索引擎用户界面的一些功能还有：对用户输入的拼写纠错提示，搜索结果的分类统计，根据用户搜索词返回相关搜索词，在搜索结果中再次查找，记录和统计搜索日志。

4.9.1 拼写检查

因为用户的查询本身是一个符合查询语法的字符串，所以不能把用户的查询本身直接输入给拼写检查模块，而要通过一个 didYouMeanParser 给出这个提示。在 Lucene 中使用拼写检查的例子：

```
String indexDir = "D:/lg/work/nextag/indexdir";
Directory directory = FSDirectory.open(new File(indexDir));
IndexSearcher searcher = new IndexSearcher(directory, true); //只读方式打开索引

String defaultField = "title";
String queryString ="贷款";
Analyzer analyzer = new StandardAnalyzer();
QueryParser parser = new QueryParser(defaultField,
                                     analyzer);

Query query = parser.parse(queryString);
ScoreDoc[] hits = searcher.search(query, 1000).scoreDocs;   //原始查询
searcher.close();

String suggestedQueryString = null; //提示查询的字符串

int minimumHits=5;
float minimumScore=0.6f;

//如果搜索返回结果的数量小于阈值或者匹配第一个结果的分值小于最小值就查找提示词
if (hits.length < minimumHits || hits[0].score < minimumScore) {
    CompositeDidYouMeanParser didYouMeanParser =
        new CompositeDidYouMeanParser(defaultField);
```

```
        Query didYouMean = didYouMeanParser.suggest(queryString); //调用拼写检查算法
        if (didYouMean != null) {
            suggestedQueryString = didYouMean.toString(defaultField);
        }
    }
}
System.out.println("您是不是要找: "+suggestedQueryString);
```

实现一个复杂的 DidYouMeanParser:

```
public class CompositeDidYouMeanParser implements DidYouMeanParser {

    private class QuerySuggester extends QueryParser {
        private boolean suggestedQuery = false; //是否有推荐的查询对象

        public QuerySuggester(String field, Analyzer analyzer) {
            super(field, analyzer);
        }

        protected Query getFieldQuery(String field, String queryText, boolean quoted)
                throws ParseException {
            TokenStream source = getAnalyzer().tokenStream(field,
                    new StringReader(queryText));
            CharTermAttribute termAtt = source
                    .getAttribute(CharTermAttribute.class);
            Vector<String> v = new Vector<String>();

            boolean hasMoreTokens = false;
            try {
                hasMoreTokens = source.incrementToken();
                while (hasMoreTokens) {
                    String term = termAtt.toString();
                    v.addElement(term);
                    hasMoreTokens = source.incrementToken();
                }
            } catch (IOException e) {
            }

            try {
                source.close();
            } catch (IOException e) {
                // ignore
            }

            if (v.size() == 0)
                return null;
            else if (v.size() == 1)
```

```java
            return new TermQuery(getTerm(field, v.elementAt(0)));
        else {
            PhraseQuery q = new PhraseQuery();
            q.setSlop(getPhraseSlop());
            for (int i = 0; i < v.size(); i++) {
                q.add(getTerm(field, (String) v.elementAt(i)));
            }
            return q;
        }
    }

    private Term getTerm(String field, String queryText)
            throws ParseException {
        String aux = spellCheck.findMostSimilar(queryText);
        if (aux == null || aux.equals(queryText)) {
            return new Term(field, queryText);
        }
        suggestedQuery = true;
        return new Term(field, aux);
    }

    public boolean hasSuggestedQuery() {
        return suggestedQuery;
    }
}

private String defaultField; //缺省搜索列
public static SpellChecker spellCheck = null; //拼写检查类

public CompositeDidYouMeanParser(String defaultField,
                                  String dicFile, //正确词典类
                                  String commonMisspellingsFile //正误词表
                                  ) throws Exception {
    this.defaultField = defaultField;
    spellCheck = new SpellChecker();
    spellCheck.initialize(dicFile, commonMisspellingsFile);
}

public Query parse(String queryString) throws ParseException {
    QueryParser queryParser = new QueryParser(defaultField,
                    new WhitespaceAnalyzer());
    queryParser.setDefaultOperator(QueryParser.AND_OPERATOR);
    return queryParser.parse(queryString);
}

public Query suggest(String queryString) throws ParseException {
    QuerySuggester querySuggester = new QuerySuggester(defaultField,
```

```
            new WhitespaceAnalyzer());
    querySuggester.setDefaultOperator(QueryParser.AND_OPERATOR);
    Query query = querySuggester.parse(queryString);
    return querySuggester.hasSuggestedQuery() ? query : null;
    }
}
```

设计一个供界面调用的 Bean，根据输入的查询结果返回缺省查询列和查询字符串，判断是否给出提示词。

```
public class DidYouMean {
    static String defaultField;
    static int minimumHits=5;
    static float minimumScore=0.6f;

    public DidYouMean(String f){
        defaultField = f;
    }

    public String getSuggest(TopDocs hits,String queryString) throws Exception {
        String suggestedQueryString = null; //提示查询的字符串

        //如果搜索返回结果的数量小于阈值
        //或者匹配第一个结果的分值小于最小值就查找提示词
        if (hits.totalHits < minimumHits || hits.scoreDocs[0].score < minimumScore) {
            CompositeDidYouMeanParser didYouMeanParser = new CompositeDidYouMeanParser(defaultField);
            Query didYouMean = didYouMeanParser.suggest(queryString); //调用拼写检查算法
            if (didYouMean != null) {
                suggestedQueryString = didYouMean.toString(defaultField);
            }
        }

        return suggestedQueryString;
    }
}
```

在 JSP 页面使用这个类：

```
<jsp:useBean class="com.lietu.didyoumean.DidYouMean" id="suggestBean" scope="application">
  <% suggestBean = new DidYouMean("title"); %>
</jsp:useBean>
您是不是要找: <%= suggestBean.getSuggest(hits,queryString) %>
```

如果返回结果数量很少，可以直接把提示词的搜索结果放在原查询词返回结果的下面。

4.9.2 分类统计

首先定义分类统计信息类:

```java
public class CatInf implements Comparable<CatInf> {
  public String name;// 分类名
  public int count;   //类别数量

  public int compareTo(CatInf obj){
        return (int)(obj.count - this.count);
  }
}
```

为了保证分类统计和查询的结果一致性,需要共用一个查询对象。实现分类统计的方法声明如下:

```java
ArrayList<CatInf> factedCounter(IndexSearcher searcher, Query q)
```

二级子树展开的效果图如图 4-8 所示。

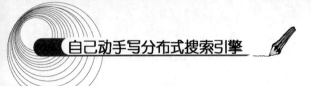

图 4-8　二级子树的展开效果

节点类定义如下:

```java
public class CatNode {
   public int no;   //编码
   public String name; //节点名
   public boolean isLeaf;   //是否为叶节点
   public List<CatNode> children = null;  //孩子节点
   public CatNode parent;   //父节点
   public int level;   //级别

   public CatNode(int no, String name,CatNode parentNo,int l, boolean isLeaf) {
      this.no = no;
      this.name = name;
      this.parent = parentNo;
      this.level = l;
      this.isLeaf = isLeaf;
      if (!isLeaf) {
         children = new ArrayList<CatNode>(5);
      }
   }
}
```

```java
    public void addChildren(CatNode node) throws Exception  {
        if (isLeaf)   {
            throw new Exception("add child error to leaf node:" + no);
        }
        children.add(node);
    }

    public String toString()  {
        String temp = this.name;
        for (CatNode child : children)  {
            temp += "\n"+ "child:"+ child.no +":"+child.name ;
        }
        temp += "\n";
        return temp;
    }
}
```

用于查找父子节点的映射表：

```java
public class CategoryMap extends HashMap<Integer, CatNode> {
    public CategoryMap(){ //构造方法
        String sql =
          "SELECT ID,isnull(FATHER_ID,0) FATHER_ID,CAT_NAME,CAT_LEVEL "+
          " FROM DP_DISPLAY order by ID" ;
        PreparedStatement stmt = con.prepareStatement(sql);
        ResultSet rs = stmt.executeQuery();

        CatNode thisNode = new CatNode(0,"ROOT",null,0,false);
        this.put(0, thisNode );

        while(rs.next()) {
            int code = rs.getInt("ID");
            String name = rs.getString("CAT_NAME");
            int fatherID = rs.getInt("FATHER_ID");
            int level = rs.getInt("CAT_LEVEL");

            boolean isLeaf = true;

            CatNode parentNode = this.get(fatherID);

            thisNode = new CatNode(code,name,parentNode,level,isLeaf);

            if(parentNode.isLeaf)   {
                parentNode.isLeaf = false;
                parentNode.children = new ArrayList<CatNode>(5);
            }
            parentNode.children.add(thisNode);
            this.put(code, thisNode );
```

```
        }
    }
}
```

搜索页面用于计数的类:

```java
public class CountNode {
    public int no;// 分类号
    public String name;// 分类名
    public boolean isLeaf;// 是否为末级
    public List<CountNode> children = null;
    public CountNode parent;
    public int count;

    public CountNode(int no, String name,CountNode parentNo,boolean isLeaf) {
        this.no = no;
        this.name = name;
        this.parent = parentNo;

        this.isLeaf = isLeaf;
        if (!isLeaf) {
            children = new ArrayList<CountNode>(5);
        }
    }

    @Override
    public boolean equals(Object o)  {
        if(o instanceof CountNode)  {
            CountNode t = (CountNode)o;
            return (t.no == this.no);
        }
        return false;
    }

    @Override
    public int hashCode(){
        return this.no;
    }
}
```

二级子树展开的实现:

```java
//搜索形成的二级子树
HashMap<Integer, CountNode> cat1Set = new HashMap<Integer, CountNode>();
for (Count c : facetCounts) {
    Integer cat2Id = Integer.parseInt(c.getName());
    CatNode cat2Node = catMap.get(cat2Id);
    CountNode newParen = cat1Set.get(cat2Node.parent.no);
```

```
    if (limitCat > 0) {
        if (cat2Node.parent.no != limitCat) {
            continue;
        }
    }

    if (newParen == null) {
        newParen = new CountNode(cat2Node.parent.no,
                cat2Node.parent.name, null, false);
        CountNode childNode = new CountNode(cat2Node.no, cat2Node.name,
                newParen, true);
        childNode.count = c.getCount();
        newParen.children.add(childNode);
        cat1Set.put(newParen.no, newParen);
    } else {
        CountNode childNode = new CountNode(cat2Node.no, cat2Node.name,
                newParen, true);
        childNode.count = c.getCount();
        newParen.children.add(childNode);
    }
}
```

使用 MatchAllDocsQuery 返回所有的记录，然后再分类统计，这样可以实现一个分类导航的页面。

有个专门实现分类统计功能的项目 bobo-browse(http://code.google.com/p/bobo-browse/)。不过它需要集成 Spring 框架，用起来麻烦一点。

然后在页面执行搜索时执行如下过程：

```
/**
 *  @param cat 当前类别编号
 *  @param q 当前查询
 *  @return 分类统计列表
 *  @throws Exception
 */
public List<CatInf> catCounter(int cat,Query q) throws Exception {
    CategoryNode thisNode = ListContainer.catMap.get(String.valueOf(cat));
    if(thisNode.isLeaf()) {
        //已经到达最后一级，不能再展开统计
        return null;
    }

    List<CategoryNode> children = thisNode.children;
    if(children == null) {
        return null;
    }
    ArrayList<CatInf> catList =new ArrayList<CatInf>( children.size() );
    DocSetHitCollector all = new DocSetHitCollector(reader.maxDoc());
```

```
searcher.search(q, all );

DocSet allDocSet = all.getDocSet();

String termField = null;//层次列
if (cat<=0) {
    termField = "hs1";//第一层的ID号存储在hs1列
} else if (cat<100) {
    termField = "hs2";//第二层的ID号存储在hs2列
}
//如果还有后续层,则按前缀匹配来搜
for (int i=0;i<children.size();++i)    {
        //统计每个子类的搜索结果数
        CategoryNode currentNode = children.get(i);
        DocSetHitCollector these = new DocSetHitCollector(reader.maxDoc());

        searcher.search(
            new TermQuery(new Term(termField, currentNode.no)),
            these );
        int count = these.getDocSet().intersectionSize(allDocSet);
        if(count>0) {
            CatInf catInf = new CatInf();
            catInf.name = currentNode.name;
            catInf.no = currentNode.no;
            catInf.count = count;
            catList.add(catInf);
        }
    }
    Collections.sort(catList);//对分类统计结果排序后输出
    return catList;
}
```

可以通过类别编码来控制是否按类别查找,但不推荐这样实现。为了实现 REST 风格的链接,可以直接根据类名按条件查询,例如 cat=Music。有些类名可能包含特殊的符号,例如:"Health & Beauty"。可以使用 URLEncoder 类对类名中的特殊符号转义,例如:把空格转换成加号。代码如下:

```
URLEncoder.encode(catName,"UTF-8");
```

在 Taglib 中输出在<table>标签中的分类统计结果:

```
public String getCatView() {
   if(_catList == null)
       return "";
   StringBuffer output = new StringBuffer();
   output.append("<table width=\"100%\" border=\"0\" cellspacing=\"0\" cellpadding=\"2\">");
   output.append("<tr> ");
```

```
    int count =0 ;
    for(CatInf e : _catList) {
     output.append("<td><a href=\""); //URL 地址
     output.append(_url);
     output.append("?query=");
     output.append(_query);    //查询词
     output.append("&");
     output.append(InitTag.CAT_KEY); //在 URL 中增加一个分类参数用于进一步导航
     output.append("=");
     output.append(e.no);    //类别编号
     output.append("\" class=\"m\">");
     output.append(e.name);    //类别名
     output.append("</a>(");
     output.append(e.count);   //该类别下的文档数量
     output.append(")</td>");
     if(count%3 ==2)  {
        output.append("</tr><tr>");
     }
     count++;
    }
    output.append(" </tr></table>");

    return output.toString();
}
```

最后 JSP 界面通过 Tag 调用 getCatView：

```
<list:prop property="catView"/>
```

4.9.3 相关搜索

一种方法是从搜索日志中挖掘字面相似的词作为相关搜索词列表。首先从一个给定的词语挖掘多个相关搜索词，可以用以编辑距离为主的方法查找一个词的字面相似词，如果候选的相关搜索词很多，就要筛选出最相关的 10 个词。下面是利用 Lucene 筛选最相关词的方法。

```
private static final String TEXT_FIELD = "text";

/**
 *
 * @param words 候选相关词列表
 * @param word 要找相关搜索词的种子词
 * @return
 * @throws IOException
 * @throws ParseException
 */
static String[] filterRelated(HashSet<String> words, String word) {
```

```
    StringBuilder sb = new StringBuilder();

    for(int i=0;i<word.length();++i){
        sb.append(word.charAt(i));
        sb.append(" ");
    }

    RAMDirectory store = new RAMDirectory();
    IndexWriter writer = new IndexWriter(store, new StandardAnalyzer(), true);

    for(String text:words)  {
        Document document = new Document();
        Field textField =
new Field(TEXT_FIELD, text, Field.Store.YES, Field.Index.TOKENIZED);
        document.add(textField);
        writer.addDocument(document);
    }
    writer.close();

    IndexSearcher searcher = new IndexSearcher(store);
QueryParser queryParser = new QueryParser(TEXT_FIELD,
 new StandardAnalyzer());
    Query query = queryParser.parse(sb.toString());

    Hits hits = searcher.search(query);
    int maxRet = Math.min(10, hits.length());

    String[] relatedWords = new String[maxRet];
    for (int i = 0; i < maxRet ; i++) {
      Document document = hits.doc(i);
      String text = document.get(TEXT_FIELD);
      System.out.println(text);
      relatedWords[i]=text;
    }
    searcher.close();
    store.close();

    return relatedWords;
}
```

整理出这样的相关词表,第一列是关键词,后续是 10 个以内的相关搜索词:
集福轩婚礼%集福轩
手机定位跟踪系统%手机定位系统%手机定位%手机定位仪器
喷绘材料卖店电话%我要喷绘材料卖店电话
厦门房产%厦门租房%厦门新闻%厦门桑拿%房产%青岛房产%厦门%恒雄房产
送水果%送水%水果

三星传真机%三星手机

另外一种方法，可以把多个用户共同查询的词看成相关搜索词，需要有记录用户 IP 的搜索日志才能实现。类似于推荐系统。超市把尿布与啤酒放在一起卖。因为这是关联规则挖掘出的结果。

对这个结果的解释：在美国，一些年轻的父亲下班后经常要到超市去买婴儿尿布，而他们中有 30%～40%的人同时也为自己买一些啤酒。产生这一现象的原因是：美国的太太们常叮嘱她们的丈夫下班后为小孩买尿布，而丈夫们在买尿布后又随手带回了他们喜欢的啤酒。

用户搜索"啤酒"的时候，提示他是否还要找"尿布"。

ARtool 是一个挖掘关联规则的算法工具集。

然后通过 RelatedEngine 类查找某个关键词的相关词。

```java
public static void main(String[] args) throws Exception {
    RelatedEngine re =new RelatedEngine(new File("D:/dic/relatedwords.txt"));
    String word = "徐家汇";
    String[] relatedWords = re.getRelated(word);
    for(String w : relatedWords) {
        System.out.println(w);
    }
}
```

输出相关搜索词如下：

```
上海徐家汇
徐汇
徐家汇价格是
上房徐家汇路附近有吗
```

最后通过自定义的<tag>标签 RelatedTag 在 JSP 页面显示出相关搜索词。

在标签库描述符中定义<tag>：

```
    <tag>
      <name>relatedWords</name>
      <tag-class>com.bitmechanic.listlib.RelatedTag</tag-class>
      <description></description>

      <attribute>
         <name>index</name>
         <required>false</required>
         <rtexprvalue>true</rtexprvalue>
      </attribute>

      <attribute>
         <name>url</name>
         <required>false</required>
         <rtexprvalue>true</rtexprvalue>
```

```
    </attribute>

    <attribute>
      <name>query</name>
      <required>true</required>
      <rtexprvalue>true</rtexprvalue>
    </attribute>

  </tag>
```

最后在 JSP 页面中引用标签：

```
<list:relatedWords index="D:/search/related" url="Search.jsp"
query="<%=query%>"/>
```

4.9.4 再次查找

经常需要从结果中缩小范围再次查找信息。一个实现方法是通过"+"连接符连接上次查询和当前查询。例如：inputstr 记录了上次查询词，queryString 记录当前查询词，实现代码如下：

```
if (refind) //如果需要再次查找
    queryString = " + ("+queryString+") + ("+inputstr+")";
```

使用这个新的查询词就可以实现再次搜索的功能。

4.9.5 搜索日志

搜索日志是用来分析用户搜索行为和信息需求的重要依据。一般记录如下信息。
- 搜索关键字；
- 用户来源 IP；
- 本次搜索返回结果数量；
- 搜索时间；
- 其他需要记录的应用相关信息。

IP 地址是最容易获取的信息，但其局限性也较为明显：伪 IP、代理、动态 IP、局域网共享同一公网 IP 出口……这些情况都会影响基于 IP 来识别用户的准确性，所以 IP 识别用户的准确性比较低，目前一般不会直接采用 IP 来识别用户。

可以通过 Cookie 记录用户 ID。Cookie 是从用户端存放的 Cookie 文件记录中获取的，这个文件里面一般在包含一个 Cookie ID 的同时也会记下用户在该网站的 User ID(如果你的网站需要注册登录并且该用户曾经登录过你的网站且 Cookie 未被删除)，所以在记录日志文件中 Cookie 项的时候可以优先去查询 Cookie 中是否含有用户 ID 类的信息，如果存在则将用户 ID 写到日志的 Cookie 项，如果不存在则查找是否有 Cookie ID，如果有则记录，没有则记为"-"，这样，日志中的 Cookie 就可以直接作为最有效的用户唯一标识符被用于统计。当然这里需要注意：该方法只有网站本身才能够实现，因为用户 ID 作为用户隐私信息只有该网站才知道其在 Cookie 的设置及存放位置，第三方统计工具一般很难获取。

通过以上的方法实现用户身份的唯一标识后，我们可以通过一些途径来采集用户的基础信息、特征信息及行为信息，然后为每位用户建立起详细的 Profile，具体途径有：

- 用户注册时填写的用户注册信息及基本资料；
- 从网站日志中得到的用户浏览行为数据；
- 从数据库中获取的用户网站业务应用数据；
- 基于用户历史数据的推导和预测；
- 通过直接联系用户或者用户调研的途径获得的用户数据；
- 有第三方服务机构提供的用户数据。

通过用户身份识别及用户基本信息的采集，我们可以通过网站分析的各种方法在网站实现一些有价值的应用：

- 基于用户特征信息的用户细分；
- 基于用户的个性化页面设置；
- 基于用户行为数据的关联推荐；
- 基于用户兴趣的定向营销。

为了不影响即时搜索的速度，一般不把搜索日志记录直接记录在数据库中，而是写在文本文件中。推荐使用 Logback(http://logback.qos.ch/)的日志功能实现。Logback 提供了 3 个 jar 包 core、classic、access。其中 core 是基础，其他两个包依赖于这个包。logback-classic 是 SLF4J 原生的实现。所以可以用其他 logging 系统去替换它。当然 logback-classic 依赖于 slf4j-api。logback-access 与 Servlet 容器集成。提供 http-access 的 log 功能。SLF4J(http://www.slf4j.org/)几乎已经称为业界日志的统一接口。

这里的项目一共需要 3 个包：slf4j-api-1.6.1.jar、logback-classic-0.9.21.jar 和 logback-core-0.9.21.jar。Logback 通过 logback.xml 进行配置。

这里把当前日志写到 D:/logs/log 文件中，新的一天日志开始的时候，昨天的日志生成一个新文件。

在搜索类中初始化日志类：

```
private static Logger logger = LoggerFactory.getLogger(SearchBbs.class);
```

然后当用户执行一次搜索时，记录查询词、返回结果数量、用户 IP 以及查询时间等。

```
logger.info(_query+"|"+desc.count+"|"+"bbs"+"|"+ip);
```

日志文件 log.txt 记录的结果例子如下：

```
什么是新生儿|37|topic|124.1.0.0|2007-11-21 12:25:36
什么是新生儿|28|bbs|124.1.0.0|2007-11-21 12:25:42
怀孕|18|topic|124.1.0.0|2007-11-21 12:26:05
怀孕|2|shangjia|124.1.0.0|2007-11-21 12:26:05
怀孕|145|bbs|124.1.0.0|2007-11-21 12:26:06
怀孕|18|topic|124.1.0.0|2007-11-21 12:30:33
```

这里，第 1 列是用户搜索词，第 2 列是搜索返回结果数量，第 3 列是搜索类别，第 4

列是 IP 地址，第 5 列是搜索的时间。

然后定义搜索日志统计表，例如我们需要统计搜索最多的词，可以把搜索最多的词放在 keywordAnalysis 表中：

```
CREATE TABLE [keywordAnalysis] (
    [searchTerms] [varchar] (50)  NOT NULL ,--搜索词
    [AccessCount] [int] NULL ,   --搜索计数
    [Result] [int] NULL    --该词返回结果数
)
```

4.10 查询分析

因为搜索关键词是用户输入的文本信息，所以可以从搜索日志中了解用户使用搜索的意图。有人把 Google 的搜索关键词排行榜称为人类意图数据库。这来源于对搜索日志的分析。

4.10.1 历史搜索词记录

可以通过 Cookie 记录用户经常搜索的关键字，然后就可以从用户经常搜索的关键字来判断用户的兴趣。先看一下怎么设置用户查询词。Cookie 在用户电脑中是以一种类似于 Map 的方式存放，且只能存放字符串类型的对象。通过 Response 对象增加 Cookie，代码如下：

```
Cookie cookie = new Cookie("query", query);
cookie.setMaxAge(60*60*24*30);   //设置 Cookie 的存放时间(单位是秒)
//然后通过 Response 对象的 addCookie()方法添加 Cookie 才能生效
response.addCookie(cookie);
```

通过 Request 对象的 getCookies()方法得到一个包含所有 Cookie 的数组。

```
Cookie[] cookies = request.getCookies();
//然后遍历这个数组就能得到记录查询词的 Cookie
String query = null;
for (int i = 0; i < cookies.length; i++) {
        Cookie c = cookies[i];
        if (c.getName().equals("query")) {
                query = c.getValue();
        }
}
```

上面的例子显示了如何设置并获取名为 query 的 Cookie。如果要记录 5 个查询词，则需要设置 5 个不同的 Cookie。如果要在 Web 界面显示历史查询词，则需要把这些关键词去重，然后再显示出来。

4.10.2 日志信息过滤

公开的搜索会有很多爬虫的访问。搜索日志中包括大量的 Google 爬虫信息，需要把它

和普通用户的搜索区分出来。

可以从请求的信息中判断出是哪一种爬虫。例如，Baidu 爬虫的"User-Agent"信息：

```
Baiduspider+(+http://help.baidu.jp/system/05.html)
```

Goolge 爬虫的"User-Agent"信息：

```
Mozilla/5.0 (compatible; Googlebot/2.1; +http://www.google.com/bot.html)
```

下面是程序实现：

```
String userAgent = request.getHeader( "User-Agent" );

    public static String[] getBotName(String userAgent) {
        userAgent = userAgent.toLowerCase();
        int pos=0;
        String res=null;
    if ((pos=userAgent.indexOf("google/"))>-1) {
        res= "Google";
        pos+=7;
    } else
    if ((pos=userAgent.indexOf("msnbot/"))>-1) {
        res= "MSNBot";
        pos+=7;
    } else
    if ((pos=userAgent.indexOf("googlebot/"))>-1) {
        res= "Google";
        pos+=10;
    } else
    if ((pos=userAgent.indexOf("webcrawler/"))>-1) {
        res= "WebCrawler";
        pos+=11;
    } else
    if ((pos=userAgent.indexOf("inktomi"))>-1) {
        res= "Inktomi";
        pos=-1;
    } else
    if ((pos=userAgent.indexOf("teoma"))>-1) {
        res= "Teoma";
        pos=-1;
    }
    if (res==null) return null;
    return getArray(res,res,res + getVersionNumber(userAgent,pos));
}
```

4.10.3 信息统计

可以按搜索次数排列搜索热词，此外还可以按地区来源统计搜索次数。

按地区统计搜索词需要从用户的访问 IP 查询出对应的地址。如下所示，IP 地址表记录了一个地区的 IP 范围，见表 4-3。

表 4-3 IP 地址表

ip1	ip2	country	city	countryNo	provinceNo
3740518268	3740518268	湖南省永州市	金鹰网吧	1	12
3740518269	3740518400	湖南省永州市	电信	1	12
3740518401	3740518401	湖南省永州市祁阳县	怡心苑网吧	1	12
3740518402	3740518417	湖南省永州市	电信	1	12

下面的例子代码返回用户 IP 所在省：

```java
String ipin[] = getIp.split("\\.");
long ipinfo[] = new long[4];
for (int i = 0; i < ipinfo.length; i++) {
    //从 String 类型的 IP 地址到 long 型的转换可以用查表法更快地实现
    ipinfo[i] = Integer.parseInt(ipin[i]);
}
long num=ipinfo[0] * 256 * 256 * 256 + ipinfo[1] * 256 * 256 + ipinfo[2] * 256 + ipinfo[3] - 1;
String sql = "select DIC_Province.caption from ip,DIC_Province where ip.provinceNo = DIC_Province.id and ip.ip1<=" + num + " and ip.ip2>= " + num + "";
st = con.createStatement();
rs = st.executeQuery(sql);
if (rs.next()) {
    area = rs.getString(1);
}
```

首先建立搜索日志统计表：

```sql
CREATE TABLE SC_SEARCH_STAT (
    ID NUMERIC(12 , 0) IDENTITY ,  // 自增长 ID
    SEARCH_WORD VARCHAR(90) NULL,   // 搜索词
    SEARCH_NUM INT NULL,            // 搜索次数
    SEARCH_DATE DATE NULL,          // 搜索日期
    SEARCH_RESULT INT NULL          // 搜索词的返回结果数量
)
```

搜索统计表每天更新一次。每次把上一天的搜索日志文件中的数据统计后写入该表。搜索次数 SEARCH_NUM 指一天内搜索 SEARCH_WORD 这个词的独立 IP 的统计数量,同日同地址同搜索词只算一次。搜索统计用到的主要数据结构有：

```java
HashMap<String,HashSet<String>> word2IP =
        new HashMap<String,HashSet<String>>();//搜索词到 IP 的映射
HashMap<String,Integer> word2ResultNum =
```

```
            new HashMap<String,Integer>();//搜索词到搜索结果数的映射
…
//统计信息
HashSet<String> ips = word2IP.get(key);//如果搜索词已存在
if (ips!=null) {
         ips.add(strIP);//增加当前IP
         word2ResultNum.put(key, resultNum);
} else if (resultNum > 0) {//如果搜索返回结果数大于零
         ips = new HashSet<String>();
         ips.add(strIP);
         word2IP.put(key, ips);
         word2ResultNum.put(key, resultNum);
}
…
//统计信息写入统计表
String sql =
"insert into SC_SEARCH_STAT
(SEARCH_WORD,SEARCH_NUM,SEARCH_DATE,SEARCH_RESULT) values(?,?,?,?)";
PreparedStatement pstmt = con.prepareStatement(sql);

for (Entry<String, HashSet<String>> e : word2IP.entrySet()) {
      pstmt.setString(1, e.getKey());
      pstmt.setInt(2, e.getValue().size());
      pstmt.setString(3, yesDate);
      pstmt.setInt(4, word2ResultNum.get(e.getKey()));
      pstmt.executeUpdate();
}
```

4.10.4 挖掘日志信息

可以从不同的角度挖掘搜索日志。例如：

- 挖掘单个词。可以挖掘出用户对查询语法的使用情况，例如：filetype、site 等查询语法。可以统计用户搜索中使用空格分开多个词来搜索的比例。
- 或者按用户会话(session)挖掘词。有用户先搜索"工程电磁学基础(第6版)"，没有结果返回，几秒钟后他换了一个搜索词"工程电磁学基础"，这次有22条结果返回。因此可以把用户对查询词的修改分为4种情况：减少查询词、增加查询词、部分替换查询词、完全更换查询词。另外，还可以统计用户搜索词之间的词序，考虑上一个词到下一个词之间的转移概率。
- 根据用户与词之间的关联可以挖掘出词与词之间的关联或者用户与用户之间的关联。观察到一个用户搜索"眉笔"后，又搜索了"粉底液"和"紧肤水"，这些词都是与美容相关的商品名，所以考虑使用关联规则挖掘出相关词。可以根据用户的搜索词对用户分类，计算出用户对每个类别的隶属度。

4.10.5 查询词意图分析

用户输入 138777，也许是想找以这个数字串开始的手机号码。用户输入"张"，也许是想找所有姓"张"的人。

4.11 部署网站

服务器端操作系统推荐采用 Linux。Linux 有各种版本，这里以免费的 CentOS 为例。用静态 IP 地址配置一个网络连接的 IPv4 属性。例如静态 IP 地址是 201.147.214.149。eth0 配置文件的内容如下：

```
# cat /etc/sysconfig/network-scripts/ifcfg-eth0
TYPE=Ethernet
DEVICE=eth0
HWADDR=00:2g:fc:1b:c3:9e
ONBOOT=yes
USERCTL=no
IPV6INIT=no
PEERDNS=yes
NETMASK=255.255.255.128
IPADDR=201.147.214.149
GATEWAY=201.147.214.254
```

重新启动网络服务：

```
#service network restart
```

如果托管的机器要换机柜。可以远程先修改好网卡的配置，然后让机房人员挪机器，重启机器。同时修改 DNS 中的 IP 地址。

4.11.1 部署到 Web 服务器

配置 Java 环境，设置环境变量 JAVA_HOME 和 PATH 的值。修改脚本文件/etc/bashrc：

```
#vi /etc/bashrc
```

增加如下行：

```
export JAVA_HOME=/usr/local/jdk1.6.0_21
export PATH=$JAVA_HOME/bin:$PATH
```

从 Tomcat 官方网站 http://tomcat.apache.org/ 下载 tar.gz 包。

```
# wget http://www.fayea.com/apache-mirror/tomcat/tomcat-7/v7.0.33/bin/apache-tomcat-7.0.33.tar.gz
```

解压缩这个文件：

```
#tar -xf apache-tomcat-7.0.33.tar.gz
```

然后增加 Tomcat 所使用的内存。修改配置文件 catalina.sh：

```
#vi /usr/local/apache-tomcat-6.0.32/bin/catalina.sh
```

在文件 catalina.sh 的开始位置增加如下行：

```
JAVA_OPTS=-Xmx1024m
```

修改 Tomcat 配置文件 server.xml，把监听端口号由 8080 改为 80，并且支持 UTF-8 编码：

```
#vi /usr/local/apache-tomcat-6.0.32/conf/server.xml
```

增加配置：

```
useBodyEncodingForURI="true" URIEncoding="UTF-8"
```

可以把 Web 应用打一个 war 包，然后传到服务器上的 webapps/子路径下，会自动解压缩 war 包中的 Web 应用。

也可以压缩开发环境中的文件：

```
#tar -cjf price.tar.bz2 ./price
```

在正式环境中下载压缩好的文件 price.tar.bz2，然后解压缩文件：

```
#tar -xjf price.tar.bz2
```

启动 Tomcat：

```
#startup.sh
```

查看 Tomcat 是否已经起来了：

```
#pgrep -l java
```

或者使用命令：

```
#ps -ef |grep java
```

虚拟主机服务器提供商可能提供这样的服务：当 Tomcat 服务停止的时候，会自动起来一个 Apache 显示错误信息。启动 Tomcat 之前，先停止 Apache 服务：

```
#httpd -k stop
```

查看 Apache 服务是否已经停止了：

```
#pgrep httpd
```

如果需要更好的性能，可以使用 Resin。处理静态页面，Resin 比 Nginx 或者 Apache 都快。从 http://www.caucho.com/download/下载 Resin 免费版本。

在 bin 目录下面，用 vi 命令新建一个名为 startResin.sh 的文件。

```
vi ./startResin.sh
```

在文件内输入如下信息：

```
export LC_ALL=zh_CN.GB18030
export LANG=zh_CN.GB18030
nohup ./httpd.sh & -Xms512M -Xmx1024M
```

备份到目录/home/webbak/ROOT2012：

```
cp -r ./ROOT/  /home/webbak/ROOT2012
```

启动 Resin 4：

```
resin.sh start
```

如果需要，可以给网站买一个好记的域名。向域名供应商购买域名后，增加 DNS 信息。如果修改 DNS 信息后，要清空本地的 DNS 缓存信息，在 Windows 下可使用如下的命令：

```
>ipconfig /flushdns
```

查询一个域名的 A 记录：

```
>nslookup www.lietu.com 198.153.192.1
```

4.11.2 防止攻击

当站点无法访问了，可能是被攻击了。一种常见的攻击称为分布式拒绝服务攻击，英文叫作 DDoS(Distributed denial of service)。检查服务器是否在 DDoS 状态的一个快速而有用的命令是：

```
#netstat -anp |grep 'tcp\|udp' | awk '{print $5}' | cut -d: -f1 | sort | uniq -c | sort -n
```

这会列出花最多的连接到服务器的 IP。但是要记住，DDoS 正在变得更复杂，它可能用更多的 IP 地址，每个 IP 地址使用更少的连接。例如使用代理 IP。如果是这样，即使在 DDoS 下，仍然只有很少数量的连接。这叫作 CC 攻击。

另外一个非常重要的事情是看有多少正在处理的活跃的连接。

```
#netstat -n | grep :80 |wc -l
```

将显示活跃的连接数。许多攻击通常从一个到服务器的连接开始，然后不发送任何答复，让服务器等待到超时。活动连接的数量会有很大的不同，但如果是 500 以上，就可能有问题。

```
#netstat -n | grep :80 | grep SYN |wc -l
```

如果超过 100，就有 SYN 攻击的麻烦。大量的 SYN 请求导致未连接队列被塞满，使正常的 TCP 连接无法顺利完成 3 次握手，通过增大未连接队列空间可以缓解这种压力。

Linux 用变量 tcp_max_syn_backlog 定义 backlog 队列容纳的最大半连接数。在 Redhat AS 中，这个值默认是 1024。这个值是远远不够的，一次强度不大的 SYN 攻击就能使半连接队列占满。可以通过以下命令修改此变量的值：

```
# sysctl -w net.ipv4.tcp_max_syn_backlog="2048"
```

在 Linux 下可以通过修改/etc/sysctl.conf，添加下列选项达到效果。

```
# add by geminis for syn crack
net.ipv4.tcp_syncookies = 1
net.ipv4.tcp_max_syn_backlog=2048
net.ipv4.tcp_synack_retries=1
```

Linux 有一个很好的工具来拒绝为不怀好意的 IP 提供服务，称为 iptables。

很多管理员害怕使用 iptables。阻塞 IP 会阻塞这个 IP 访问任何服务器资源，不仅仅是 Web 服务器，还包括 FTP、telnet 等。不幸的是，很多系统管理员和网站站长害怕使用 iptables，因此没有经验。如果直接编辑 iptables 的配置文件，可能会导致宕机。例如，一个语法错误，可能会阻止你访问 SSH、FTP、HTTP 和任何其他的服务。因此，不要直接编辑配置文件 iptables-config。最好从 Linux 命令行进入 iptables 命令。如果有语法错误，命令行接口会拒绝这个命令。下面是一些常用的例子：

阻塞 120.60.0.0～120.60.255.255 范围内的 IP。

```
#iptables -I INPUT -m iprange --src-range 120.60.0.0-120.60.255.255 -j DROP
```

要阻止一个 IP，例如 120.60.43.201，使用下面的命令：

```
#iptables -A INPUT -s 120.60.43.201 -j DROP
```

显示当前的 iptables 文件，而不编辑它，使用下面的命令：

```
#iptables -L
```

iptables 不能在自动屏蔽恶意 IP，只能手动屏蔽。一个轻量级的脚本 DDoS deflate 能够自动屏蔽 DDoS 攻击者的 IP。

可以配置白名单的 IP 地址，通过配置：/usr/local/ddos/ignore.ip.list。

IP 地址被封时间是预先设定的，默认 600 秒后自动解除封锁。通过配置文件，脚本可以定时周期性运行(默认是 1 分钟)。有 IP 地址被封锁时，可以为指定的邮箱接收电子邮件警报。

这些都可以写在配置文件/usr/local/ddos/ddos.conf 中。

安装 DDoS deflate：

```
# wget http://www.inetbase.com/scripts/ddos/install.sh
# chmod 0700 install.sh
# ./install.sh
```

下面解释一下 DDoS deflate 脚本主要配置文件 ddos.conf：

```
##### Paths of the script and other files
PROGDIR="/usr/local/ddos"              //文件存放目录
PROG="/usr/local/ddos/ddos.sh"         //主要功能脚本
IGNORE_IP_LIST="/usr/local/ddos/ignore.ip.list"     //白名单地址列表
CRON="/etc/cron.d/ddos.cron"           //crond 定时任务脚本
```

```
APF="/etc/apf/apf"
IPT="/sbin/iptables"

##### frequency in minutes for running the script
##### Caution: Every time this setting is changed, run the script with --cron
#####          option so that the new frequency takes effect
FREQ=1         //间隔多久检查一次，默认1分钟

##### How many connections define a bad IP? Indicate that below.
NO_OF_CONNECTIONS=150       //最大连接数设置，超过这个数字的IP就会被屏蔽，默认即可

##### APF_BAN=1 (Make sure your APF version is atleast 0.96)
##### APF_BAN=0 (Uses iptables for banning ips instead of APF)
APF_BAN=0      //1：使用APF，0：使用iptables，推荐使用iptables

##### KILL=0 (Bad IPs are'nt banned, good for interactive execution of script)
##### KILL=1 (Recommended setting)
KILL=1     //是否屏蔽IP，默认即可

##### An email is sent to the following address when an IP is banned.
##### Blank would suppress sending of mails
EMAIL_TO="root"          //发送电子邮件报警的邮箱地址，换成自己使用的邮箱即可

##### Number of seconds the banned ip should remain in blacklist.
BAN_PERIOD=600      //屏蔽IP的时间，根据情况调整
//最后开启系统crond服务即可
//执行uninstall.ddos来卸载脚本：
# wget http://www.inetbase.com/scripts/ddos/uninstall.ddos
# chmod 0700 uninstall.ddos
# ./uninstall.ddos
```

更好的方法是安装CSF (Config security firewall)，它可以在极大程度上保护服务器安全。CSF可以免费使用的，是基于iptables的防火墙，很容易集成到CPanel。CPanel是为网站所有者设计的一套Web形式的控制系统，网站所有者甚至可以直接把它当作网站后台Windows操作系统。

银行为了防止黑客暴力破解持卡人的密码，采用连续3次输入密码错误，就锁定该账户的方法。CSF防火墙为了防止暴力破解密码，也会自动屏蔽连续登录失败的IP。

可以通过它管理网络端口，只开放必要的端口。可以免疫小流量的DDoS和CC攻击。

CentOS下需要先安装CSF依赖包：

```
yum install perl-libwww-perl perl iptables
wget http://www.configserver.com/free/csf.tgz
tar zxf csf.tar.gz
```

如果和Apache服务器配合使用，则执行：

```
sh ./csf/install.sh
```

如果直接管理防火墙,则执行:

```
sh ./csf/install.directadmin.sh
```

按照安装程序的指示装好后,可以运行测试程序:

```
perl /etc/csf/csftest.pl
```

如果没问题就可以启动防火墙。

```
csf -s
```

重新启动防火墙

```
csf -r
```

刷新规则,或者停止防火墙

```
csf -f
```

4.12 手机搜索界面

可以开发 WML 或 XHTML 格式的手机搜索界面。
用户可以选择搜索"网页""图片"或者"移动互联网"。

```xml
<?xml version="1.0" encoding="GBK"?>
<!DOCTYPE html PUBLIC "-//WAPFORUM//DTD XHTML Mobile 1.0//EN"
"http://www.wapforum.org/DTD/xhtml-mobile10.dtd">

<html xmlns="http://www.w3.org/1999/xhtml">
  <head>
    <meta http-equiv="Content-Type" content="text/html; charset=GBK"/>
    <meta http-equiv="Cache-Control" content="no-cache"/>
    <style type="text/css">
      a:link {color:#0000cc}
      a:visited {color:#0000cc}
      a:hover {color:#0000cc}
      a:active {color:#0000cc}
      .c0 {vertical-align:middle;}
    </style>
    <title>
      Lietu
    </title>
  </head>
  <body>
    <div><img class="c0" src="../images/logo.gif" alt="Lietu"/>
    </div>
    <form action="/pda/search">
      <div><input type="hidden" name="mrestrict" value="xhtml"/>
```

```html
            <input type="text" name="q" size="15" maxlength="2048"/>
            <br/>
            <input type="submit" name="btnG" value="搜索"/>
            <br/>
            <input type="radio" name="site" value="search" checked="checked"/>
            网页<br/>
            <input type="radio" name="site" value="images"/>
            图片<br/>
            <input type="radio" name="site" value="mobile"/>
            移动互联网<br/>
        </div>
    </form>
    <div><a href="/pda?pref=d"
        >使用偏好</a>
    <br/><a href="/pda/help"
        >帮助及使用条款</a>
    <br/><a href="/pda?hl=en">English</a>
    <br/>
    <br/><a href="http://www.lietu.com/pda"
        >lietu.com</a>
    <br/>&copy;2012 Lietu
    <br/>
    </div>
  </body>
</html>
```

4.13 本章小结

本章介绍了使用 Java 框架 Struts2 实现的搜索界面。并且介绍了很多重要而且基本的搜索功能界面实现,例如复杂条件搜索界面和用户输入提示词、分类查找界面等。

因为不需要重复输入,所以在浏览器的输入框中提交查询词到一个搜索引擎,需要花费的时间更少一些,但仍然需要搜索结果有一定用处。

因为有的浏览器禁用 JavaScript,所以尽量用普通的 HTML 标签,只是在必要的时候才使用 JavaScript。

在 2010 年之前使用 Script.aculo.us 实现自动完成功能。当 JQuery 开始流行后,依赖 Prototype 的 Script.aculo.us 不再流行。

Struts2 并没有退出历史舞台,它的版本仍然在更新中。除了可以使用 Struts2 来实现搜索界面,还可以使用 Spark 网站开发框架(http://sparkjava.com/)。

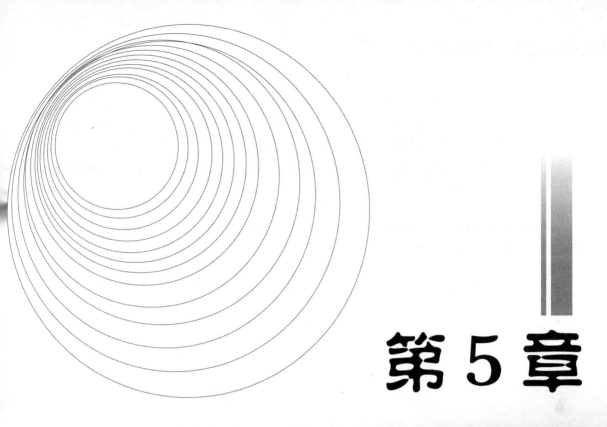

第 5 章

Solr 分布式搜索引擎

Lucene 仅仅是一个全文检索包，不是一个独立的搜索服务。如果在异构环境直接调用 Lucene 可能会导致系统不稳定，例如开发 Web 界面的工具采用 PHP 时，采用桥接方式调用 Lucene 容易导致内存泄露。本章介绍的 Solr 搜索服务器就是一个支持多种前台开发语言的搜索框架。Solr 提供了 XML 接口的全文检索服务，支持 ASP.NET、JSP、PHP、Python 等多种前台开发语言。

Solr 自身由服务器端和客户端组成。服务器端提供搜索服务，客户端提供 Web 界面开发支持。经常使用 JSP 开发 Web 界面。支持 JSP 开发的 Solr 客户端叫作 SolrJ。使用 Solr 的软件架构如图 5-1 所示。

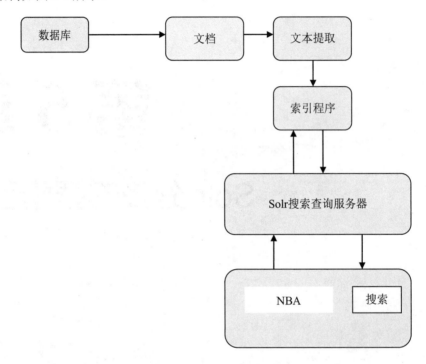

图 5-1 应用 Solr 的软件架构

Solr 服务器类似一个全文检索数据库服务器。SolrJ 类似 JDBC 驱动程序。JDBC 驱动程序往往使用私有的协议，而 SolrJ 则可以使用调试方便的 HTTP 协议。

5.1 Solr 简介

如果说 Lucene 是半自动步枪，则 Solr 是全自动步枪。Solr 把对 Lucene 索引的调用和管理做了一个 REST 风格的封装。它是一个 Web 方式的索引服务器。需要搜索的应用可以通过 HTTP 请求调用 Solr 索引服务。

可以把 Solr 看成类似 MySQL 一样的数据库，在配置文件 schema.xml 中定义表结构。和用 JDBC 访问数据库不同，通过 HTTP 协议的 POST 方式发送需要增加的数据给 Solr，通过 HTTP 协议的 GET 方式查询 Solr 管理的索引库。外部程序首先把 HTTP 请求提交给类似

Tomcat 或者 Resin 这样的 Web 服务器，然后 Solr 内部包含 Servlet 来写数据到索引库或者查询索引库。Solr 的内部结构如图 5-2 所示。

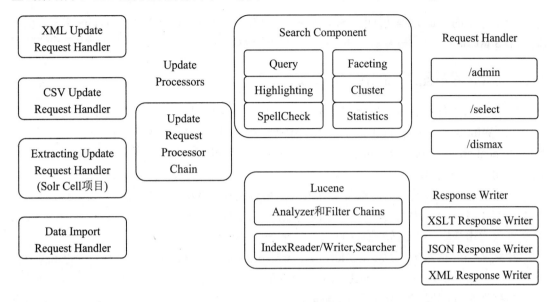

图 5-2　Solr 内部结构图

5.2　Solr 基本用法

Solr 自身由服务器端和客户端组成。服务器端提供搜索服务，客户端提供 Web 界面开发支持。

本书中介绍的 Solr 是 6.4 版本，从 Solr 的官方网站 http://lucene.apache.org/solr/可以下载到这个版本和当前最新的版本。准备一个比较新的 JDK，例如：Solr 6 需要 JDK1.8 或以上版本。

Solr 并不能独立运行，它只是一个 Web 项目。需要把它放在一个 Java Web 服务器中，启动 Web 服务器后才能使用它。

下载 solr-6.4.0.tgz 后可以看到其中包括一个 example 目录。这是一个完整的 Solr 例子服务器。其中包括 Jetty Servlet 引擎，也就是一个 Java Web 服务器。example 目录中包括一些例子数据和一个 Solr 配置例子。

使用 Solr 最简单的方式，就是运行 Jetty 中的 Solr，也就是/solr-6.4.0/server/start.jar。在命令行执行：

```
>java -jar start.jar
```

然后打开网址 http://localhost:8983/solr，可以看到一个指向 Solr 管理界面的链接。

在配置文件中来定义索引库的结构。配置文件位于 solr-4.2.0\example\solr\collection1\conf\目录下。

另外一种新的运行方法是使用 solr.cmd 命令脚本：

```
>bin/solr start              # 启动 solr
>bin/solr create -c demo     # 这将创建一个称为"demo"的文档集合
```

一个完整的 Lucene 索引叫作 SolrCore。一个 Solr 实例可以管理多个 Lucene 索引，也就是多个 SolrCore。

5.2.1 Solr 服务器端的配置与中文支持

Web 应用服务器通常可以选用 Tomcat 或者 Resin。这里以 Tomcat 为例说明安装过程。

在 Windows 下运行的 Tomcat 中部署 Solr：首先从 Tomcat 官方网站 http://tomcat.apache.org/ 下载 zip 包。然后在命令行运行 startup.bat。如果要停止 Tomcat，只需要关闭有 Java 标志的控制台窗口。

从分布路径上来说，Solr 由两部分组成：后台路径和 Web 管理界面。后台路径可以放在 Web 应用服务器的根路径下的 solr 路径下，其中的 conf 路径存放配置文件，其中的 data 路径存放索引数据。Web 管理界面存放在 webapps 目录下。

首先从 Solr 的官方网站 http://lucene.apache.org/solr/ 下载 6.4 版本。解压缩 solr-6.4.0.tgz 文件后，把 solr-6.4.0\example\webapps 下的 solr.war 文件复制到 Tomcat 的 webapps 目录下。

直接把 Solr 配置路径复制到 Tomcat 的子目录下就可以。也就是把 solr-6.4.0\example\solr 复制到 tomcat\ 目录下。需要复制与日志相关的文件到 tomcat\lib。首先把 solr-6.4.0\example\lib\ext 路径下的 jar 文件复制到 tomcat\lib。然后把 solr-6.4.0\example\resources 下的配置文件也复制到 tomcat\lib。

Solr 运行所需要的文件准备好以后，在 Tomcat 根目录下运行 .\bin\startup.bat。这样做是为了把当前路径设置到 Tomcat 目录，而不是其 bin 子目录。在浏览器中输入网址 http://localhost:8080/solr，打开 Solr 的控制界面，检查 Solr 是否正常运行。初次配置 Solr 的时候，建议在命令行运行 Tomcat，这样容易发现错误。

另外也可以通过 Java 虚拟机的系统属性 solr.solr.home 来指定索引库路径。

```
#export JAVA_OPTS="$JAVA_OPTS -Dsolr.solr.home=/my/custom/solr/home/dir/"
```

或者把 solr/home 参数的值写到配置文件中。也就是通过 JNDI 中的 solr/home 来指定 Solr 索引库的路径。例如创建文件 apache-tomcat-6.0.32\conf\Catalina\localhost\solr.xml。内容如下：

```
<?xml version="1.0" encoding="utf-8"?>
<Context docBase="E:\solr-app\apache-solr-6.0.32.war" debug="0"
crossContext="true">
<Environment name="solr/home" type="java.lang.String"
value="E:\lucene\apache-solr-6.0.32\example\solr" override="true"/>
</Context>
```

可以在 Solr 的控制面板中看到 Java 虚拟机的内存使用量。通过设置 JAVA_OPTS 增加 Web 服务器的内存使用量。

在文件 bin/catalina.bat 的开始位置增加如下行：

第 5 章　Solr 分布式搜索引擎

```
set JAVA_OPTS=-Xms1024m -Xmx1024m -XX:PermSize=256m -XX:MaxPermSize=128m
```

有时候需要建立好几个不同格式的索引。例如垂直行业搜索中需要分别搜索产品和公司，需要把产品信息和公司信息放在不同的索引格式中。

例如建立两个核：usercore1 和 usercore2。solr.xml 中的配置信息：

```xml
<solr persistent="true" sharedLib="lib">
 <cores adminPath="/admin/cores">
  <core name="usercore1" instanceDir="usercore1" />
  <core name="usercore2" instanceDir="usercore1" />
 </cores>
</solr>
```

查询 usercore1 的 URL 如下：

```
http://host.com/solr/usercore1/select?q=test
```

查询 usercore2 的 URL 如下：

```
http://host.com/solr/usercore2/select?q=test
```

可以通过 defaultCoreName 指定缺省的核心。例如：

```xml
<cores adminPath="/admin/cores" defaultCoreName="collection1"
host="${host:}" hostPort="${jetty.port:}" hostContext="${hostContext:}"
zkClientTimeout="${zkClientTimeout:15000}">
    <core name="collection1" instanceDir="collection1" />
     <core name="collection2" instanceDir="collection2" />
 </cores>
```

可以在同一个 Tomcat 下可以配置多个 Solr 实例来对应不同的索引格式。例如要配置两个 Solr 实例：company 和 product。通过两个不同的 URL 访问：

```
http://localhost/product/
http://localhost/company/
```

可以在 /usr/local/tomcat55/conf/Catalina/localhost 路径下建立两个配置文件：company.xml 和 product.xml，分别对应两个 Web 应用，其中 product.xml 的内容如下：

```xml
#cat ./product.xml
<Context docBase="" debug="0" crossContext="true" >
    <Environment name="solr/home" type="java.lang.String"
        value="/usr/local/product" override="true" />
</Context>
```

和部署于同一个 Web 服务器中的多个 Web 应用不同，Solr 中的多核(Multi Core)功能指在同一个 Web 应用中管理多个索引。在并发量很少的情况下，多核在性能上有优势，但当并发量增加后，多核响应速度反而比标准配置慢。

多核允许一个 Solr 实例有不同的配置和索引模式。也就是说一个 Solr 的 Web 管理界面可以同时管理多个不同结构的索引。可以用一个 SolrCore 代替另外一个 SolrCore 而不需要重新启动 Servlet 容器。Solr 5 以上的版本可以自动发现位于 Solr 根目录下的子目录下的核。

为了支持中文，Tomcat 中的相关配置如下：在 server.xml 中增加对 UTF-8 的处理，这样是为了支持 HTTP 的 GET 方法：

```
<Connector port="80"  useBodyEncodingForURI="true"  URIEncoding="UTF-8" />
```

server.xml 位于 E:\lucene\apache-tomcat-6.0.32\conf\server.xml 这样的路径下。useBodyEncodingForURI="true"，设置这个值的意思是用页面的编码去处理 POST 请求。

另外在 Solr 的 Web 项目的 web.xml 配置中增加对 UTF-8 的转码：

```
    <filter>
     <filter-name>Set Character Encoding</filter-name>
     <filter-class>filters.SetCharacterEncodingFilter</filter-class>
     <init-param>
        <param-name>encoding</param-name>
        <param-value>utf-8</param-value>
     </init-param>
   </filter>

   <filter-mapping>
     <filter-name>Set Character Encoding</filter-name>
     <url-pattern>/*</url-pattern>
   </filter-mapping>
```

增加这个 Filter 是为了支持 HTTP 的 POST 方法。SetCharacterEncodingFilter 类可以在 webapps\examples\WEB-INF\classes 路径下找到。

Solr 中有两个在 conf 路径下的配置文件，分别是：

(1) solrconfig.xml：用来配置 Solr 运行的系统参数，例如缓存、插件等。和 Lucene 索引库相比，Solr 的好处是实现了缓存预加载。

(2) schema.xml：主要定义与索引结构相关的信息，例如，types、fields 和其他的一些缺省设置。

然后需要调整的是 solrconfig.xml。

```
<maxWarmingSearchers>4</maxWarmingSearchers>
```

自动加载缓存的线程数缺省是 4，可以调整小一点，最好少于 CPU 核数量。否则可能使一个双 CPU 双核的机器的 CPU 使用率达到近 100%。

此外，还可以设置默认的 ResponseWriter 为 javabin 格式。

schema.xml 配置文件定义了 Solr 中索引库的结构，可以把它类比为数据库中的一张表。可以通过 field 来定义列，fieldType 定义列类型，uniqueKey 指定索引库的唯一标识列（类似数据库表中的主键），defaultSearchField 指定默认搜索列。

```
<uniqueKey>id</uniqueKey>
```

schema.xml 使用的基本的数据类型如下：

```
<fieldtype name="sint" class="solr.SortableIntField" />   //整型
<fieldtype name="slong" class="solr.SortableLongField"/>   //长整型
```

```
<fieldtype name="sfloat" class="solr.SortableFloatField"/>   //浮点数
<fieldtype name="sdouble" class="solr.SortableDoubleField" />  //双精度
<fieldtype name="date" class="solr.DateField" />   //日期
```

和 Lucene 比起来,这些数据类型有更优化的存储方式。

不要把 _version_ 列删除了。这列是 Solr 内部用于分布式处理的系统保留字段。

在 schema.xml 中定义了索引库的结构。可以有各种不同的 field,索引就是图书馆的目录中有这本书,存储就是能借到这本书。当原文很长,而且对性能要求很高的时候,可以考虑只索引而不存储。多值就是一列可以有很多值。例如,一个用户可以有多个标签。

下面定义 3 列,分别是唯一 id、标题和内容。

```
<field name="id" type="string" indexed="true" stored="true"
multiValued="false" />
<field name="title" type="text_ws" indexed="true" stored="true"
multiValued="false" />
<field name="body" type="text_ws" indexed="true" stored="true"
multiValued="false" />
```

在这里,multiValued="false"的意思是一列只允许存储一个值。如果需要一列存储多个值,例如一个商品属于多个类别(cat),则定义:

```
<field name="cat" type="string" indexed="true" stored="true"
multiValued="true" />
```

定义 id 为唯一标识列(id 即为上面定义的名字是"id"的列)。

```
<uniqueKey>id</uniqueKey>
```

定义 body 为默认搜索列。

```
<defaultSearchField>body</defaultSearchField>
```

为了高亮显示,必须把 stored 设置成 true。对于中文列,还需要添加 termVectors 和 termOffsets,这可以提高性能。如果进一步添加 termPositions,额外的加速也是可能的。

```
<field name="includes" type="text_general" indexed="true" stored="true"
termVectors="true" termPositions="true" termOffsets="true" />
```

它的服务器端管理界面包含 Solr 和索引库中的一些宏观的参数。相比较而言,Luke 则显示更细一些的 Document 和 Term 等信息。http://localhost/solr/admin/是一个已经安装好的 Solr 界面。可以先通过 http:// localhost/solr/admin/analysis.jsp?highlight=on 来对每个列做基本的测试,看看对定义好的 title 列是否能对中文做正确的处理。

有时候需要调试某个查询词是否能匹配上某个文档内容。可能会困惑于一个查询词不能匹配上某个文档。就好像一把钥匙打不开一把锁。所以可以把查询词和文档看成是钥匙和锁的关系,查询词是钥匙,而待查找的文档则是锁。当查询词正好匹配上文档,感觉像用正确的钥匙打开正确的锁一样。

在 Tomcat 中配置多个 Solr 实例的方法是:用 JNDI 的方法配置多个 solr.home。例如在 $CATALINA_HOME/conf/Catalina/localhost 下为每个 Solr 实例 的 webapp 创建独立的

context:

```
$ cat /tomcat55/conf/Catalina/localhost/solr1.xml
  <Context docBase="" debug="0" crossContext="true" >
    <Environment name="solr/home" type="java.lang.String"
value="f:/solr1home" override="true" />
  </Context>
$ cat /tomcat55/conf/Catalina/localhost/solr2.xml
  <Context docBase="" debug="0" crossContext="true" >
    <Environment name="solr/home" type="java.lang.String"
value="f:/solr2home" override="true" />
  </Context>
```

把每个 Solr 实例服务器端配置都放在不同的目录下，修改 solrconfig.xml 的索引数据存储路径：

```
<dataDir>${solr.data.dir:./solr1/data}</dataDir>
```

solrconfig.xml 文件记录一个 Solr 实例的配置信息。如上述索引数据存储路径，还可以配置自动加载缓存的线程数量：

```
<maxWarmingSearchers>2</maxWarmingSearchers>
```

自动加载缓存的一个线程数缺省是 4，有时候要调整小一点。否则会使一个双 CPU 双核的机器 CPU 使用率达到近 100%。

还可以在 solrconfig.xml 定义默认搜索处理类及其他 Handler 信息等。

默认搜索处理器 standard Handler 的定义：

```
<requestHandler name="standard" class="solr.SearchHandler" default="true">
    <lst name="defaults">
      <str name="echoParams">explicit</str>
    </lst>
</requestHandler>
```

有个小问题是 Solr 的日志在缺省情况下增加得很快，可以通过日志的输出级别来控制日志输出，Solr 中的配置界面参考 http://localhost/solr/admin/logging.jsp。

5.2.2 数据类型

首先定义类型，然后指定某个列使用这个分析类型。例如 text_general 类型定义如下：

```
<fieldType name="text_general" class="solr.TextField"
positionIncrementGap="100">
    <analyzer type="index">
      <tokenizer class="solr.StandardTokenizerFactory"/>
      <filter class="solr.StopFilterFactory" ignoreCase="true"
words="stopwords.txt" enablePositionIncrements="true" />
      <!-- in this example, we will only use synonyms at query time
      <filter class="solr.SynonymFilterFactory"
synonyms="index_synonyms.txt" ignoreCase="true" expand="false"/>
```

```xml
      -->
      <filter class="solr.LowerCaseFilterFactory"/>
    </analyzer>
    <analyzer type="query">
      <tokenizer class="solr.StandardTokenizerFactory"/>
      <filter class="solr.StopFilterFactory" ignoreCase="true"
words="stopwords.txt" enablePositionIncrements="true" />
      <filter class="solr.SynonymFilterFactory" synonyms="synonyms.txt"
ignoreCase="true" expand="true"/>
      <filter class="solr.LowerCaseFilterFactory"/>
    </analyzer>
</fieldType>
```

使用 text_general 类型：

```xml
<field name="content" type="text_general" indexed="true" stored="true" multiValued="true"/>
```

Solr 时间使用 UTC 格式。放入和取出都要转换。

```java
public static String toUtcDate(String dateStr) {
    SimpleDateFormat out = new SimpleDateFormat("yyyy-MM-dd'T'HH:mm:ss'Z'",
            Locale.CHINESE);
    // out.setTimeZone(TimeZone.getTimeZone("GMT-8"));
    // out.setTimeZone(TimeZone.getTimeZone("Z"));
    String[] dateFormats = { "yyyy-MM-dd", "MMM dd, yyyy hh:mm:ss Z" };
    for (String dateFormat : dateFormats) {
        try {
            return out.format(new SimpleDateFormat(dateFormat)
                    .parse(dateStr));
        } catch (ParseException ignore) {
        }
    }
    throw new IllegalArgumentException("Invalid date: " + dateStr);
}
```

将 UTC 时间转换为东 8 区时间：

```java
public static String getLocalTimeFromUTC(String UTCTime) {
    DateFormat format = new SimpleDateFormat("yyyy-MM-dd'T'HH:mm:ss'Z'",
            Locale.CHINESE);

    java.util.Date UTCDate = null;
    String localTimeStr = null;
    try {
        UTCDate = format.parse(UTCTime);
        localTimeStr = format.format(UTCDate);
    } catch (ParseException e) {
        e.printStackTrace();
    }
```

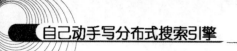

```
    return localTimeStr;
}
```

5.2.3 解析器

Solr 规定 TokenizerFactory 在 inform()方法中使用 ResourceLoader 加载词典文件这样的资源。

```
public interface ResourceLoader {
  public InputStream openResource(String resource) throws IOException;
}
```

需要加载外部资源的 TokenizerFactory 需要实现 ResourceLoaderAware。这个接口中定义的唯一一个方法是：inform()。

```
public interface ResourceLoaderAware {
  void inform(ResourceLoader loader) throws IOException;
}
```

使用 ResourceLoader 加载外部文件，例如中文分词所需要的词典文件。org.apache.lucene.util.IOUtils 这个类模拟新的 Java7 "Try-With-Resources" 语句。IOUtils.getDecodingReader()方法得到输入流。使用 FilesystemResourceLoader 加载 GBK 编码的词典文件例子如下：

```
File baseDirectory = new File("d:/eclipse/workspace/seg-Lucene4.3/dic/");
ResourceLoader loader=new FilesystemResourceLoader(baseDirectory);
String filename="unigramDict.txt";
InputStream dicStream = loader.openResource(filename);
BufferedReader reader = new BufferedReader
    (IOUtils.getDecodingReader(dicStream, Charset.forName("GBK")));

String line = null;
while ((line = reader.readLine()) != null) {
  System.out.println(line);
}
reader.close();
```

除了使用路径创建 FilesystemResourceLoader，还可以使用类通过 org.apache.lucene.analysis.util.ClasspathResourceLoader 创建 ResourceLoader。

```
ResourceLoader loader = new ClasspathResourceLoader(this.getClass());
```

CnTokenizerFactory 实现框架如下：

```
public class CnTokenizerFactory extends TokenizerFactory implements
        ResourceLoaderAware {

    @Override
```

```java
    public void inform(ResourceLoader loader) {
        // 加载资源
    }
}
```

测试 CnTokenizerFactory：

```java
CnTokenizerFactory factory = new CnTokenizerFactory(
        new HashMap<String, String>());
factory.inform(new ClasspathResourceLoader(this.getClass()));
String text = "大学生活动中心";
Reader reader = new StringReader(text);
TokenStream tokenizer = factory.create(reader);
```

生成分词 jar 文件后，把 seg.jar 复制到 webapps\solr\WEB-INF\lib 目录下。

修改配置文件 schema.xml，增加列类型 text_cn：

```xml
<fieldtype name="text_cn" class="solr.TextField"
positionIncrementGap="100" sortMissingLast="true">
    <analyzer>
        <tokenizer class="com.lietu.bigramSeg.CnTokenizerFactory"
                   dicDir="d:/ BigramSeg/dic/"/>
    </analyzer>
</fieldtype>
```

5.2.4 把数据放进 Solr

为了把数据放入 Solr，需要用 POST 方式提交要索引的数据。可以通过 post.jar 把数据放到 Solr 索引库。post.jar 读取的是 XML 文件。如果处理中文，这个 XML 文件必须是 UTF-8 编码的，否则就会出现乱码了。下面是一个 XML 文件的样例。

```xml
<add>
  <doc>
    <field name="id">126788</field>
    <field name="title">猎兔搜索</field>
    <field name="body">搜索内容</field>
  </doc>
</add>
```

curl 是一个知名的网络命令行工具。可以用来上传或者下载文件，也支持用 POST 方式提交数据。在 Linux 下缺省已经安装了这个命令行工具，但是也有 Windows 版本的 curl，下载地址为 http://www.paehl.com/open_source/?CURL_7.21.6，这是一个编译好的 curl.exe 文件。需要在 Windows 命令行下运行这个工具。增加数据：

```
>curl http://localhost:8983/solr/update?commit=true -H "Content-Type:
text/xml; charset=utf-8" --data-binary '<add><doc><field
name="id">testdoc</field></doc></add>'
```

如果提交的数据格式不对，HTTP 请求会返回一个状态码是 400 的错误。

也可以用 RESTClient(http://www.restclient.org/)代替 curl。

可以通过客户端程序 SolrJ 把数据库中的数据转换成符合 Solr 格式的 XML 数据流，然后通过 HTTP 协议发送出去。SolrJ 位于已经下载的 solr-4.3.1.zip 文件中。

除了把 solr-4.3.1\dist\solrj-lib 下相关的 jar 文件复制到项目的 lib 路径下，还需要把 log4j.properties 文件放到项目的 lib 路径下。检查能否找到日志的配置文件。

```
System.out.println(Thread.currentThread().getContextClassLoader().getResource("log4j.properties"));
```

索引文件代码如下：

```
public final static String url = "http://localhost:8080/solr/collection1/"; //solr 服务的网址

public static void main(String[] args) {
    SolrServer sever = new HttpSolrServer(url);
    Collection<SolrInputDocument> docs = new HashSet<SolrInputDocument>();
    for (int i = 0; i < 10; i++) {
        SolrInputDocument s = new SolrInputDocument();
        s.addField("id", "00"+i);
        s.addField("title", "title_0"+i);
        s.addField("content", "content_0"+i);

        docs.add(s);
    }

    UpdateResponse add = sever.add(docs);
    sever.commit(true, true);
}
```

一个完整的例子：

```
SolrServer server = new HttpSolrServer("http://localhost:8080/solr/collection1/");

//索引一些文档
Collection<SolrInputDocument> docs = new HashSet<SolrInputDocument>();
for (int i = 0; i < 10; i++) {
    SolrInputDocument doc = new SolrInputDocument();
    doc.addField("id", i);
    doc.addField("title", "Title: " + i);
    doc.addField("content", "This is the " + i +"(th|nd|rd) piece of content.");

    System.out.println("Doc[" + i + "] is " + doc);
    docs.add(doc);
}
UpdateResponse response = server.add(docs);
System.out.println("Response: " + response);
```

```
//把文档加入到 Solr 服务器后,需要提交数据,才能让前台搜索界面搜到
response =server.commit();
System.out.println("Response: " + response);
```

批量增加数据时,如果数据量太大,则会出错。

或者使用 POJO 类,定义一个 Item 类:

```
public class Item {
   @Field
   String id;
   @Field("cat")
   String[] categories;
}
```

注意这里必须要有注解@Field。

放入数据:

```
SolrServer server = getSolrServer();
Item item = new Item();
item.id = "one";
item.categories = new String[] { "aaa", "bbb", "ccc" };
server.addBean(item);
```

boost 缺省是 1.0。例如对下面这个文档加权。

```
<add>
  <doc boost="2.5">
    <field name="employeeId">05991</field>
    <field name="office" boost="2.0">Bridgewater</field>
  </doc>
</add>
```

SolrJ 中的写法:

```
doc.setDocumentBoost(0.001f);
```

数据更新以后,如果要在索引库即刻看到更新的数据,需要提交更新,也就是发送如下的内容给 Solr:

```
<Commit/>
```

在 SolrJ 中调用执行 server.commit()来更新数据。

因为索引写入硬盘花时间,可能会导致 java.net.SocketTimeoutException: Read timed out 错误。所以要延长等待时间。

```
server.setSoTimeout(120000);  // 读取 solr 响应的过期时间设置成 2 分钟
```

此外还可以修改其他的参数:

```
server.setConnectionTimeout(100);
server.setDefaultMaxConnectionsPerHost(100);
```

```
server.setMaxTotalConnections(100);
server.setFollowRedirects(false);
```

除了显式调用 commit，同时 Solr 也支持自动提交数据。可以在 solrconfig.xml 配置文件中修改自动提交的条件：

```
<!-- 当满足以下条件时执行自动提交:
        maxDocs —自从上次提交以来,更新数量超过指定次数时提交更新
        maxTime —  当超过指定时间(单位为 ms) 后提交
-->
<autoCommit>
    <maxDocs>10000</maxDocs>
    <maxTime>100000</maxTime>
</autoCommit>
```

如果不需要自动提交数据，只是通过程序显式地提交，可以把相关的值都置成负数。

```
<autoCommit>
  <maxDocs>-1</maxDocs>
  <maxTime>-1</maxTime>
</autoCommit>
```

完整的参数如下：

```
StreamingUpdateSolrServer server =
        new StreamingUpdateSolrServer( "http://localhost:8983/solr",5,5);
server.setSoTimeout(30000);
server.setConnectionTimeout(30000); // socket read timeout
server.setDefaultMaxConnectionsPerHost(100);
server.setMaxTotalConnections(100);
server.setFollowRedirects(false);  // 缺省是 false
// allowCompression 缺省是 false
// 服务器端必须支持 gzip 或 deflate 压缩才能起作用
server.setAllowCompression(true);
server.setMaxRetries(1); // 缺省是 0。不推荐采用大于 1 的值
```

可以编写一个后台独立运行的程序来同步数据库和索引，把数据库中的数据增量放入索引库中。

```
public static void main(String[] args) throws Exception,
InterruptedException {
    long sleepTime = 5000L; //5 秒
    for(;;){
        boolean findNew = indexDb();//查看数据库是否有新数据要索引

        if(findNew) {
            sleepTime = 5000L;
        }else{
            if(sleepTime < 500000L) {
                sleepTime = sleepTime*2; //没有新数据则延长检测新数据的间隔
```

```
            }
        }
        System.out.println("sleep..."+sleepTime);
        Thread.sleep(sleepTime);
    }
}
```

考虑 indexDb()如何实现,也就是如何从待索引的表中找出要索引的新数据。可以根据表中的唯一 id 列查询出要索引的新数据。

StreamingUpdateSolrServer 缓存所有的增加文档并把这些文档写到开放的 HTTP 连接。StreamingUpdateSolrServer 内部使用 HttpClient 发送文档到 Solr 服务器。StreamingUpdateSolrServer 可以高速批量更新,提高索引文档的速度。

```
StreamingUpdateSolrServer server =
 new StreamingUpdateSolrServer( "http://localhost:8983/solr",5,5);
```

此外,还可以改变日志级别,例如从 INFO 到 WARNING,这样可以加快索引速度。

可以使用 dataimport(简称 DIH)把数据通过 SQL 语句导入。只需要配置好数据源和 SQL 语句,就可以把数据库中的数据索引到 Solr。需要增加一个数据导入配置文件 data-config.xml。并且在配置文件 solrconfig.xml 中注册这个数据导入功能。

在 solrconfig.xml 中注册一个通过 URL 地址"/dataimport"访问的请求处理器。

```
<requestHandler name="/dataimport"
    class="org.apache.solr.handler.dataimport.DataImportHandler">
    <lst name="defaults">
        <str name="config">data-config.xml</str>
    </lst>
</requestHandler>
```

data-config.xml 文件内容如下:

```
<dataConfig>
  <dataSource type="JdbcDataSource"
            driver="com.mysql.jdbc.Driver"
            url="jdbc:mysql://127.0.0.1:3306/sq_xidi?useUnicode=true&amp;characterEncoding=utf-8&amp;zeroDateTimeBehavior=round"
            user="root"
            password="root"/>
  <document>
    <entity name="expert"
query="select id,company_name,company_desc,key_words,url,address,tel,create_time,substring(SUBSTRING_INDEX(url,'/',3),8) as webSite from company where status=1 and is_indexed=0">
      <field name="id" column="id"/>
<field name="companyName" column="company_name"/>
<field name="companyDesc" column="company_desc"/>
<field name="keyWords" column="key_words"/>
<field name="url" column="url"/>
```

```xml
<field name="address" column="address"/>
<field name="tel" column="tel"/>
<field name="create_time" column="create_time"/>
<field name="webSite" column="webSite"/>
    </entity>
  </document>
</dataConfig>
```

配置 deltaQuery 实现增量查询。例如，按时间增量索引。

```xml
<entity name="dbbook"
     query="select id,isbn from content"
     deltaQuery="select id,isbn from content where lastUpdateTime > '${dataimporter.last_index_time}'" />
```

5.2.5 删除数据

可以通过查询的方式删除 Solr 的索引数据。例如根据查询删除数据：

```
#curl http://192.168.10.30:8080/solr/update
 --data-binary "<delete><query>id:314685</query></delete>" -H
"Content-type:text/xml"
#curl http://localhost/gongkong/update
 --data-binary
"<delete><query>id:http\:\/\/www.aljoin.com\/news\/2007-10\/20071010154644.htm</query></delete>" -H "Content-type:text/xml"
```

或者直接通过 id 列删除：

```
#curl http://192.168.10.30:8080/solr/update --data-binary
"<delete><id>314685</id></delete>" -H "Content-type:text/xml"

#curl http://localhost/gongkong/update --data-binary
"<delete><id>http\:\/\/www.aljoin.com\/news\/2007-10\/20071010154644.htm</id></delete>" -H "Content-type:text/xml"
```

如果要删除所有标题或者内容中含有"答案"这个关键词的结果应该怎么写？

```
#curl http://192.168.10.30:8080/solr/update --data-binary
"<delete><query>title:答案 OR body:答案</query></delete>" -H
"Content-type:text/xml"
```

SolrJ 删除数据的例子：

```
String url = "http://localhost/solr/";
SolrServer server = new HttpSolrServer( url );
String q = "source:\"industrial acs\"";
UpdateResponse res = server.deleteByQuery( q );
System.out.println(res.getStatus());//取得返回状态值
server.commit( true, true );
```

第 5 章　Solr 分布式搜索引擎

如果要清空所有的数据，可以把 Solr 服务停止后直接删除 Solr 的索引路径，Solr 启动后会重建索引路径。不是清空 solr\data\index 下所有的文件，因为那样会被 Solr 误认为索引文件不完整。

5.2.6　查询语法

只查询某一列，可以使用如下的查询语法：

```
列名:查询词
```

如果是搜索缺省列，则可以省略列名。如果要返回所有文档，可以用通配符"*:*"。
可以使用布尔逻辑条件查询多列：

```
title:NBA OR body:NBA
```

配置文件 schema.xml 中的 defaultOperator 参数指定有空格时是用"AND"还是用"OR"操作逻辑，缺省是 OR 操作逻辑。

```
<solrQueryParser defaultOperator="OR"/>
```

5.3　使用 SolrJ

实际中往往使用 JSP 开发搜索界面。SolrJ 是通过 Java 访问 Solr 的客户端。
可以在 Maven 中加入依赖：

```xml
<dependency>
    <groupId>org.apache.solr</groupId>
    <artifactId>solr-solrj</artifactId>
    <version>5.4.0</version>
</dependency>
```

也可以手工添加 solr-solrj-*.jar 和 httpclient-*.jar 等 jar 包。

对于简单应用来说，Tomcat 下面有两个 Web 应用：一个作为索引服务器，往往叫作 solr；还有个作为搜索用户界面，往往放在 ROOT 目录下。搜索用户界面中包含 SolrJ 相关的一些 Jar 包。

5.3.1　Solr 客户端与搜索界面

SolrJ 是一个易于使用的客户端。下面是 SolrJ 一个简单的使用例子。

```java
String url = "http://localhost:8080/solr/";
SolrServer server = new HttpSolrServer( url );

SolrQuery query = new SolrQuery("NBA");
query.set("fl", "id,title,content"); //控制返回哪几列

QueryResponse response = server.query( query );
```

```
//取得符合查询条件的总数
long numFound = response.getResults().getNumFound();

for(SolrDocument d : response.getResults()){
    String id = (String)d.getFieldValue("id");
    System.out.println("id:"+id);
    String title = (String)d.getFieldValue("title");
    System.out.println("title:"+title);
}
```

因为 Solr 内部已经有分页控制，所以实际显示的记录会比搜索到的结果少。分页程序需要计算总共有多少页，当前在第几页。根据返回结果总数计算出总共有多少页。SolrDocumentList.getNumFound()方法得到符合查询条件的结果总数。

最多只返回 5 个文档的例子：

```
SolrQuery query = new SolrQuery().setQuery(
        "num:[20 TO 30]").setSortField("num", ORDER.desc).setRows(5);
//可以在服务器对象 SolrServer 中设置连接参数：
httpserver = new CommonsHttpSolrServer(url);
httpserver.setSoTimeout(5000);   // socket 读取超时
httpserver.setConnectionTimeout(5000);
httpserver.setDefaultMaxConnectionsPerHost(100);
httpserver.setMaxTotalConnections(100);
httpserver.setFollowRedirects(false);  //默认值为 false
httpserver.setAllowCompression(true);
httpserver.setMaxRetries(1); //默认值为 0。不推荐>1
```

下面是对"cat1id"这一列的分类统计例子：

```
String url = "http://hot.lietu.com:8080/solr/";
SolrServer server = new HttpSolrServer( url );

SolrQuery query = new SolrQuery("NBA");

query.getFacetParams().addField("cat1id");

QueryResponse response = server.query( query );
System.out.println("found:"+response.getResults().getNumFound());

List<Count> facetCounts = response.getFacetField("cat1id").getValues();
for(Count c:facetCounts ){//打印分类统计结果
    System.out.println(c);
}
```

向下缩小查询范围，可以使用 fq(filter query)过滤查询。例如：按地区分类统计，可以使用：q=QueryWord&fq=area:areaName。因为过滤器有缓存，所以比"QueryWord AND area:areaName"性能更好。

VelocityResponseWriter 可以返回从 Velocity 模板生成的内容。

按区间范围查询：

```
timestamp:[* TO NOW]
```

例如要查询年纪在 21~24 岁：

```
age:[21 TO 24]
```

5.3.2 Solr 索引库的查找

通过 Solr 的管理界面，可以直接分析索引库。还可以直接在浏览器输入 URL 地址来查询索引库。通过各种参数来指定搜索执行的方式和返回结果的格式等。Solr 中的参数分成基本参数和公共参数等类型。

Solr 中所有请求都会用到的核心查询参数有：

- qt：查询类型 (request handler)，例如 standard；
- wt：返回格式类型(response writer)，例如 XML 或 JSON。

公共的查询参数有：

- q：查询词；
- sort：排序方式；
- start：返回结果的开始行；
- rows：本次需要返回结果的行数；
- fl：需要返回的列名称；
- fq：FilterQuery，过滤查询，用于分类统计附加条件

例如：执行一个最基本的关键词搜索，搜索 "NBA" 这个词：

http://localhost:8080/solr/select/?q=NBA&version=2.2&start=0&rows=10&indent=on

这里返回最开始的前 10 行搜索结果。

通过查询参数 sort 指定排序方式。例如按价格升序，可以指定 sort=price asc。

例如，按时间升序：

http://localhost:8080/solr/select/?q=a&version=2.2&start=0&rows=10&indent=on&sort=timestamp+asc

http://localhost/solr/admin/ 管理界面可以输入查询词，自动生成这样的查询条件。例如搜索标题，可以提交这样的查询字符串：title:NBA。

如果要按 URL 地址查找一个网站的数据，则需要把 URL 地址中的字符 ":" 和 "/" 转义：

```
url:http\:\/\/2packaging*
```

可以通过 qt 参数来调用不同的 RequestHandler 处理查询请求，下面这个查询调用 StandardRequestHandler 来处理。

```
http://localhost:8983/solr/select/?q=video&fl=name+score&qt=standard
```

Solr 的分析页面是不可缺少的实验方法和故障排除工具。可以用这个工具分析不同的语句，来验证是否是期望的结果。如果查询结果和你所想要的结果不同，也可以利用这个故

障排除工具找出原因。在 Solr 的管理页面顶部有一个"ANALYSIS"的超级链接。在这个页面顶部的第一个选项表示字段的类型，这一项是必选的。既可以直接输入字段类型，也可以按字段名称来分析。但是这个工具是用来分析文本型字段类型的，而不是用来分析布尔类型、日期类型和数字类型的。这个工具能同时分析索引和(或)查询文本。如果同时做查询和索引分析并且想查看分析时索引部分的匹配情况，可以高亮显示匹配结果。

输入"Quoting,Wi-Fi"and stopword 后，分析的输出结果中最重要的行是叫"term text"的第二行。"term"是实际存储和查询的原子单位。因此，查询的分析结果必须和索引分析阶段的结果包含相同的 term 才能够匹配。请注意，位置 3 上有两个词条。在同一位置的多个词条可能是由于同义词扩展产生，但是在这里，是由 WordDelimeterFilterFactory 引入了 WiFi 这个词。Quoting 在词干化和小写化之后变成 quot。因为 and 在停用词表里，所以被 StopFilter 省略。

使用 StopFilter 的副作用是，因为有些查询中的词全部在停用词表中，所以可能完全搜索不到了。如果不想直接去掉停用词，可以用 CommonGramsFilter 来改进短语查询。CommonGramsFilter 对布尔查询没有意义。布尔查询仅仅通过 CommonGramsFilter 而不发生改变。

有两个相关的类：CommonGramsFilter 和 CommonGramsQueryFilter。在索引时使用 CommonGramsFilter，在查询时使用 CommonGramsQueryFilter。CommonGramsFilter 输出 CommonGrams 和 Unigrams，这样就可以使用布尔查询代替短语查询。

如下是一个使用 CommonGramsFilter 的 schema.xml 配置文件例子：

```xml
<fieldType name="CommonGramTest" class="solr.TextField"
positionIncrementGap="100">
 <analyzer type="index">
  <tokenizer class="solr.WhitespaceTokenizerFactory"/>
  <filter class="ISOLatin1AccentFilterFactory"/>
  <filter class="solr.PunctuationFilterFactory"/>
  <filter class="solr.LowerCaseFilterFactory"/>
  <filter class="solr.CommonGramsFilterFactory" words="new400common.txt"/>
 </analyzer>
 <analyzer type="query">
  <tokenizer class="solr.WhitespaceTokenizerFactory"/>
  <filter class="ISOLatin1AccentFilterFactory"/>
  <filter class="solr.PunctuationFilterFactory"/>
  <filter class="solr.LowerCaseFilterFactory"/>
  <filter class="solr.CommonGramsQueryFilterFactory"
words="new400common.txt"/>
 </analyzer>
</fieldType>
```

有没有办法把最满足条件的文档排在最前面，如果查找很多个关键词时，如何把都满足的或者满足得最多的排在最前面？

可以制定如下规则：如果有 3 个条件，那么这 3 个条件都必须满足，如果条件超过 3 个，只要满足 75%的条件。按照这个思路，首先重新配置 solrconfig.xml，修改 Dismax 的缺省参数配置，否则有些列名可能无效。

```xml
<requestHandler name="dismax" class="solr.DisMaxRequestHandler" >
    <lst name="defaults">
    <str name="echoParams">explicit</str>
    <float name="tie">0.01</float>
    <str name="qf">
      body^0.5 title^1.2
    </str>
    <str name="pf">
      body^0.2 title^1.5
    </str>
    <str name="fl">
      title,body
    </str>
    <str name="mm">
      75%
    </str>
    <int name="ps">100</int>
    <str name="q.alt">*:*</str>
    <!-- example highlighter config, enable per-query with hl=true -->
    <str name="hl.fl">text features name</str>
    <!-- for this field, we want no fragmenting, just highlighting -->
    <str name="f.name.hl.fragsize">0</str>
    <!-- instructs Solr to return the field itself if no query terms are
        found -->
    <str name="f.name.hl.alternateField">name</str>
    <str name="f.text.hl.fragmenter">regex</str> <!-- defined below -->
    </lst>
</requestHandler>
```

然后在浏览器中输入：

http://localhost/solr/select/?q=test&rows=0&qt=dismax&mm=75%

这里设置了查询满足条件"mm=75%"并且指定 Solr 的 DisMaxRequestHandler 类来处理查询请求。

Velocity 是一个模版系统。用 wt=velocity 参数指定使用 VelocityResponseWriter 输出返回结果。

5.3.3 分类统计

分类统计展示搜索结果在类别中的分布情况。对于容量不大的索引都可以执行分类统计搜索。Solr 已经集成了分类统计功能。通过 facet 相关的参数控制如何执行分类统计搜索。facet.field 参数指定分类统计列，分类统计列是不分词的列。在搜索结果中会增加 facet_counts 相关的信息。

索引库中已经定义了一列叫作 cat。假设用 cat 列做分类统计列，首先我们可以看一下这个索引分了几类。

```
http://localhost/solr/select/?q=*%3A*&rows=0&facet.field=cat&rows=0&face
t=true&&facet.limit=-1
```

部分返回结果如下:

```
<result name="response" numFound="4" start="0"/>
    <lst name="facet_counts">
        <lst name="facet_queries"/>
        <lst name="facet_fields">
            <lst name="cat">
                <int name="体育">1</int>
                <int name="商业">3</int>
            </lst>
        </lst>
        <lst name="facet_dates"/>
        </lst>
...
```

按商品上市时间统计最近一个月和最近 3 个月的商品。

Solr 为日期字段提供了更为方便的查询统计方式。与数值分类统计类似,日期分类统计也可以对多个字段进行分类统计。并且针对每个字段都可以单独设置参数。需要注意的是,使用日期分类统计时,必须提供字段名、起始时间、结束时间、时间间隔这 4 个参数。

需要进行日期分类统计的字段名通过 facet.date 参数指定。当然,字段的类型必须是 DateField(或其子类型)。

起始时间通过 facet.date.start 参数指定。时间的一般格式为"1995-12-31T23:59:59Z",另外还可以使用"NOW","YEAR","MONTH"。

结束时间通过 facet.date.end 参数指定。

时间间隔用 facet.date.gap 参数指定。如果 start 为 2009-1-1,end 为 2010-1-1。gap 设置为"+1MONTH"表示间隔 1 个月,那么将会把这段时间划分为 12 个间隔段。注意:"+"因为是特殊字符,所以应该用"%2B"代替。

例如,按出版日期分类统计:

http://localhost:8080/solr/select/?q=*%3A*&version=2.2&start=0&rows=10&indent=on&facet=on&facet.date=pubDate&facet.date.start=2002-1-1T0:0:0Z&facet.date.end=2004-1-1T0:0:0Z&facet.date.gap=%2B1MONTH&facet.date.other=all

facet.date.gap=+1DAY,+2DAY,+3DAY,+10DAY

创建 4 个桶大小,1 天为单位的区间、2 天为单位的区间、3 天为单位的区间和 10 天为单位的区间。

为了实现向下钻取,需要按某一列的值过滤结果。可以使用 fq 参数实现。

分类有层次结构,例如对商品分类可能有 3 层或 4 层。多层分类统计把刻面组织成树形结构。

```
Sneakers
   Men (22)
   Women (43)
```

在用户过滤其中一个后,可以使用一个新的分类统计列显示:

```
Sneakers
  Men
    Size 7 (10)
    Size 8 (11)
    Size 9 (23)
```

5.3.4 高亮

首先必须确保索引库已经保存列的原始值。然后增加如下的参数到查询:

```
hl=true
hl.fl=*
```

下面是一个示例查询:

```
http://localhost:8989/solr/select?q=britta&hl=true&hl.fl=*
```

对应的示例响应可能是:

```
<response>
?
<lst name="responseHeader">
<int name="status">0</int>
<int name="QTime">8</int>
?
<lst name="params">
<str name="q">britta</str>
<str name="hl.fl">*</str>
<str name="hl">true</str>
</lst>
</lst>
?
<result name="response" numFound="1" start="0">
?
<doc>
<str name="description">britta is awesome</str>
<str name="id">poddObject:67</str>
</doc>
</result>
?
<lst name="highlighting">
?
<lst name="poddObject:67">
?
<arr name="description">
<str><em>britta</em> is awesome</str>
</arr>
```

```
</lst>
</lst>
</response>
```

设置高亮相关的参数:

```
query.setHighlight(true); // 开启高亮组件
query.addHighlightField(FIELD_NAME);// 高亮字段
query.addHighlightField(FIELD_DESC);// 高亮字段
query.addHighlightField(FIELD_KEYWORDS);// 高亮字段
query.setHighlightSimplePre("<font color='red'>");//标记,高亮关键字前缀
query.setHighlightSimplePost("</font>");//后缀
query.setHighlightSnippets(1);//结果分片数,默认为1
query.setHighlightFragsize(1000);//每个分片的最大长度,默认为100
query.set("hl.usePhraseHighlighter", true);
query.set("hl.highlightMultiTerm", true);
```

DIY 台式机,首先写好配置,然后再拿到需要的机器。首先设置参数,然后再执行搜索。完整的流程如下:

```
//连接到服务器
String url = "http://59.151.1.71/gongkong/";
SolrServer server = new HttpSolrServer( url );

//设置查询词
SolrQuery query = new SolrQuery("NBA");

//增加高亮参数
HightlightingParams hp = query.getHighlightingParams();
hp.addField("body");

hp.setSimplePre("<font color=red>");
hp.setSimplePost("</font>");

//返回查询结果
QueryResponse response = server.query( query );
```

怎么得到高亮值?根据 id 得到一个 Map<String, List<String>>类型的对象。例如:

```
response.getHighlighting().get(id).get( "body" );
```

完整的代码:

```
for(SolrDocument d : response.getResults()){
    String id = (String)d.getFieldValue("id");
    System.out.println("id:"+id);
    //取得高亮显示的第一段
    String hl = response.getHighlighting().get(id).get( "body" ).get(0);
    System.out.println("hl:"+hl);
}
```

有时候不希望把所有的搜索词全部高亮。例如按地区分类统计查询，当导航到某个地区，可能使用如下的查询条件：

```
QueryWord AND area:areaName
```

可能不希望高亮这里的地区名。为了正确地高亮显示，这样的条件可以使用过滤查询：

```
query.addFilterQuery("area:areaName");
```

Solr API 还包含"facet 查询"的方法。不要混淆刻面查询和刻面的筛选查询。乍一看，它看起来像刻面查询的概念，提供向下钻取的可能性。但实际不是这样的。

刻面查询是一种动态的刻面列，只适用于对数字或日期范围分类项目。例如：如果项目需要分为价格范围，像[100～200 元]，[200～300 元]等。然后刻面查询必须用来"获取所有价格大于 100 元并且小于 200 元的物品数量"。只是指定价格字段作为一个刻面字段是不会有用的，因为它只在搜索结果中返回所有不同的价格列表。在这个例子中，真正提供钻取功能的是刻面查询的概念。

使用语法列声明刻面查询：[start TO end]。在 URL 中，它应该以编码格式存在：

```
facet.query=age:[20+TO+22]
```

在 API 中，它指定为：

```
solrQuery.addFacetQuery("age:[20 TO 22]");
```

URL 中输入：

```
&facet=on
&facet.query=date:[2009-1-1T0:0:0Z TO 2009-2-1T0:0:0Z]
&facet.query=date:[2009-4-1T0:0:0Z TO 2009-5-1T0:0:0Z]
```

返回结果：

```
<lst name="facet_queries">
        <int name="date:[2009-1-1T0:0:0Z TO 2009-2-1T0:0:0Z]">5</int>
        <int name="date:[2009-4-1T0:0:0Z TO 2009-5-1T0:0:0Z]">3</int>
</lst>
```

完整的代码示例：

```
String url = "http://59.50.104.93:8983/solr/docs/";
SolrServer searchServer = new CommonsHttpSolrServer( url ); //连接到服务器

SolrQuery query = new SolrQuery("content:中国"); //设置查询词

// 设置高亮
query.setHighlight(true);
query.addHighlightField("title");
query.addHighlightField("content");
query.setHighlightSimplePre("<font color='red'>");// 高亮标记
query.setHighlightSimplePost("</font>");// 高亮标记
```

```java
QueryResponse queryResponse = searchServer.query(query);

SolrDocumentList docs = queryResponse.getResults();
// 取出值
if (docs == null)
    return;

long recordCount = docs.getNumFound();
System.out.println("found:" + recordCount);

// 高亮显示
Map<String, Map<String, List<String>>> highlightMap = queryResponse
        .getHighlighting();

for (SolrDocument solrDocument : docs) {
    String id = solrDocument.getFieldValue("id").toString();
    System.out.println("id:"+id);
    if(id != null){
        Map<String, List<String>> lighters = highlightMap.get(id);
        for (String key : lighters.keySet()) {
            if(key.equals("title")){
                System.out.println("title:" + lighters.get(key).get(0));
            }
            if(key.equals("content")){
                System.out.println("content:" + lighters.get(key).get(0));
            }
        }
    }
}
```

5.3.5 同义词

使用 SOLR 加入中文同义词需要把 synonyms.txt 的默认编码改成与自己系统使用的编码一致。一般来说，系统的编码都是用 UTF-8，那么就要把 synonyms.txt 这个文件的编码格式转换成 UTF-8。

5.3.6 嵌入式 Solr

Embedded Solr 提供了不使用 HTTP 连接的索引和搜索接口。但是并不灵活，扩展性也不好。仅在必要的时候才用。

可以用嵌入式 Solr 来做集成测试。Solr 结构有点复杂，所以不用 Tomcat 直接在 Eclipse 中测试。

首先增加和 Solr 相关的 jar 引用。具体来说，就是在 Solr 的 Web 应用中的 jar 包。

```java
System.setProperty("solr.solr.home", "/home/example/solr");
CoreContainer.Initializer initializer = new CoreContainer.Initializer();
```

```
CoreContainer coreContainer = initializer.initialize();
EmbeddedSolrServer server = new EmbeddedSolrServer(coreContainer, "");
```

增加文档:

```
SolrInputDocument doc1 = new SolrInputDocument();
doc1.addField( "id", "1", 1.0f );
doc1.addField( "name", "玛瑙宝石真漂亮", 1.0f );
doc1.addField( "price", 10 );
SolrInputDocument doc2 = new SolrInputDocument();
doc2.addField( "id", "2", 1.0f );
doc2.addField( "name", "MM 也很漂亮, 天地真爱可见", 1.0f );
doc2.addField( "date", "2011-8-21 14:02:41", 1.0f );
doc2.addField( "price", 20 );
SolrInputDocument doc3 = new SolrInputDocument();
doc3.addField( "id", "3", 1.0f );
doc3.addField( "name", "simple is beauty 风领导好", 1.0f );
doc3.addField( "date", "2011-8-29 14:03:01", 1.0f );
doc3.addField( "price", 1000 );
Collection<SolrInputDocument> docs = new ArrayList<SolrInputDocument>();
docs.add( doc1 );
docs.add( doc2 );
docs.add( doc3 );
server.deleteByQuery( "*:*" );// delete everything!
server.add(docs);
server.commit();
```

5.3.7 Spring 实现的搜索界面

通过 Spring 应用程序上下文配置 HttpSolrServer 最简单的方式是定义一个 SolrServer 的 Bean。然后把它注入到想要使用的类中。Spring 配置文件例子如下:

```
<bean id="solrServer"
    class="org.apache.solr.client.solrj.impl.HttpSolrServer" >
    <constructor-arg>
        <value>http://localhost:8080/solr/core0</value>
    </constructor-arg>
</bean>
```

更完整的设置 solrServer 参数的例子:

```
<bean id="solrServer"
class="org.apache.solr.client.solrj.impl.HttpSolrServer">
<constructor-arg value="${solr.serverUrl}"/>
<property name="connectionTimeout" value="${solr.connectionTimeout}"/>
<property name="defaultMaxConnectionsPerHost"
value="${solr.defaultMaxConnectionsPerHost}"/>
```

```
<property name="maxTotalConnections"
value="${solr.maxTotalConnections}"/>
</bean>
```

在服务层使用这个叫作"solrServer"的 Spring Bean。这里使用 Spring IOC 机制，创建 org.apache.solr.client.solrj.SolrServer 对象作为服务中的一个成员变量。使用如下的代码和 Solr 打交道：

查询关键词：

```
SolrQuery query = new SolrQuery();
query.setQuery(search);   //查询词是 search
QueryResponse qr = solrServer.query(query);
return qr.getBeans(SearchItem.class);
```

提交项目：

```
solrServer.addBean(item);
solrServer.commit();
```

删除项目：

```
solrServer.deleteById(id);
solrServer.commit();
```

也可以使用 Spring Data Solr 项目。这个项目的地址是：https://github.com/SpringSource/spring-data-solr/。Spring Data Solr 是 Spring Data 的子项目。Spring Data Solr 实现了 Spring Data 访问 Solr 存储并提供了 Spring Data JPA(Java 持久化 API)模型的访问方式。此外，Spring Data Solr 提供了一个更底层的 SolrTemplate，以方便启动嵌入式的 Solr 服务器。SolrTemplate 是 Solr 操作的核心支持类。

Spring Data 提供了一套数据访问层(DAO)的解决方案，致力于减少数据访问层的开发量。它使用一个叫作 Repository 的接口类作为基础。Repository 定义如下：

```
public interface Repository<T, ID extends Serializable> {
}
```

Repository 是访问底层数据模型的超级接口。而对于某种具体的数据访问操作，则在其子接口中定义。例如，Spring Data Solr 项目中定义了访问 Solr 中数据的 SolrRepository 接口。

所有继承 Repository 接口的界面都由 Spring 管理，此接口作为标识接口，功能就是用来控制领域模型。Spring Data 可以让我们只定义接口，只要遵循 Spring Data 的规范，就无须写实现类。Spring 可以根据接口中定义的方法名实现 Repository。

用一个简单的实例说明如何使用 Spring Data Solr。创建一个叫作 HttpSolrContext 的类并用@Configuration 注解标注这个类。用@EnableSolrRepositories 注解标注这个类来启用 Spring Data Solr 存储，并且配置 Solr 存储的根包。

```
@EnableSolrRepositories("com.lietu.spring.datasolr.todo.repository.solr")
```

使用@PropertySource 注解标注这个类，并把值设置成"classpath:application.properties"。用这个值配置属性文件的位置，并且增加一个 PropertySource 到 Spring 的环境。

第 5 章 Solr 分布式搜索引擎

加一个 Environment 列到这个类，并用@Resource 标注该列。注入的 Environment 用来取得加到属性文件的属性。

```
@Resource
private Environment environment;
```

创建一个叫作 solrServerFactoryBean()的方法，并且用@Bean 注解标注这个方法。这个方法的实现创建一个新的 HttpSolrServerFactoryBean 对象，然后设置 Solr server url 的值，并返回这个创建的对象。

创建一个叫作 solrTemplate()的方法，并且用@Bean 注解标注这个方法。此方法的实现创建一个新的 SolrTemplate 对象，并传递使用过的 SolrServer 实现作为构造函数的参数。

```
@Bean
public SolrTemplate solrTemplate() throws Exception {
    return new SolrTemplate(solrServerFactoryBean().getObject());
}
```

HttpSolrContext 完整的源代码如下：

```
import org.springframework.context.annotation.Bean;
import org.springframework.context.annotation.Configuration;
import org.springframework.context.annotation.Profile;
import org.springframework.core.env.Environment;
import org.springframework.data.solr.core.SolrTemplate;
import org.springframework.data.solr.repository.config.EnableSolrRepositories;
import org.springframework.data.solr.server.support.HttpSolrServerFactoryBean;

import javax.annotation.Resource;

@Configuration
@EnableSolrRepositories("com.lietu.spring.datasolr.todo.repository.solr")
@PropertySource("classpath:application.properties")
public class HttpSolrContext {

    @Resource
    private Environment environment;

    @Bean
    public HttpSolrServerFactoryBean solrServerFactoryBean() {
        HttpSolrServerFactoryBean factory = new HttpSolrServerFactoryBean();

        factory.setUrl(environment.getRequiredProperty("solr.server.url"));

        return factory;
    }

    @Bean
```

```
    public SolrTemplate solrTemplate() throws Exception {
        return new SolrTemplate(solrServerFactoryBean().getObject());
    }
}
```

通过如下步骤为 HTTP Solr 服务器创建一个 XML 配置文件：
- 通过使用上下文命名空间的属性占位符元素配置使用的属性文件；
- 启用 Solr 的索引库，并使用 Solr 命名空间的存储元素配置 Solr 索引库的基础包；
- 利用 SOLR 命名空间的 solr-server 元素配置 HTTP Solr 服务器 Bean。设置 Solr 服务器的 URL；
- 配置 Solr 模板 Bean。设置配置好的 HTTP Solr 服务器 Bean 作为构造函数的参数。

exampleApplicationContext-solr.xml 文件内容如下：

```
<?xml version="1.0" encoding="UTF-8"?>
<beans xmlns="http://www.springframework.org/schema/beans"
    xmlns:xsi="http://www.w3.org/2001/XMLSchema-instance"
    xmlns:context="http://www.springframework.org/schema/context"
    xmlns:solr="http://www.springframework.org/schema/data/solr"
    xsi:schemaLocation="http://www.springframework.org/schema/beans
http://www.springframework.org/schema/beans/spring-beans.xsd
    http://www.springframework.org/schema/context
http://www.springframework.org/schema/context/spring-context-3.2.xsd
http://www.springframework.org/schema/data/solr
http://www.springframework.org/schema/data/solr/spring-solr.xsd">

    <context:property-placeholder location="classpath:application.properties"/>

    <!-- Enable Solr repositories and configure repository base package -->
    <solr:repositories base-package="com.lietu.spring.datasolr.todo.repository.solr"/>

    <!-- Bean definitions -->
    <beans>
        <!-- Configures HTTP Solr server -->
        <solr:solr-server id="solrServer" url="${solr.server.url}"/>

        <!-- Configures Solr template -->
        <bean id="solrTemplate" class="org.springframework.data.solr.core.SolrTemplate">
            <constructor-arg index="0" ref="solrServer"/>
        </bean>
    </beans>
</beans>
```

为了把文档加入到 Solr 索引，接下来创建一个文档类。文档类基本上是一个按照如下规则实现的POJO：
- @Field注解用于在POJO列和Solr文档列之间建立一个连接；
- 如果Bean的列名称不等于文档的列名称，则文档列的名称必须作为@Field注解的值给出；
- 可以用@Field注解标注一个字段或setter()方法。

Spring Data Solr假设文档默认id字段的名称是"id"。可以使用@Id注解标注id字段来覆盖此设置。创建一个叫作TodoDocument的类来索引要完成的任务条目。添加id字段到TodoDocument类，并用@Field注解标注该列。添加description字段到TodoDocument类，并用@Field注解标注该列。添加title字段到TodoDocument类，并用@Field注解标注该列。给TodoDocument类的字段创建getter()方法。

创建一个叫作Builder的静态内部类用来构建新的TodoDocument对象。添加一个静态getBuilder()方法到TodoDocument类。此方法的实现返回一个新的TodoDocument.Builder对象。

TodoDocument类的源代码如下所示：

```java
import org.apache.solr.client.solrj.beans.Field;
import org.springframework.data.annotation.Id;

public class TodoDocument {

    @Id
    @Field
    private String id;

    @Field
    private String description;

    @Field
    private String title;

    public TodoDocument() {

    }

    public static Builder getBuilder(Long id, String title) {
        return new Builder(id, title);
    }

    //忽略了Getters()方法

    public static class Builder {
        private TodoDocument build;

        public Builder(Long id, String title) {
```

```
            build = new TodoDocument();
            build.id = id.toString();
            build.title = title;
        }

        public Builder description(String description) {
            build.description = description;
            return this;
        }

        public TodoDocument build() {
            return build;
        }
    }
}
```

接下来创建存储接口。Spring Data Solr 存储的基本接口是 SolrCrudRepository<T, ID>接口，并且每个存储接口必须扩展这个接口。

当扩展 SolrCrudRepository<T, ID>接口时，必须给出 T 和 ID 两个类型参数。其中，T 类型参数表示文档类的类型；ID 类型参数表示文档 id 的类型。

可以按照以下步骤创建存储接口：

- 创建一个叫作 TodoDocumentRepository 的接口；
- 扩展 SolrCrudRepository 接口，并用文档类类型和它的 ID 类型(字符串)作为类型参数。

TodoDocumentRepository 接口的源代码如下所示：

```
import org.springframework.data.solr.repository.SolrCrudRepository;

public interface TodoDocumentRepository
        extends SolrCrudRepository<TodoDocument, String> {
}
```

下一步是使用创建的 Solr 存储来创建服务。首先创建一个叫作 TodoIndexService 的服务接口，然后实现这个接口。TodoIndexService 接口的源代码如下所示：

```
public interface TodoIndexService {

    public void addToIndex(Todo todoEntry);    //增加文档到索引

    public void deleteFromIndex(Long id);    //从索引删除文档
}
```

接下来，实现创建出来的接口。通过以下步骤实现服务接口：

- 创建一个服务类的框架实现；
- 实现把文档添加到 Solr 索引的方法；
- 实现从 Solr 的索引中删除文档的方法。

第 5 章 Solr 分布式搜索引擎

接下来详细讲解如何创建一个服务类的框架实现。

创建一个叫作 RepositoryTodoIndexService 的类，并用@Service 注解标注这个类。这个注解把这个类标记成为一个服务，并确保在类路径扫描中检测到这个类。

增加一个 TodoDocumentRepository 字段到 RepositoryTodoIndexService 类，并用 @Resource 注解标注这个字段。这个注解指示 Spring IoC 容器把实际的存储实现注入到服务的存储列。

虚拟服务实现的源代码如下所示：

```java
import org.springframework.stereotype.Service;
import org.springframework.transaction.annotation.Transactional;

import javax.annotation.Resource;

@Service
public class RepositoryTodoIndexService implements TodoIndexService {

    @Resource
    private TodoDocumentRepository repository;

    //Add methods here
}
```

创建把文档添加到 Solr 索引的方法。

在 RepositoryTodoIndexService 类中添加 addToIndex()方法，然后使用@Transactional 注解标记这个方法。这将确保 Spring Data Solr 参与 Spring 的事务管理。通过调用 TodoDocumentRepository 接口的 save()方法添加文档到 Solr 索引。创建方法的源代码如下所示：

```java
import org.springframework.stereotype.Service;
import org.springframework.transaction.annotation.Transactional;

import javax.annotation.Resource;

@Service
public class RepositoryTodoIndexService implements TodoIndexService {

    @Resource
    private TodoDocumentRepository repository;

    @Transactional
    @Override
    public void addToIndex(Todo todoEntry) {
        TodoDocument document = TodoDocument.getBuilder(todoEntry.getId(),
                todoEntry.getTitle())
                .description(todoEntry.getDescription())
                .build();
```

```
        repository.save(document);
    }

    //Add deleteFromIndex() method here
}
```

使用如下步骤创建从 Solr 索引删除文档的方法：
- 添加 deleteFromIndex() 方法到 RepositoryTodoDocumentService 类，并用 @Transactional 注解标记此方法。这将确保 Spring Data Solr 存储参与 Spring 的事务管理。
- 通过调用 TodoDocumentRepository 接口的 delete()方法从 Solr 的索引中删除文件。

创建的方法源代码如下所示：

```
import org.springframework.stereotype.Service;
import org.springframework.transaction.annotation.Transactional;

import javax.annotation.Resource;

@Service
public class RepositoryTodoIndexService implements TodoIndexService {

    @Resource
    private TodoDocumentRepository repository;

    //Add addToIndex() method here

    @Transactional
    @Override
    public void deleteFromIndex(Long id) {
        repository.delete(id.toString());
    }
}
```

通过使用查询方法实现搜索功能。可以使用以下技术创建 Spring Data Solr 的查询方法：
- 根据方法名生成查询；
- 命名查询；
- @Query 注解。

这 3 种方法中，根据方法名生成查询是最简单的一种方法。这里先只介绍这种方法，其他方法可以查看 Spring Data Solr 的说明文档。从方法名生成查询是一个查询生成策略，从查询方法的名称解析出来要执行的查询。

查询方法的名称必须用一个特殊的前缀标识查询方法。这些前缀是：find, findBy, get, getBy, read 和 readByreadBy。当解析要执行的查询时，从方法名称中剥离出这个前缀。

用属性表达式指出文档类的属性。可以组合多个属性表达式，方法是：在它们之间加入 And 或 Or 关键字。

查询方法的参数数量必须与在名称中使用的属性表达式的数量相等。例如：TodoDocumentRepository 接口的源代码中的 findByTitleContainsOrDescriptionContains()方法包括 title 和 description 两个参数。

TodoDocumentRepository 接口的源代码如下所示：

```
import org.springframework.data.solr.repository.SolrCrudRepository;
import java.util.List;

public interface TodoDocumentRepository
        extends SolrCrudRepository<TodoDocument, String> {

    public List<TodoDocument>
findByTitleContainsOrDescriptionContains(String title,
                                         String description);
}
```

注意：如果搜索词包含多个单词，那么这个查询方法就行不通了。

使用已经创建好的查询方法。首先在 TodoIndexService 接口中声明 search()方法，然后添加 search()方法的实现到 RepositoryTodoIndexService 类。

调用 TodoDocumentRepository 接口的 findByTitleContainsOrDescriptionContains()方法返回一个 TodoDocument 对象组成的列表。

RepositoryTodoIndexService 类的相关部分如下所示：

```
import org.springframework.stereotype.Service;
import javax.annotation.Resource;
import java.util.List;

@Service
public class RepositoryTodoIndexService implements TodoIndexService {

    @Resource
    private TodoDocumentRepository repository;

    @Override
    public List<TodoDocument> search(String searchTerm) {
        return
repository.findByTitleContainsOrDescriptionContains(searchTerm,
searchTerm);
    }
}
```

5.3.8 索引分发

为了支持更多的搜索访问量或者避免单点失败，使索引服务有更高的可用性，需要分发索引到多个服务器上来同时提供服务。利用 rsync 压入和弹出索引快照不会对服务器性能有大的影响。下面介绍如果有一个已经存在的 Lucene 索引，如何把这个索引从主节点复制到从节点。

在每个 Linux 服务器上创建路径：

```
/usr/local/solr/
```

这个路径下包含一些重要的可执行脚本：
主节点需要的脚本如下：

- rsyncd-enable；
- rsyncd-disable；
- rsyncd-start；
- rsyncd-stop；
- snapshooter；
- snapcleaner。

从节点需要的脚本如下：

- snappuller-enable；
- snappuller-disable；
- snappuller；
- snapinstaller；
- snapcleaner。

然后在 Solr 文件夹下面创建一个 conf 目录，conf 目录下放一个配置文件：scripts.conf。
主节点里面的 scripts.conf 的配置文件类似这样：

```
solr_hostname=localhost
solr_port=8983
rsyncd_port=18983
data_dir=<lucene 索引路径>
webapp_name=solr
master_host=
master_data_dir=
master_status_dir=
```

注意：如果 index 的目录路径为：/usr/local/lucene/search/index/，则配置文件里 data_dir 路径应该为：/usr/local/lucene/search/。

从节点里面的 scripts.conf 的配置文件类似这样：

```
user=
solr_hostname=<主节点 ip 地址>
solr_port=8993
rsyncd_port=18993
data_dir=<从节点的 lucene 索引路径>
webapp_name=solr
master_host=<主节点 ip 地址>
master_data_dir=<主节点的 lucene 索引路径>
master_status_dir=<主节点的 solr 路径>/logs/clients/
```

例如这里 master_status_dir 可能是：/usr/local/solr/logs/clients/。

通过如下的步骤即可开始复制索引：
(1) 在主节点里，运行 solr/bin 目录下面的 Linux 命令：

```
#rsyncd-enable
#rsyncd-start
```

执行这两个命令后在 solr/logs 目录下将会创建一个 pid 文件和一个 log 文件。
(2) 为了做一个索引的快照，只需要在主节点运行：

```
#snapshooter
```

现在可以在 crontab 里进行配置 snapshooter 命令，指定在一天的某几个时间运行。当新创建一个快照时，索引必须是完整的，也就意味着如果正在写入索引时，不能运行 snapshooter 命令。

如果运行 snapshooter 时，增加-c 参数，只有当最后一个快照改变时，才创建一个新的 snapshooter。为了防止磁盘被快照塞满，可以经常运行 snapcleaner 命令。运行 snapshooter –N 2 将会移除所有旧的快照，只保留两个最新的快照。

(3) 在从节点，需要执行如下的脚本拉回并且安装快照：

```
#snappuller-enable
#snappuller -P 18983
#snapinstaller
```

注意：snappuller 后面的端口号是在主节点配置的 rsyncd_port。同时在从节点应该调度执行 snapcleaner。

(4) 现在在从节点调度执行 snappuller 和 snapinstaller 的问题就剩下到主节点里做身份认证了。为了解决这个问题，可以添加 ssh-keys 到主节点的认证文件中。

在从节点执行：

```
#ssh-keygen -t dsa
```

执行 ssh-keygen 命令后，认证文件生成到 home 目录下的.ssh 目录。
用 scp 命令把从节点上的 id_dsa.pub 文件复制到主节点：

```
#scp id_dsa.pub usrname@masternode:./id_dsa.pub
```

在主节点的.ssh 目录执行：

```
#touch authorized_keys
#chmod 600 authorized_keys
#cat ../id_dsa.pub >> authorized_keys
```

5.3.9　Solr 搜索优化

如果要在做索引时把不同的列设置不同的权重：

```
<field name="myBoostedField" boost="7.0">value1</field>
<field name="myBoostedField" boost="8.0">value2</field>
<field name="myBoostedField" boost="4.0">value3</field>
```

发送数据的写法：

```
XML.writeXML(xmlContent, "field", value1, "name",
                          " myBoostedField","boost","7.0");
```

也可以在搜索的时候动态加权。对于标准的请求处理器，对标题列加权的例子：q=title:superman^2 subject:superman。

使用 dismax 请求处理器，可以在参数 qf 中对指定的列申明加权，例如：q=superman&qf=title^2 subject。

使用函数式查询实现日期和相关度混合排序：

```
queryWord += "
AND _val_:\"linear(recip(rord(timestamp),1,10000,10000),10000000,0)\"";
```

在索引中出现"头疼 药"和"头疼药"这样的内容，其分词都是"头疼"和"药"，前者有搜索结果，后者无搜索结果。这样的原因是 Solr 的查询词解析类使用的是短语匹配的方式。"头疼药"只按照"短语匹配"搜索是不恰当的，应该是先短语匹配，后按照分词进行垂直搜索。为了实现这样的效果，修改 SolrQueryParser 类。

```
protected Query getFieldQuery(String field, String queryText) throws
ParseException {
    //如果碰到"-"，则把查询当成内部函数处理
    if (field.equals("_val_")) {
      return QueryParsing.parseFunction(queryText, schema);
    }
    //缺省执行普通的字段查询
    TokenStream source = this.getAnalyzer().tokenStream(field, new
StringReader(queryText));
    ArrayList<Token> v = new ArrayList<Token>(10);
    Token t;
    while (true) {
       try {
          t = source.next();
       }
       catch (IOException e) {
          t = null;
       }
       if (t == null)
          break;
       v.add(t);
    }
    try {
       source.close();
    }
    catch (IOException e) {
       // 省略
    }
    if (v.size() == 0)
```

```
            return null;
    else if (v.size() == 1)
        return new TermQuery(new Term(field, ((Token)v.get(0)).termText()));
    else {
        PhraseQuery q = new PhraseQuery();
        BooleanQuery b = new BooleanQuery();
        q.setBoost(2048.0f);
        b.setBoost(0.001f);
        for (int i = 0; i < v.size(); i++) {
            Token token = v.get(i);
            q.add(new Term(field, token.termText()));
            TermQuery tmp = new TermQuery(new Term(field, token.termText()));
            tmp.setBoost(0.01f);
            b.add(tmp, BooleanClause.Occur.MUST);
        }
        BooleanQuery bQuery = new BooleanQuery();
        // 用OR条件合并两个查询
        bQuery.add(q,BooleanClause.Occur.SHOULD);
        bQuery.add(b,BooleanClause.Occur.SHOULD);
        return bQuery;
    }
}
```

如果需要限制这里的 BooleanQuery bQuery 的查询范围还可以采用：

```
if(!stopwords.contains(token.termText())){
    b.add(tmp, BooleanClause.Occur.MUST);//OR 连接
    System.out.println("add tmp:"+tmp);
}
```

在通常的搜索中用户可以按内容相关度排序，或者按照日期逆序排序。

```
queryWord;createtime desc
```

还可以按 sort 参数排序：

```
inStock desc, price asc
```

SolrJ 中按时间列排序的例子：

```
query.addSortField("timestamp", ORDER.desc);
```

经常通过一个选项来切换这两种排序方式。为了更加简化搜索界面，可以综合内容相关度排序和日期排序。基本的方法是设置时间加权，在考虑到搜索词相关性的同时还考虑到搜索结果的时间。例如：索引库中有 3 个字段：标题、内容、日期。然后组合搜索"内容+标题"，然后按时间排序。这样存在一个问题：比如搜索"手机"，其结果如果不按日期排序，那结果的相关性很好；如果按日期排序，有些商情的内容里面会留些联系人的手机号这种字样，但是其实这条商情并不是卖手机的，有可能是卖衣服的。可这条商情的权重还是会排到卖手机的上面。Solr 通过函数查询来实现时间加权排序。

- OrdFieldSource 实现了 ord(myfield)函数。

- ReverseOrdFieldSource 实现了 rord(myfield)函数。
- LinearFloatFunction 实现了数值列的 linear(myfield,1,2) 函数。
- MaxFloatFunction 实现了数值列或常量的 max(linear(myfield,1,2),100) 函数。
- ReciprocalFloatFunction 实现了数值列的 recip(myfield,1,2,3) 函数。

实际搜索"NBA"这个词的时候，使用时间加权的例子：

```
+_val_:"linear\(recip\(rord\(timestamp\),1,10000,10000\),10000000,0\)" NBA
```

其基本原理是用索引中"timestamp"列的值来影响排序结果。

在优化索引阶段，Solr 需要大约 2 倍于索引大小的临时空间。因此需要大约 400 GB 的硬盘来容纳 200 GB 的索引。

5.4 从 FAST Search 移植到 Solr

微软在 2008 年收购 FAST 搜索公司。FAST 搜索作为一个索引服务运行。因为 FAST 搜索的 Linux 版本已经停止开发，所以有必要移植到 Solr。为了移植到 Solr，需要修改提供数据端和查询端接口。

FAST 有丰富的特征集合(语言，导航，连接器等)，并不是所有的特征都在 Lucene/Solr 中已经实现，或者实现的方式不一样。不同于 Solr 的以 Field 为中心的分析，FAST 搜索的文档处理器在索引文档前转换文档。

例如下面的特征在 Solr 中没有很好地实现：

- 编码归一化，语言识别；
- 文本抽取(HTML、PDF、MS Office 等)；
- 分词，原型化(lemmatization)，实体提取；

两者的术语对照见表 5-1。

表 5-1 术语对照表

Lucene/Solr	FAST
Replica	Search Row
Shard	Column
Facet	Navigator
Spellcheck	Did you mean
Update processor	Document processor
Request Handler	Query Transformer(QT)
Response Writer	Result Processor(RP)/TWM
Schema	Index profile
Index segment	Index partition
Multi core	Multi cluster
用同样的方式处理的文档	Collection

第 5 章　Solr 分布式搜索引擎

移植的步骤如下。

(1) 评估一下当前的特征和体系结构：保留所有的特征还是要增加新的特征？

(2) 安装 Solr 并且迭代式开发：编写 solrconfig.xml 和 schema.xml，导出并索引一些真实的数据，测试各种查询。

(3) 设计覆盖所有移植领域的说明：模式、内容、装载数据并分析；前端界面，查询和 API；管理和操作性的功能。

(4) 实现移植到 Solr。

从 FAST 搜索的索引配置到 Solr 的 schema.xml 移植的例子。FAST 列的例子：

```
<field name="postcode">
<field name="postalplace">
<field name="iprice" type="int32" fullsort="yes"/>
<field name="viewprice" type="int32"/>
```

Solr 中的等价声明：

```
<field name="postcode" type="string" indexed="true" stored="true">
<field name="postalplace" type="text_ws" indexed="true" stored="true">
<field name="iprice" type="int" indexed="true" stored="true"/>
<field name="viewprice" type="int" indexed="true" stored="true"/>
```

Solr 缺少成熟的处理管线。可以使用 https://issues.apache.org/jira/browse/SOLR-1725 中的更新处理器或者 OpenPipeline(http://www.openpipeline.org/)来模拟。

需要特别注意的是，FAST 搜索更好地支持多种自然语言，大量地使用了文档处理，包括实体提取。

5.5　Solr 扩展与定制

Solr 的实现很灵活，可以定制输入输出或者增加内部功能。org.apache.solr.common.params.CommonParams 类中定义了参数的名字。例如取得查询参数可以通过下面这行代码：

```
String q = params.get( CommonParams.Q );
```

5.5.1　缺省查询

无查询词的时候，返回一个该类别下的文章列表。q=*:*,在过滤参数中设置文档类别。当索引量大的时候，会消耗 Solr 的缓存。解决方法是：绕过 Solr，直接查询 Lucene，这样不会占用 Solr 缓存。

SolrIndexSearcher searcher = req.getSearcher();
IndexReader reader = searcher.getReader();
然后调用 searcher.search(query, collector)。

5.5.2 插件

由于 Solr 默认的 Query Parser 生成的 Query 一般是 PhraseQuery，导致只有很精确的结果才被搜索出来。大多时候我们需要分词后的 BooleanQuery。这时候，可以创造一个查询解析器插件。查询解析器插件都是 QParserPlugin 的子类。如果有自定义解析需求，可能需要扩展这个类来创建自己的查询解析器。

插件代码加载到 Solr 的方法是：将类放到一个 JAR 文件，然后配置 Solr，让它知道如何找到这些类。可以把你的 jar 文件放在 SolrCore 的 instanceDir 的 lib 目录。在示例程序中，位置是：example/solr/lib。在发布包中并不存在这个目录，所以需要先做 mkdir。

例如：JoinQParser 扩展 QParserPlugin。它允许用连接操作规范文档之间的关系。

这和在关系数据库中的 join 是不同的概念，因为并没有真正的连接信息。一个适当的 SQL 类比是一个"内部查询"。例如：

找到所有含有"ipod"这个词的产品，连接制造商的文档并返回制造商的名单：

```
{!join from=manu_id_s to=id}ipod
```

5.5.3 Solr 中字词混合索引

在 Lucene 的介绍中已经提到了实现字词混合索引的方法。在此基础上增加 FilterFactory。

```
public class SingleFilterFactory extends BaseTokenFilterFactory {
    public SingleFilter create(TokenStream input) {
            return new SingleFilter(input);
        }
}
```

在 schema.xml 中定义 text 列类型如下：

```xml
<fieldType name="text" class="solr.TextField" positionIncrementGap="100">
    <analyzer type="index">
      <tokenizer class="CnTokenizerFactory"/>
      <filter class="solr.SingleFilterFactory"/>
    </analyzer>
    <analyzer type="query">
      <tokenizer class="CnTokenizerFactory"/>
      <filter class="solr.SingleFilterFactory"/>
    </analyzer>
</fieldType>
```

修改 org.apache.solr.search 包中的 SolrQueryParser 类：

```
//缺省执行普通的字段查询
TokenStream source = this.getAnalyzer().tokenStream(field,
                                    new StringReader(queryText));
```

```java
ArrayList<Token> v = new ArrayList<Token>(10);
Token t;
while (true) {
   try {
       t = source.next();
   }
   catch (IOException e) {
       t = null;
   }
   if (t == null)
      break;
   v.add(t);
}
try {
   source.close();
}
catch (IOException e) {
   //忽略
}

if (v.size() == 0)
   return null;
else if (v.size() == 1)
   return new TermQuery(new Term(field, ((Token)v.get(0)).termText()));
else {
   PhraseQuery q = new PhraseQuery();
   q.setBoost(2048.0f);
   ArrayList<SpanQuery>s=new ArrayList<SpanQuery>(v.size());
   for (int i = 0; i < v.size(); i++) {
      Token token = v.get(i);
      if(token.getPositionIncrement()>0) {
          q.add(new Term(field, token.termText()));
      }
      if(token.termText().length()==1) {
         SpanTermQuery tmp =
             new TermQuery(new Term(field, token.termText()));
         s.add(tmp);
      }
   }

   BooleanQuery bQuery = new BooleanQuery();
   // 用 OR 条件合并两个查询
   bQuery.add(q,BooleanClause.Occur.SHOULD);
   if(s.size()>0) {
     SpanNearQuery nearQuery =
         new SpanNearQuery(s.toArray(new SpanQuery[s.size()]),s.size(),true);
```

```
        nearQuery.setBoost(0.001f);
        bQuery.add(nearQuery,BooleanClause.Occur.SHOULD);
    }
    return bQuery;
}
```

如果要向词表中增加词，则旧的按词切分会不准确。解决方法是：先查找出包含相关词的文档，然后删除文档，最后增加文档。

5.5.4 相关检索

经常需要实现相似检索的功能。例如输入一篇文章，返回相关的 5 篇文章。Solr 中包括了一个 MoreLikeThis 的处理模块。在 solrconfig.xml 中包括 MoreLikeThisHandler 的定义：

```xml
<requestHandler name="/mlt" class="solr.MoreLikeThisHandler">
  <lst name="defaults">
    <str name="mlt.fl">title,abstract</str>
    <int name="mlt.mindf">1</int>
  </lst>
</requestHandler>
```

mlt.fl 中包括提取关键词的缺省列。

搜索的时候输入类似：

```
http://www.lietu.com:8080/solr/select/?q=%E6%B1%BD%E8%BD%A6&version=2.2&start=0&rows=1&mlt=true&mlt.mindf=1&mlt.mintf=1&mlt.fl=title,abstract
```

返回的结果后面包括了相似匹配的结果：

```xml
<lst name="moreLikeThis">
        <result name="521838" numFound="6586103" start="0">
    <doc>
    <str name="abstract">
本文通过分析世界汽车工业的发展趋势，总结出 21 世纪世界汽车工业发展的八大趋势，主要表现在汽车产业组织、汽车生产方式、汽车产品及造型和汽车可持续发展战略上。
</str>
<str name="cn_institution">武汉理工大学</str>
<str name="en_abstract"/>
        <str name="en_title">
Eight Developmont Trends of The World Automobile Industry in The 21st Century
</str>
<str name="id">5141078</str>
<str name="pin_yin_name">杨瑞海,韩雄辉</str>
<str name="title">21 世纪世界汽车工业发展的八大趋势</str>
<str name="user_real_name">杨瑞海,韩雄辉</str>
</doc>
…
</result>
</lst>
```

这个缺省的相关检索仍然有待改进，因为 MoreLikeThis 查询的准确性依赖于提取关键词的准确性，为了提取关键词，首先要做的工作是去掉 StopWord。MoreLikeThis 类缺省使用的是英文的 StopWord。下面我们修改 MoreLikeThisHandler，让它从外部读取 StopWord.txt。

```xml
<requestHandler name="/mlt" class="solr.MoreLikeThisHandler">
        <lst name="defaults">
            <str name="mlt.fl">manu,cat</str>
            <int name="mlt.mindf">1</int>
        </lst>
<str name="stopWordFile">stopwords.txt</str>
</requestHandler>
```

MoreLikeThisHandler 类的实现代码修改如下：

```java
private static Set stopWords = null;

    public void init(NamedList args) {
   super.init(args);
    SolrParams p = SolrParams.toSolrParams(args);
    String stopWordFile = p.get("stopWordFile");
 if(stopWordFile == null)    {
   stopWords = StopFilter.makeStopSet(StandardAnalyzer.STOP_WORDS);
       return;
   }

    List<String> wlist;
 try {
        wlist = Config.getLines(stopWordFile);
   } catch (IOException e) {
       throw new SolrException( SolrException.ErrorCode.NOT_FOUND,
           "MoreLikeThis requires a stop word list " );
   }
    stopWords = StopFilter.makeStopSet((String[])wlist.toArray(new String[0]), true);
}
…
    mlt.setStopWords(stopWords);
…
```

5.5.5 搜索结果去重

折叠对于一个给定列的相同或相似值的搜索结果叫作 collapsing。例如对同一站点搜索结果的折叠，经常也会在这个搜索结果上加上"……中还有几条相关信息"。如图 5-3 所示。这需要我们给 Solr 增加折叠的功能。首先我们假定折叠的列是未分词的。

https://issues.apache.org/jira/browse/SOLR-236 正是关于这个问题的解决方法。先取得这个版本：

```
# svn export -r592129 http://svn.apache.org/repos/asf/lucene/solr/
# patch -u -p0 < field-collapsing-extended-592129.patch
```

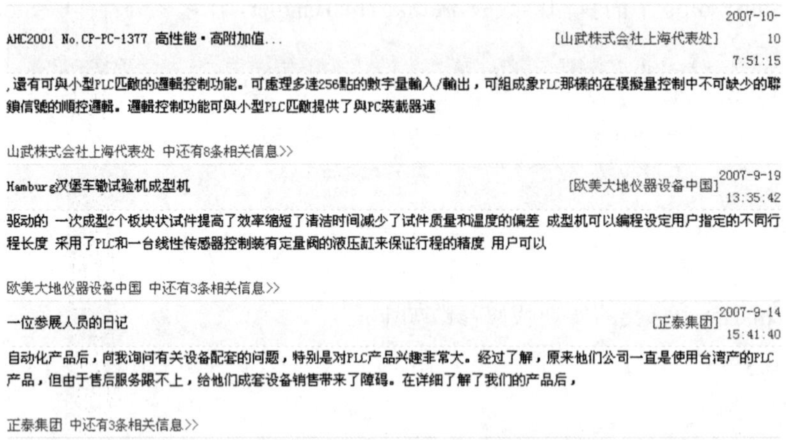
AHC2001 No.CP-PC-1377 高性能・高附加值... [山武株式会社上海代表处] 2007-10-10 7:51:15

,還有可與小型PLC匹敵的邏輯控制功能。可處理多達256點的數字量輸入/輸出，可組成象PLC那樣的在模擬量控制中不可缺少的聯鎖信號的順控邏輯。邏輯控制功能可與小型PLC匹敵提供了與PC裝載器連

山武株式会社上海代表处 中还有8条相关信息>>

Hamburg汉堡车辙试验机成型机 [欧美大地仪器设备中国] 2007-9-19 13:35:42

驱动的 一次成型2个板块状试件提高了效率缩短了清洁时间减少了试件质量和温度的偏差 成型机可以编程设定用户指定的不同行程长度 采用了PLC和一台线性传感器控制装有定量阀的液压缸来保证行程的精度 用户可以

欧美大地仪器设备中国 中还有3条相关信息>>

一位参展人员的日记 [正泰集团] 2007-9-14 15:41:40

自动化产品后，向我询问有关设备配套的问题，特别是对PLC产品兴趣非常大。经过了解，原来他们公司一直是使用台湾产的PLC产品，但由于售后服务跟不上，给他们成套设备销售带来了障碍。在详细了解了我们的产品后，

正泰集团 中还有3条相关信息>>

图 5-3 搜索结果去重效果图

搜索测试：

```
http://localhost:8080/select/?q=words_t%3AApple&version=2.2&start=0&rows
=10&indent=on&collapse.field=t_s& collapse.threshold=1
```

返回的结果是：

```
<?xml version="1.0" encoding="UTF-8"?>
<response>

<lst name="responseHeader">
      <int name="status">0</int>
 <int name="QTime">0</int>
 <lst name="params">
  <str name="start">0</str>
   <str name="collapse.max">1</str>
  <str name="indent">on</str>
  <str name="q">words_t:Apple</str>
      <str name="version">2.2</str>
  <str name="rows">10</str>
  <str name="collapse.field">t_s</str>
      </lst>
</lst>
<result name="response" numFound="2" start="0">
 <doc>
  <int name="c_i">3</int>
  <str name="id">1</str>
     <int name="popularity">0</int>
  <str name="sku">1</str>
  <str name="t_s">movie</str>
```

```xml
      <date name="timestamp">2007-12-21T15:22:42.211Z</date>
      <str name="words_t">Apple Orange</str>
    </doc>
<doc>
    <int name="c_i">4</int>
  <str name="id">3</str>
    <int name="popularity">0</int>
  <str name="sku">3</str>
    <str name="t_s">book</str>
    <date name="timestamp">2007-12-22T02:01:53.328Z</date>
    <str name="words_t">Apple Orange</str>
  </doc>
</result>
<lst name="collapse_counts">
    <str name="field">t_s</str>
 <lst name="doc">
    <int name="1">1</int>
 </lst>
 <lst name="count">
    <int name="movie">1</int>
 </lst>
 <str name="debug">HashDocSet(1) Time(ms): 0/0/0/0</str>
</lst>
</response>
```

这样的返回结果表示：对于列"t_s"，还有一个同名列"t_s"的值是 movie。

主要参数介绍：

- collapse=true：启用 collapse 组件；
- collapse.facet = before|after 参数：控制 faceting 是在 collapsing 前或后发生；
- collapse.threshold 参数：控制在多少重复后开始折叠结果。例如：collapse.threshold =1 就是只显示一条结果；
- collapse.maxdocs 参数：控制折叠执行时最多考虑的文档数量。当结果集很大的时候，增加此参数能缩短执行时间；
- collapse.info.doc 和 collapse.info.count 参数：控制 collapse_counts 返回结果内容。
- collapse.field=popularity：用哪个字段来作为重复内容折叠的依据。

实际应用场景设想：一个 Solr 索引库包含很多新闻故事，这些故事来源于许多报纸或有线电视。

一条新闻故事可能来源于多种不同的报纸，例如《人民日报》或一些地方小报等。每种报纸对同一个新闻起上不同的标题，并截成不同的长度。

需要检测并把这些重复的故事分组在一起显示。假设每个故事都由一个 Hash 整数代表(也就是语义指纹)，例如取故事的几个关键词作为"similarity_hash"，把这个值索引并存储起来作为检测重复故事的依据。

我们可以基于这个"similarity_hash"值来折叠搜索结果，这样同一个新闻故事的多次出现就折叠到了一起。

而且，用户可能更愿意读到更加权威的版本，需要把这个结果优先显示到搜索结果中，当然也会有个重复新闻的计数和连接。权威性的取值可能是：

(1) 表示有限电视新闻网；
(2) 表示国家级报纸；
(3) 表示地区报纸；
(4) 表示地方小报。

可以把权威性的取值索引和存储成一个整数权值——authority。然后，可以显示给用户：

```
"人咬狗"
**日报，连接可见其他 77 个重复新闻
```

通过折叠"similarity_hash"列实现这个功能，选择基于另外一列"authority"的值返回显示的新闻中的一条。

这样需要进一步修改这个折叠列的实现，增加一个参数：

```
collapse.authority=[field] //索引列，用来控制折叠后的组中返回哪一个值
```

以前 CollapseFilter 只对一列排序，然后在排序后的结果中发现重复，实现折叠功能。现在改为实现多个列排序，把 collapse.authority 也加到排序列中。我们修改它的主要实现文件 src/java/org/apache/solr/search/CollapseFilter.java：

```
if (collapseType == CollapseType.NORMAL) {
    if(sort!=null)  {
        SortField[] ofields = sort.getSort();
        SortField[] nfields = new SortField[ofields.length+1];
        for(int k=0;k<ofields.length;++k)   {
            nfields[k] = ofields[k];
        }
        nfields[nfields.length-1] = new SortField(collapseField);
        sort.setSort(nfields);
    } else    {
        sort = new Sort(new SortField(collapseField));
    }
}
```

5.5.6　定制输入输出

Solr 的输入和输出格式是固定的，有时候为了方便其他系统调用，需要把输入的 GET 请求改成 POST 请求。POST 请求可以用如下表单模拟：

```
<form action="Seach" method="post">
   <input type="text" name=" ers_word" />
   <input type="submit" value="submit" />
</form>
```

搜索结果需要返回的格式规定见表 5-2。

第 5 章 Solr 分布式搜索引擎

表 5-2 返回结果格式

变量名称	变量命名	说 明
检索结果基本信息	baseinfo	ers_word、ers_id、ers_s、ers_tc、ers_nc 的父节点
检索词	ers_word	字符型字符串 最大长度 120 字节
检索 ID	ers_id	数字型字符串 最大长度 20 字节
检索状态	ers_s	字符型字符串
检索耗时	ers_tc	数字型字符串
检索结果数量	ers_nc	数字型字符串
检索结果内容	r	ers_n、ers_pn 父节点
检索结果顺序码	ers_n	数字型字符串
检索结果对应所在电子书中页码	ers_pn	数字型字符串，非负

例如下面这个搜索结果：

```
<root>
  <baseinfo ers_word="软件" ers_id="123123" ers_s="000000" ers_tc="31" ers_nc="1299" />
  <r ers_n="1" ers_pn="">软件开发</r>
  <r ers_n="2" ers_pn="">懂 Protel 等软件，熟悉电子线路</r>
  <r ers_n="3" ers_pn="">一般办公室庶务，文件整理，进销存软件操作</r>
  <r ers_n="4" ers_pn="">计算机软件相关专业，本科以上，英语四级以上；有视频会议系统、流媒体软件、网络通信、图像处理、数据库等相关软件开发经验。</r>
</root>
```

在服务器端定义一个 Search 类(Servlet)来处理查询请求。

```
Public class Search {
    public void doPost(HttpServletRequest request, HttpServletResponse response)
            throws ServletException, IOException {
//1. 接收用户查询请求参数：
        String ers_word = request.getParameter("ers_word");
        if(ers_word==null||ers_word.length()<=0){
            response.sendError( 400, "检索词不能为空" );
            return ;
        }

        final SolrCore core = SolrCore.getSolrCore();
        SolrServletRequest solrReq = new SolrServletRequest(core, request);
//2. 将参数传递到 Solr 内部：
        SolrParams spold = solrReq.getParams();
```

```
            NamedList nl = spold.toNamedList();
            nl.add("q", ers_word);
            nl.add("start", 0);
            nl.add("rows", 20);
            Map <String, String []> map  = spold.toMultiMap(nl);
            SolrParams sp = new  ServletSolrParams(map);
            solrReq.setParams(sp);

//经过以上处理Solr已经基本完成查询配置。
            SolrQueryResponse solrRsp = new SolrQueryResponse();
            try {
              SolrRequestHandler handler 
                    = core.getRequestHandler(solrReq.getQueryType());
              if (handler==null) {
                log.warn("Unknown Request Handler '" + solrReq.getQueryType()
+"' :" + solrReq);
                throw new
SolrException(SolrException.ErrorCode.BAD_REQUEST,"Unknown Request
Handler '" + solrReq.getQueryType() + "'", true);
              }

//3. Solr执行查询
            core.execute(handler, solrReq, solrRsp );
            if (solrRsp.getException() == null) {

//4. 调用自定义输出类进行输出：
            MyXMLWriter responseWriter = new MyXMLWriter();
                response.setContentType(responseWriter.getContentType(solrReq,
solrRsp));
                PrintWriter out = response.getWriter();
                responseWriter.rewrite(out, solrReq, solrRsp);

            } else {
              Exception e = solrRsp.getException();
              int rc=500;
              if (e instanceof SolrException) {
                rc=((SolrException)e).code();
              }
              sendErr(rc, SolrException.toStr(e), request, response);
            }
            } catch (SolrException e) {
              if (!e.logged) SolrException.log(log,e);
              sendErr(e.code(), SolrException.toStr(e), request, response);
            } catch (Throwable e) {
              SolrException.log(log,e);
              sendErr(500, SolrException.toStr(e), request, response);
            } finally {
```

```
            // This releases the IndexReader associated with the request
            solrReq.close();
        }
    }
}
```

自定义 MyXMLWriter 和 MyWriter 类。MyXMLWriter 继承 XMLResponseWriter，而 MyWriter 继承 XMLWriter。在 Servlet 里调用 MyXMLWriter 输出搜索结果。

```
public class MyXMLWriter extends XMLResponseWriter {
public void rewrite(Writer writer, SolrQueryRequest req, SolrQueryResponse
rsp) throws IOException {
//MyXMLWriter 调用 MyWriter 进行输出:
    MyWriter.rewriter(writer,req,rsp);
}
}

public class MyWriter extends XMLWriter {
public static float CURRENT_VERSION=2.2f;
private static final char[] XML_START="<?xml version=\"1.0\"
encoding=\"UTF-8\"?>\n".toCharArray();

private static final char[] XML_START_NOSCHEMA=("<root>\n").toCharArray();

public MyWriter(Writer writer, IndexSchema schema, SolrQueryRequest
req,String version) {
        super(writer,schema,req,version);
}

public static void rewriter(Writer writer, SolrQueryRequest req,
SolrQueryResponse rsp) throws IOException{
        String ver = req.getParams().get(CommonParams.VERSION);

        MyWriter xw = new MyWriter(writer, req.getSchema(), req, ver);
        xw.defaultFieldList = rsp.getReturnFields();

        String ers_word = req.getParam("ers_word");
        //获取查询结果集
        NamedList lst = rsp.getValues();

        StringBuilder sb = new StringBuilder("");
        //添加 XML 头信息
        sb.append(XML_START).append(XML_START_NOSCHEMA);

        //添加 XML 结果状态信息
        NamedList response = rsp.getResponseHeader();
        if(response!=null){
         String ers_s = (String)response.get("ers_s");
```

```
            String ers_tc = (String)response.get("ers_tc");
            String ers_nc = (String)response.get("ers_nc");

        XML.writeXML(sb, "beseinfo", "",
"ers_word",ers_word,,"ers_s",ers_s,"ers_tc",ers_tc,"ers_nc",ers_nc);
        }
        //添加 XML 详细结果信息
        Object result = lst.get("result");
        if(result!=null){
            DocList docs = (DocList)result;
            SolrIndexSearcher searcher = req.getSearcher();
          DocIterator iterator = docs.iterator();
          int ss = docs.size();
          for (int i=0; i<ss; i++) {
            int id = iterator.nextDoc();
            Document doc = searcher.doc(id,defaultFieldList);
            Fieldable ers_pn = doc.getFieldable("ers_pn");
            Fieldable val = doc.getFieldable("describe");
             XML.writeXML(sb, "r", val==null ?"":val.stringValue(),
                "ers_n",i+1,"ers_pn",ers_pn==null?"":ers_pn.stringValue());
          }
        }
        //添加 XML 结尾
        sb.append("\n</root>\n");

        String temp = sb.toString();
        //输出数据
writer.write(temp);
    }
    private static Set<String> defaultFieldList;
}
```

5.5.7 聚类

Carrot2 是一个开源的类聚工具，在 Solr 中可以很方便地集成 Carrot2 实现的聚类。

首先运行 contrib/clustering 目录下的 build.xml。

在%solr_home%/lib 目录下添加扩展包：

从下载的 Solr 项目中将 dist/apache-solr-clustering-*.jar，contrib/clustering 目录下的所有 jar 包，contrib/clustering/downloads 目录下的所有 jar 包加入到%solr_home%/lib。

在 solrconfig.xml 中添加搜索组件：clusteringComponent。

最后运行 solr：http://localhost:8080/solr/clustering?q=*:*&rows=10，需要分词优化。

5.5.8 分布式搜索

大容量分布式搜索的一种实现方式是：不同的文档索引到不同机器上。每台机器都有一个相对独立的索引。另外一种实现方法是：把某个词的所有文档编号存到一台机器上。

这样每台机器的索引都不是独立的。

单台机器的计算能力有限，可以采用 Solr 搭建多机集群的分布式搜索来实现高负载和高可用性。一个完整的分布结构如图 5-4 所示。

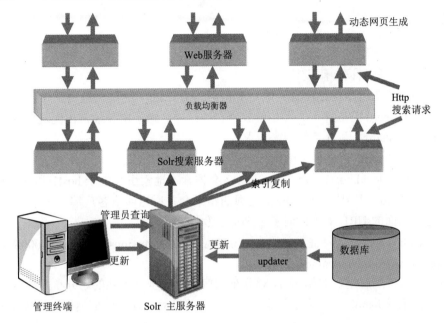

图 5-4 Solr 的完整分布结构图

当一个索引的大小超过一个机器的处理能力的时候，就需要把索引分布到多台机器上了。Solr 直接用 HTTP 协议来实现分布式搜索。待搜索的文档分到不同的索引，每个机器的索引都是独立和完整的，叫作一个 shard。可以查询和合并多个 shard 中的结果。每个机器的索引的并集组成整个可搜索的索引。唯一的列在所有的 shard 中都必须是唯一的。各 shard 的 schema 也必须一致。

同时发送查询请求给所有活跃的 shard，然后异步等待 Shard 返回检索结果。为了返回最相关的 n 个结果，用根据多个查询条件的优先队列存储查询结果。优先队列的长度是：[0, 1, …, start + rows]。

首先在 shard 机器上分别做 top N 查询，最后有个合并的 top N 队列。合并的 top N 队列中包括全局打分得到的最相关的 N 个文档。

评分在 shard 之间有可比性吗？显然通常没有可比性。即使对同样大小的 shard 来说，也往往没有可比性。分布式索引的打分方法是：词频除以全局性的文档频率。

向量空间检索模型中需要得到查询中每个词的 IDF。全局评分需要全局的 IDF。可以采用两轮查询的过程：

(1) 提交查询到每个 shard，以便得到查询中每个词的每 shard 的 IDF；

(2) 收集和汇总 IDF 从 shard 成为全局的 IDF；

(3) 提交查询和全局的 IDF。修改 IndexSearcher，让它能够使用汇总的 IDF 来产生反映全局 IDF 的权重。

但这样存在的问题是查询速度慢。估计的全局 IDF 只用查询一轮就能得到近似的结果。根据词频的分布规律，约有一半的词出现次数少于 10。IDF 不必精确，只要在 shard 之间一致即可。

分布的压缩表示方法如下：

(1) 在列表中记录前 1000 词+ 频率；

(2) 用布隆过滤器记录其他频率大于 5 的词；

(3) 跳过其他任何词，假设频率是 1。

定期广播这个压缩结构给所有其他 shard 服务器。例如，经过大量更新导致 IDF 显著改变后，如果有的 shard 响应速度太慢，就放弃它的结果。为了避免经常出现这样的情况，可以复制负担重的 shard，并且负载均衡请求。

如果"shards"出现在请求参数中，分布式搜索就开始起作用了，否则搜索的是本地索引库。例如，shards=local,host1:port1,host2:port2 会搜索本地索引和位于 host1 和 host2 的两个远程索引。协调 Solr 实例把 3 个 shard 返回的结果合并后返回给客户端。

可以通过下面的 URL 地址访问多个 Solr 实例：

http://localhost/select/?q=%E5%8C%BB%E7%94%9F&version=2.2&start=1000&rows=20&indent=on&shards=local,localhost:8080

在这里，客户端应用，例如 SolrJ 对分布式搜索是完全不可知的。分布式搜索的处理和结果合并都在请求 handler 内部处理了。在结果合并以后，保持响应返回的结果格式不变。SolrJ 把请求发送给协调 Solr 实例。

分布式搜索的具体实现方法是：一个新的叫作 MultiSearchRequestHandler 的 RequestHandler 执行对多个子搜索的分布式搜索(子搜索服务叫作"shard")。handleRequestBody()方法被分成查询构建和执行两个方法。为了增加分布式搜索功能，所有的搜索请求 handler 都扩展这个 MultiSearchRequestHandler。标准的 StandardRequestHandler 和 DisMaxRequestHandler 改成继承这个类。

在 shard 集合上的搜索请求按 6 步执行，其中有两个查询阶段：在第一个查询阶段找到包含查询词的文档唯一列；在第二个查询阶段根据合并的文档唯一列查询高亮信息和 MoreLikeThis 信息。处理流程如下。

第 1 步：构建查询，提取搜索词。通过请求所有的 shard 并求和计算全局的文档数量和文档频率。

第 2 步：在第一个查询阶段查询所有的 shard。把全局的文档数量和文档频率作为参数传递。所有的文档列都不请求，仅仅请求文档唯一列和排序列。

第 3 步：基于"sort", "start"和"rows" 参数合并从第一个查询阶段返回的请求。收集合并的文档唯一列和排序列。同时也合并其他的信息，例如 facet 和调试信息。

第 4 步：在第二个查询阶段合并文档唯一列和排序列按 shard 分组。向在组中的所有的 shard 查询合并的文档的唯一列(第一个查询阶段查询)、高亮信息和 MoreLikeThis 信息。

第 5 步：合并从所有的 shard 的第二个查询阶段的响应。

第 6 步：从第二个查询阶段得到的文档列、高亮和 moreLikeThis 信息合并进第一个查询阶段查询的响应。

分布式搜索系统整体结构如图 5-5 所示。其中，需要考虑如何在多个 shard 中分布文档，也就是分布文档算法。一个简单的分布文档方法是根据文档的唯一编号分布到不同的 shard：uniqueId.hashCode() % numShards。如果重建索引很容易，或者 shard 数量固定不变，这样是可以的。在 shard 数量变化的情况下，可考虑采用一致性散列(Consistent Hashing)。设想删除或者增加一个 shard，不会造成文档到 shard 映射的剧烈改变，这就是一致性散列要达到的效果。一致性散列的结果不只是一个数，而是一个完整的探查序列的情形。想象有一个年级的同学分在几个班，把其中一个班解散，这个班的学生分到其他的班，而其他班的学生仍旧在原来的班。

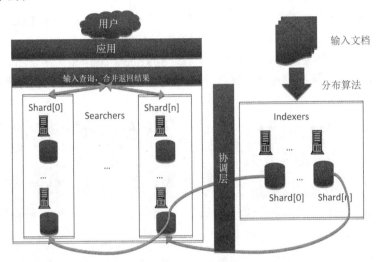

图 5-5　分布式搜索系统整体结构图

索引分布到多台机器后，提高了单点失败的可能性，也就是一台搜索服务器无法访问后，导致整个搜索结果都不可用了。可以利用虚拟 IP 地址(VIP)来避免单点失败。VIP 是一个不连接到一个指定的计算机或者计算机上的网卡的 IP 地址。送到 VIP 地址的输入包被重定向到物理网络接口。分布式搜索通过 HTTP 网络协议实现，因此可以查询 shard 的 VIP。通过 VIP 来实现负载均衡和容错的结构如图 5-6 所示。

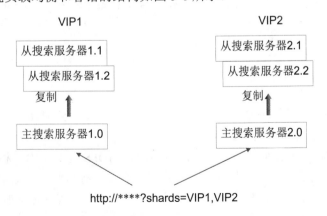

图 5-6　分布式搜索与索引复制结构图

为了集群配置和调度，Solr 包含并且使用 Zookeeper 作为存储器。把 Zookeeper 看成一个包含所有的 Solr 服务器的分布式文件系统。Zookeeper 可以实现负载均衡和容错。

5.5.9 分布式索引

使用 SOLR-1301 可以实现用 Hadoop 做索引。因为索引很多文档往往比较耗时，这个方法用于索引到 N 个主服务器也需要花费很多时间的情况。

有数 TB 索引时，就会觉得索引速度很慢了。采用每个 Hadoop 节点对应一个 Shard，并行的方式更改索引列可能要花费几分钟到几小时，而不是几天。

实现 SolrDocumentConverter 类，把对象从固定格式转换成 SolrInputDocument，然后可以索引进 Solr。SolrRecordWriter 实例化一个构建 Solr 索引的 RecordWriter。Hadoop 中的 RecordWriter 类实现把任务的输出写到文件系统。SolrRecordWriter 调用了 BatchWriter 来构建索引。BatchWriter 类按批增加文档到一个嵌入的 SolrServer，这样完成写索引。

SolrMapper 把输入的值写入到索引库。

```java
public static class SolrMapper extends Mapper<LongWritable, Text, Text, IntWritable>{
  public void map(LongWritable key, Text value, Context context)
                  throws IOException, InterruptedException {
    String line = value.toString();
    StringTokenizer lineStringTokenizer = new StringTokenizer(line, "\t");
    int fieldId = 0;
    SolrInputDocument rawOccurrenceRecordDocument = new SolrInputDocument();
    while(lineStringTokenizer.hasMoreTokens()){
        String token = lineStringTokenizer.nextToken();
        rawOccurrenceRecordDocument.setField(recordFileds[fieldId], token);
        fieldId+=1;
    }
    try {
        getServerInstance().add(rawOccurrenceRecordDocument);
    } catch (SolrServerException e) {
        log.equals(e);
    }
  }
}
```

HDFS 不支持足够多的 POSIX 来直接把 Lucene 索引直接写到 HDFS，所以在每个节点的本地存储空间创建索引。关闭索引后复制到 HDFS。

Nutch 使用的的确是 Lucene 索引，不过将索引放在 HDFS 上面是为了利用 Hadoop 平台的计算性能对索引进行合并等一些操作。在 Hadoop 平台上进行这些操作比单机处理强很多。处理完成之后，可以将索引下载到本地进行访问，并不是提供搜索服务的时候也是在 HDFS 上面的。

使用 MapFile 是为了利用 Hadoop 平台处理数据，最终生成索引。MapFile 并不是用来存储索引，而是存储原始数据的，比如网页快照。

Hadoop 实现了一个分布式文件系统(Hadoop Distributed File System，HDFS)，HDFS 有着高容错性的特点，即使一些数据节点失败了，集群仍然能工作。可以自动检测到数据节点失败。自动采取行动来重新平衡存在失败的数据节点上的数据块到集群中的其他节点。并且，如果可能的话，重新执行失败节点上的任务。最后，Hadoop 集群的用户甚至不会注意到一个数据节点的失败。所以可以使用 Hadoop 集群来备份索引。

5.5.10　SolrJ 查询分析器

为了支持像 AND(与)、OR(或)、NOT(非)这样的高级查询语法。Lucene 使用 JavaCC 生成的 QueryParser 类实现用户查询串的解析。SolrJ 自身没有这样的实现，下面实现一个和 Lucene 查询语法兼容的查询语法解析器。

查询分析一般用两步实现：词法分析和语法分析。词法分析阶段根据用户输入返回单词符号序列，而语法分析阶段则根据单词符号序列返回需要的查询串。

词法分析的功能是从左到右扫描用户输入查询串，从而识别出标识符、保留字、整数、浮点数、算符、界符等单词符号，把识别结果返回到语法分析器，以供语法分析器使用。这一部分的输入是用户查询串，输出是单词符号串的识别结果。例如，对如下的输入片断：

```
title:car site:http://www.sina.com
```

词法分析的输出可能是：

```
TREM title
COLON :
TREM car
TREM site
COLON :
TREM http://www.sina.com
```

词法分析可以采用 JFlex 这样的工具生成，也可以手工编写。因为查询词法比较简单，所以这里采用手工实现一个词法分析。语法分析采用 YACC 的 Java 版本 BYACC/J(http://byaccj.sourceforge.net/)。BYACC/J 根据 YACC 源文件生成 Java 源代码。YACC 推导的返回类型 Query 定义如下：

```
public interface Query {
    public String getQueryType();  //取得查询类型,对应Solr的Request Handler
        public String getQuery();  //取得查询串
}
```

语义值存储在一个叫作 ParserVal 的类中，因此修改 ParserVal 中的属性 obj 的类型定义。

```
public class ParserVal{
…
/**
 * 联合体对象的值
```

```
*/
public Query obj;
…
```

定义 Token 的类型有如下几种：

```
%token AND OR NOT PLUS MINUS LPAREN RPAREN COLON TREM SKIP RANGEIN_START RANGEIN_TO RANGEIN_END
```

下面是词法分析器的实现：

```java
public class Yylex {
    private Parser yyparser; //解析器
    private String buffer;  //查询串缓存
    private int tokPos = 0; //Token 的当前位置
    private int tokLen = 0; //Token 的长度

    /**
     * 词法分析器的构造函数
     *
     * @param r 输入查询串
     * @param yyparser 解析器对象
     */
    public Yylex(String r, Parser yyparser) {
        buffer = r;
        this.yyparser = yyparser;
    }

    /**
     * 遍历扫描直到匹配正则表达式
     * 如果结束，返回 0
     *
     * @返回下一个 Token
     */
    public int yylex() {
        tokPos += tokLen;
        if (tokPos >= buffer.length()) {
            //输入串解析结束
            return 0;
        }

        char ch = buffer.charAt(tokPos);
        switch (ch) {
            case '+':
                tokLen = 1;
                return Parser.PLUS;
            case '-':
                tokLen = 1;
                return Parser.MINUS;
```

```
        case '(':
        case '(':
            tokLen = 1;
            return Parser.LPAREN;
        case ')':
        case ')':
            tokLen = 1;
            return Parser.RPAREN;
        case '[':
            tokLen = 1;
            return Parser.RANGEIN_START;
        case ']':
            tokLen = 1;
            return Parser.RANGEIN_END;
        case ':':
        case ': ':
            tokLen = 1;
            return Parser.COLON;
        case '|':
            if (tokPos + 1 < buffer.length() && buffer.charAt(tokPos + 1) == '|') {
                tokLen = 2;
                return Parser.OR;
            }
            tokLen = 1;
            return Parser.TREM;

        case '&':
            if (tokPos + 1 < buffer.length() && buffer.charAt(tokPos + 1) == '&') {
                tokLen = 2;
                return Parser.AND;
            }
            tokLen = 1;
            return Parser.TREM;
        case ' ':
        case '\t':
        case ' ':
            tokLen = 1;

            return yylex();
        case '"':
            tokLen = 1;
            while (tokPos + tokLen < buffer.length())    {
                char chTerm = buffer.charAt(tokPos + tokLen);
                if (chTerm != '"') {
                    tokLen++;
                } else {
                    tokLen++;
```

```
                break;
            }
        }
        yyparser.yylval=new ParserVal(buffer.substring(tokPos,
tokPos+tokLen));

        return Parser.TREM;
    default:
        tokLen = 1;
        while (tokPos + tokLen < buffer.length()) {
            char chTerm = buffer.charAt(tokPos + tokLen);
            if (chTerm != ' ' &&
                chTerm != '\t' &&
                chTerm != ' ' &&
                chTerm != '(' &&
                chTerm != ')' &&
                chTerm != ')' &&
                chTerm != '(' &&
                chTerm != ':' &&
                chTerm != ']') {
                if( chTerm == ':' )   {
                    if("http".equals(buffer.substring(tokPos,
tokPos+tokLen)))   {
                        tokLen++;
                    } else {
                        break;
                    }
                } else {
                    tokLen++;
                }
            } else {
                break;
            }
        }
        String cur = buffer.substring(tokPos, tokPos+tokLen);
        if (cur.equals("AND")) {
            return Parser.AND;
        }
        if (cur.equals("OR")) {
            return Parser.OR;
        }
        if (cur.equals("TO")) {
            return Parser.RANGEIN_TO;
        }
        yyparser.yylval = new ParserVal(cur);
        return Parser.TREM;
}
```

可以测试一下这个词法分析程序,看是否能返回正确的 Token 类型:

```java
public static void main(String[] args) {
    String i = "title:car link:http://www.sina.com"; //输入字符串
    Parser yyparser = new Parser();
    Yylex lexer = new Yylex(i, yyparser);
    
    int type = 1;
    while (type!=0){
        type = lexer.yylex(); //得到 Token 类型
        String result = String.valueOf(type);
        if (yyparser!=null && yyparser.yylval!=null ){
            result+= " "+ yyparser.yylval.sval; //得到 Token 值
        }
        System.out.println(result);
    }
}
```

YACC 源文件由 3 部分组成:

第一部分是 DECLARATIONS 区域,在这里定义 Token 和过程等;
第二部分是 ACTIONS 区域,在这里定义文法和行为;
第三部分是 CODE 区域,在这里定义用户方法。
三个部分由一个 "%%" 行分隔开。格式是:

```
DECLARATIONS
%%
ACTIONS
%%
CODE
```

生成 Solr 查询串的 YACC 语法文件 queryparser.y 的内容如下:

```
%start querystr
%token AND OR NOT PLUS MINUS LPAREN RPAREN COLON TREM SKIP RANGEIN_START RANGEIN_TO RANGEIN_END
%left OR
%left AND
%left NOT
%left PLUS
%left MINUS
%%
querystr    : query { $$ = $1; }
    ;

query   : /*empty */
        | query clause {
```

```
                    if ($1 == null)
                    {
                        $$ = $2;
                    }
                    else
                    {
                        BooleanQuery bq = new BooleanQuery();

                        bq.Add($1.obj, BooleanClause.Occur.SHOULD);
                        if ($2.obj instanceof BooleanQuery)
                        {
                            BooleanQuery clause = (BooleanQuery)$2.obj;
                            if (clause.plus)
                            {
                                bq.Add($2.obj, BooleanClause.Occur.MUST);
                            }
                            else if (clause.minus )
                            {
                                bq.Add($2.obj, BooleanClause.Occur.MUST_NOT);
                            }
                            else
                            {
                                bq.Add($2.obj, BooleanClause.Occur.SHOULD);
                            }
                        }
                        else
                        {
                            bq.Add($2.obj, BooleanClause.Occur.SHOULD);
                        }
                        $$ = new ParserVal(bq);
                    }
                }
        ;
clause   : clause OR clause  {
            BooleanQuery bq = new BooleanQuery();
                bq.Add($1.obj, BooleanClause.Occur.SHOULD);
                bq.Add($3.obj, BooleanClause.Occur.SHOULD);
                $$ = new ParserVal( bq );
    }
        |   clause AND clause  {
                BooleanQuery bq = new BooleanQuery();
                bq.Add($1.obj, BooleanClause.Occur.MUST);
                bq.Add($3.obj, BooleanClause.Occur.MUST);
                $$ = new ParserVal( bq );
    }
       |   LPAREN clause RPAREN  {
         $$ = new ParserVal( $2.obj );
    }
```

```
        |   PLUS clause {
        BooleanQuery bq = new BooleanQuery();
        bq.Add($2.obj, BooleanClause.Occur.SHOULD);
            bq.plus = true;
        $$ = new ParserVal( bq );
    }
        |   MINUS clause {
                    BooleanQuery bq = new BooleanQuery();
                    bq.Add($2.obj, BooleanClause.Occur.SHOULD);
                    bq.minus = true;
                    $$ = new ParserVal( bq );
    }
        |   NOT clause {
                    BooleanQuery bq = new BooleanQuery();
                    bq.Add($2.obj, BooleanClause.Occur.SHOULD);
                    bq.minus = true;
                    $$ = new ParserVal( bq );
    }
        |   TREM {
                    String termStr = $1.sval;
                    BooleanQuery bq = new BooleanQuery();
                    for (int i = 0; i < defaultSearchFields.length; ++i)
                    {
                        bq.Add(new TermQuery(defaultSearchFields[i],
termStr), BooleanClause.Occur.SHOULD);
                    }
                    $$ = new ParserVal(bq);
    }
        |   TREM COLON TREM {
            String fieldStr = $1.sval;
                    String termStr = $3.sval;
                    $$ = new ParserVal( new TermQuery(fieldStr, termStr) );
    }
            |   TREM COLON RANGEIN_START TREM RANGEIN_TO TREM RANGEIN_END {
            String fieldStr = $1.sval;
                    String fromStr = $4.sval;
                    String toStr = $6.sval;
                    $$ = new ParserVal( new RangeQuery(fieldStr, fromStr,toStr) );
    }
        ;
%%
 private Yylex lexer;
 public String[] defaultSearchFields = {"title","body"}; //缺省搜索列
 private int yylex () {
    int yyl_return = lexer.yylex();
    return yyl_return;
 }
```

```
public void yyerror (String error) {
    System.err.println ("Error: " + error);
}
public Parser(String i) {
  lexer = new Yylex(i, this);
}
public static void main(String args[]) { //测试方法
      String input = "title:中国 OR 北京";
      Parser parser = new Parser(input);
      parser.yyparse();
      System.out.println(parser.val_peek(0).obj.ToString());
}
```

用 BYACC 生成 Java 源代码：

```
$yacc -J queryparser.y
```

5.5.11 扩展 SolrJ

SolrJ 通过请求类和响应类提供了方便的接口用来扩展。例如实现从 URL 地址 http://localhost/admin/stats.jsp#core 读取索引状态，如图 5-7 所示。

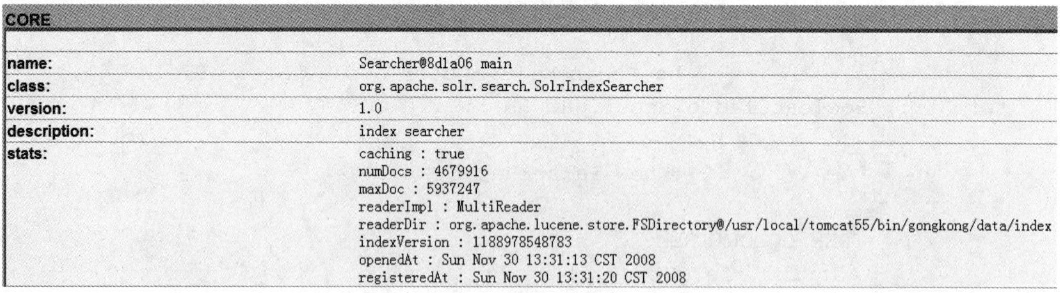

CORE	
name:	Searcher@8d1a06 main
class:	org.apache.solr.search.SolrIndexSearcher
version:	1.0
description:	index searcher
stats:	caching : true numDocs : 4679916 maxDoc : 5937247 readerImpl : MultiReader readerDir : org.apache.lucene.store.FSDirectory@/usr/local/tomcat55/bin/gongkong/data/index indexVersion : 1188978548783 openedAt : Sun Nov 30 13:31:13 CST 2008 registeredAt : Sun Nov 30 13:31:20 CST 2008

图 5-7 索引状态

首先定义请求类：

```
public CoreRequest() {
      super( METHOD.GET, "/admin/stats.jsp#core" );
}
```

然后定义响应类：

```
public CoreResponse(NamedList<Object> res) {
      super(res);
      indexInfo = res;
}
```

以及返回 XML 格式内容的解析类 CoreResponseParser：

```
protected NamedList<Object> readNamedList( XMLStreamReader parser ) {
      NamedList<Object> nl = new NamedList<Object>();
```

```
      int status;
      while( parser.hasNext()) {
         status = parser.next();
         if (status==XMLStreamConstants.START_ELEMENT) {
           String n = parser.getLocalName();
         if("stat".equals(n))  {
            //取得属性名
            n = parser.getAttributeValue(0);
            parser.next();
            //取得属性值
            String v = parser.getText().trim();
            nl.add(n, v);
          }
         }
        }
        return nl;
}
```

5.5.12 扩展 Solr

Solr 本身是 REST 风格的,它通过 Servlet 响应用户请求。以 MoreLikeThis handler 为例,在 solrconfig.xml 中包含:

```
<requestHandler name="/mlt" class="solr.MoreLikeThisHandler">
   <lst name="defaults">
     <str name="mlt.fl">manu,cat</str>
     <int name="mlt.mindf">1</int>
   </lst>
 </requestHandler>
```

MoreLikeThisHandler 对应的请求方法:

```
http://localhost:8983/solr/mlt?q=id:UTF8TEST&mlt.fl=manu,cat&mlt.mindf=1
&mlt.mintf=1
```

编写 Solr 的 handler()方法需要经过以下几个步骤的操作。

(1) 将 Solr 源文件解压,在 Eclipse 中新建工程,将源文件以及相关的 jar 文件导入。

(2) 在 org.apache.solr.handler 包下(一般在此包下进行扩展)新建 Java 类。

(3) 新建的类需要继承 RequestHandlerBase 类,并且实现其中的 handleRequestBody(SolrQueryRequest req, SolrQueryResponse rsp)方法。同时需要覆盖 getVersion()、getDescription()、getSourceId()、getSource()、getDocs()等方法。

(4) handleRequestBody(SolrQueryRequest req, SolrQueryResponse rsp)方法中,req 表示传入的参数对象,rsp 表示经过处理后得到的需要显示的对象。

(5) 根据需要编写业务处理逻辑。

(6) 在 Solr 中,常量一般在 org.apache.solr.common.params 包下的接口 CommonParams 中定义。在本项目中,需要在 CommonParams 中添加新的常量,如:

```
public static final String URI = "uri";//URI 的值表示访问的参数
```

(7) 以上工作实现后，对 Solr 重新打包，然后加入到 Web 项目目录\WEB-INF\lib 中。

(8) 打开 Solr 文件夹(对应 resin 文件夹下的 solr 文件夹)，打开 conf 文件夹下的 solrconfig.xml 文件，在<config>……</config>标签中，添加如下内容：

```
<requestHandler name="/urlinfo" class="solr.URItoWordsHandler">
</requestHandler>
```

其中，name 属性决定了访问的路径，class 属性决定了处理类。

(9) 打开 IE 浏览器，输入"http://localhost:8081/index/urlinfo?url=? wt=? "，其中，"localhost:8081/index/"表示该 Web 应用程序的访问地址；"urlinfo"对应第(8)条中的<requestHandler>标签中的 name 属性的值；"url"对应传入的查询网页地址，对应第(6)条中添加的常量的值(uri)；"wt"表示要输出的格式，值一般为 xml 和 json，表示输出的格式为 XML 格式或者 JSON 格式。

另外一个应用场景：前台用 PHP 实现的系统需要知道网页的类别。采用 PHP 调用 Solr 服务器中提供的网页分类服务。

(1) 在合适的包下编写所需要的 Java 类，本例中类名为 URItoWordsHandler，该类继承 RequestHandlerBase 类。其中的方法 handleRequestBody()实现如下：

```
public void handleRequestBody(SolrQueryRequest req, SolrQueryResponse rsp)
      throws Exception {
   // 获取传递的参数集合，并且进行相应的业务处理
   SolrParams params = req.getParams();

   // 从参数集合中得到想要的参数
   String uri = params.get(CommonParams.URI);
   if (uri == null || uri.trim().length() < 1) {
      rsp.add("type", "其他");
      rsp.add("weburl", "");
      return;
   }

   // 进行业务处理，在这里是对网页分类
   InputStream is =
this.getClass().getResourceAsStream("/spider.properties");
   try {
      properties.load(is);
      is.close();
   } catch (IOException e1) {
      e1.printStackTrace();
   }
   String modelDir = properties.getProperty("modelDir");   //分类模型所在路径
   Classifier theClassifier = new Classifier(modelDir+"/model.prj"); //文本分类器

   String catName=null;
```

```java
if(uri.startsWith("http")) {  //对网页分类
    ContentExtractor ce = new ContentExtractor();
    String body = ce.processURL(uri,null).toString();  //提取正文
    catName = theClassifier.getCategoryName(body);    //根据正文分类
    if (catName==null){
        catName="其他";
    }
} else{  //对本地文件分类
    FilterFile file=new FilterFile();
    String body=file.filter(uri); //提取正文
    catName = theClassifier.getCategoryName(body);    //根据正文分类
    if (catName==null){
        catName="其他";
    }
}

/**
 * 将业务处理结果添加到 rsp, 并且输出
 */
rsp.add("type", catName);
rsp.add("weburl", uri);
}
```

(2) 在 solrconfig.xml 中<config>标签下添加相应的配置信息：

```xml
<requestHandler name="/urlinfo" class="solr.URItoWordsHandler">
        <!-- /urlinfo 表示访问路径   class 表示对应的处理类   -->
        <!-- 以下为其他说明，可根据需要修改或删除,此信息将会在网页显示   -->
<lst name="其他说明">
        <int name="名称">参数</int>
    </lst>
</requestHandler>
```

(3) 启动服务器，在浏览器地址栏输入："http://localhost:8081/index/urlinfo?url=d&wt=xml"。其中，"urlinfo"对应访问路径；"url"表示需要传递的参数；"wt"表示返回的结果格式；"wt=xml"时的显示结果为 XML 格式，wt=json 时，指定返回结果格式为 JSON。

要对网页 http://tech.163.com/digi/09/0328/13/55GBJMQT00161MAH.html 分类，可以在浏览器地址栏输入：http://localhost:8081/index/urlinfo?uri=http://tech.163.com/digi/09/0328/13/55GBJMQT00161MAH.html&wt=xml，即可实现对该网址内容的分类：

```xml
<response>
    <lst name="responseHeader">
        <int name="status">0</int>
        <int name="QTime">9688</int>
    </lst>
    <str name="type">新闻</str>
    <str name="weburl">
```

```
            http://tech.163.com/digi/09/0328/13/55GBJMQT00161MAH.html
        </str>
</response>
```

要对本地文件进行分类,例如:D:\zhexue\zhexue\08325\1.txt,可以在浏览器地址栏输入:http://localhost:8081/index/urlinfo?uri=D:\zhexue\zhexue\08325\1.txt&wt=xml,返回结果是:

```
<response>
    <lst name="responseHeader">
        <int name="status">0</int>
        <int name="QTime">8062</int>
    </lst>
    <str name="type">新闻</str>
    <str name="weburl">D:\zhexue\zhexue\08325\1.txt</str>
</response>
```

5.5.13 日文搜索

日文分词有一个 Kuromoji(https://github.com/atilika/kuromoji)的项目。Kuromoji 把词典存成 FST。Solr 中已经集成了 "text_ja" 类型。"text_ja" 的定义如下:

```
<fieldType name="text_ja" class="solr.TextField">
 <analyzer>
  <tokenizer class="solr.JapaneseTokenizerFactory"
    mode=NORMAL
    userDictionary=user.txt
    userDictionaryEncoding=UTF-8
  />
  <filter class="solr.JapaneseBaseFormFilterFactory"/>
 </analyzer>
</fieldType>
```

在索引定义中使用 text_ja:

```
<schema name="example" version="1.1">
  <types>
    <fieldtype name="text" class="solr.TextField">
      <analyzer class="org.apache.lucene.analysis.ja.JapaneseAnalyzer"/>
    </fieldtype>
    <fieldtype name="integer" class="solr.IntField"/>
  </types>
  <fields>
    <field name="id" type="integer" indexed="true" stored="true"/>
    <field name="content" type="text_ja" indexed="true" stored="true"/>
  </fields>
  <uniqueKey>id</uniqueKey>
  <defaultSearchField>content</defaultSearchField>
</schema>
```

5.5.14 查询 Web 图

为了支持在 Solr 中以 "link:" 语法查找链接网页，写一个 LinkToThisHandler 的 Solr 插件调用 WebGraph。

```java
    public final static String PREFIX = "lt.";
    public final static String URL_FIELD   = PREFIX + "fl";
    private static final String INDEX_DIR = PREFIX + "dbDir";

    private WebGraph webGraph = null;

    @Override
    public void init(NamedList args) {
    super.init(args);
    SolrParams p = SolrParams.toSolrParams(args);
    String dir = p.get(INDEX_DIR);
try {
//读入 WebGraph
        webGraph = new WebGraph(dir);
    } catch (DatabaseException e) {
        throw new RuntimeException("Cannot open WebGraph database", e);
    }
  }

  @Override
  public void handleRequestBody(SolrQueryRequest req, SolrQueryResponse rsp)
throws Exception    {
    SolrParams params = req.getParams();
    SolrIndexSearcher searcher = req.getSearcher();
    String q = params.get( CommonParams.Q );

SolrParams required = params.required();
//取得 URL 列的名称
    String field = required.get(LinkToThisHandler.URL_FIELD) ;

    // 需要返回的字段
    String fl = params.get(CommonParams.FL);
    int flags = 0;
    if (fl != null) {
      flags |= SolrPluginUtils.setReturnFields(fl, rsp);
    }

    int start = params.getInt( CommonParams.START, 0 );
    int rows = params.getInt( CommonParams.ROWS, 10 );

    DocList docList = null;
```

```java
    if( q != null ) {

        // 找到一个基础的匹配
        Query query = QueryParsing.parseQuery(q, params.get(CommonParams.DF),
params, req.getSchema());
        //仅仅得到第一个结果
        DocList match = searcher.getDocList(query, null, null, 0, 1, flags );

        // DocIterator 是一个迭代器,但这里只处理第一个匹配
        DocIterator iterator = match.iterator();
        if( iterator.hasNext() ) {
           // 在结果中的每一个文档里都做这样一个请求处理
           int id = iterator.nextDoc();
           IndexReader reader = searcher.getReader();
           Document doc = reader.document(id);
           String toURL = doc.getField(field).stringValue();
           String[] fromURLs = webGraph.inLinks(toURL);
           ArrayList<Integer> docIds = new ArrayList<Integer>(fromURLs.length);

           for(int i = 0;i<fromURLs.length;++i)    {
               TermQuery tq = new TermQuery(new Term(field,fromURLs[i]));
               Hits hits = searcher.search(tq);
               if(hits.length()>=1)    {
               int docId = hits.id(0);
               docIds.add(docId);
            }
           }
           int[] ids = new int[docIds.size()];
           for(int i=0;i<ids.length;++i)    {
            ids[i]=docIds.get(i);
           }
           DocSlice result = new DocSlice(0,ids.length,ids,null,ids.length,0);
           docList = result.subset(start, rows);
        }
    }
    else {
      throw new SolrException( SolrException.ErrorCode.BAD_REQUEST,
         "MoreLikeThis requires either a query (?q=) or text to find similar
documents." );
    }
    if( docList == null ) {
     docList = new DocSlice(0,0,null,null,0,0); // avoid NPE
    }
    rsp.add( "response", docList );
}
    <!-- 在SolrConfig.xml 中配置LinkToThisHandler: -->
  <requestHandler name="/lt" class="solr.LinkToThisHandler">
```

```
    <lst name="defaults">
      <str name="lt.fl">url</str>
    </lst>

    <!-- Main init params for handler -->

    <!-- 下面写入 webgraph 数据库所在的路径 -->
      <str name="lt.dbDir">webgraph</str>

  </requestHandler>
```

然后我们就可以在浏览器中测试效果了:

```
http://localhost/solr/lt/?q=url%3Ahttp%5C%3A%5C%2F%5C%2Fwww.baotron.com%
5C%2Findex.asp&version=2.2&start=100&rows=10&indent=on
```

Solr 内部调用 LinkToThisHandler，查询哪些网页指向了"http://www.baotron.com/Findex.asp"。

5.6 SolrNet

如果整个网站采用 ASP.NET 开发，则可以考虑使用 Solr 的.NET 客户端开发搜索。SolrNet (http://code.google.com/p/solrnet/)是 Solr 最流行的.NET 客户端。它不是 SolrJ 的翻译版本，所以和 SolrJ 的用法不一样。

5.6.1 使用 SolrNet 实现全文搜索

假设要对招投标信息实现全文搜索。要搜标题、内容列、日期、URL 地址等。开发流程是：首先在配置文件 schema.xml 中定义索引格式，然后使用 SolrNet 开发 Windows 控制台程序测试搜索，最后开发 ASP.NET 搜索界面。

在 Solr 中的 schema.xml 定义如下这些列：

```
<fields>
  <field name="id" type="int" indexed="true" stored="true" required="true"/>
  <field name="title" type="text" indexed="true" stored="true" required="true"/>
  <field name="body" type="text" indexed="true" stored="true" required="true"/>
  <field name="adddate" type="date" indexed="true" stored="true" default="NOW" multiValued="false"/>
</fields>

<defaultSearchField>title</defaultSearchField>
<uniqueKey>id</uniqueKey>
```

在 Windows 控制台项目中增加对 SolrNet.dll 和 Microsoft.Practices.ServiceLocation.dll 的引用。然后创建 Article 类，映射 Solr 文档到 Article 类。

```csharp
public class Article{
   [SolrUniqueKey("id")]
   public int id { get; set; }

   [SolrField("title")]
   public string title { get; set; }

   [SolrField("area")]
   public string area { get; set; }

   [SolrField("industry")]
   public string industry { get; set; }

   [SolrField("body")]
   public string body { get; set; }

   [SolrField("adddate")]
   public DateTime adddate { get; set; }

   [SolrField("columns")]
   public string columns { get; set; }

   [SolrField("url")]
   public string url { get; set; }
}
```

它只是一个简单的 POCO(Plain Old CLR Object)类。SolrField 映射属性到一个 Solr 列，而 SolrUniqueKey 映射一个属性到一个 Solr 唯一标识列。

定义好模型后，现在可以连接到 Solr 并且保存一些文章。下面的代码定位 Solr 实例(这里运行在本地的 8080 端口)，创建一些文档并提交到索引。

```csharp
//初始化，指定找 solr 服务的 URL 地址和对应的类
Startup.Init<Article>("http://localhost:8080/solr");
ISolrOperations<Article> solr =
    ServiceLocator.Current.GetInstance<ISolrOperations<Article>>();

// 形成一些文章
solr.Add(new Article(){
   ID = 1,
   Title = "my laptop",
   Content = "my laptop is a portable power station",
   Tags = new List<string>() {
      "laptop",
      "computer",
```

```
        "device"
    }
});

solr.Add(new Article(){
    ID = 2,
    Title = "my iphone",
    Content = "my iphone consumes power",
    Tags = new List<string>() {
        "phone",
        "apple",
        "device"
    }
});

solr.Add(new Article(){
    ID = 3,
    Title = "your blackberry",
    Content = "your blackberry has an alt key",
    Tags = new List<string>() {
        "phone",
        "rim",
        "device"
    }
});

// 提交到索引
solr.Commit();
```

如下的例子执行一个全文搜索与"电力"相关的招标信息:

```
//查询并返回结果
ISolrQueryResults<Article> powerArticles = solr.Query(new SolrQuery("电力"));

foreach (Article article in powerArticles) { //遍历 Article 类型的对象
    Console.WriteLine(string.Format("{0}: {1}", article.ID,
article.Title));
}
```

和按标题搜索等价:

```
// 按标题搜索"电力"
ISolrQueryResults<Article> phoneTaggedArticles =
            solr.Query(new SolrQuery("title:电力"));

foreach (Article article in phoneTaggedArticles){
    Console.WriteLine(string.Format("{0}: {1}", article.ID, article.Title));
}
```

设置权重：

```
var q = new SolrQuery("name:desc").Boost(2); // (name:desc)^2
```

权重有什么用？加或不加"^2"搜出来的一样。如果有好几个查询词，就能显示出词之间的相对重要度了。

SolrNet 支持分类统计搜索。通过 QueryOptions 对象指定分类统计列。如下的例子显示按地区分类统计匹配"大桥"的例子：

```
Console.WriteLine("大桥相关的招标信息:");
ISolrQueryResults<Article> articles = solr.Query(new SolrQuery("title:大桥"),
 new QueryOptions() {
    Facet = new FacetParameters {
        // 设置分类统计列
        Queries = new[] { new SolrFacetFieldQuery("area") }
    }
});

foreach (Article article in articles){
    Console.WriteLine(string.Format("{0}: {1}", article.ID, article.Title));
}

Console.WriteLine("\n按地域分类统计:");

foreach (var facet in articles.FacetFields["area"]){
    Console.WriteLine("{0}: {1}", facet.Key, facet.Value); //地名和对应的结果数
}
```

可以通过 QueryOptions 指定排序方式。

```
QueryOptions options = new QueryOptions();
//按"Popularity"列排序
options.AddOrder(new SolrNet.SortOrder("Popularity", Order.DESC));
options.Rows = intPageSize;    //返回结果行数
options.Start = 0;             //开始位置
```

可以使用 QueryOptions 的 ExtraParams 属性来增加参数到 Solr 查询字符串。例如：

```
var results = solr.Query("myquery", new QueryOptions {
  ExtraParams = new Dictionary<string, string> {
    {"bf", "recip(rord(myfield),1,2,3)^1.5"}
  }
});
```

5.6.2 实现原理

客户端的基本原理是：发送 HTTP 请求，返回 XML 格式的数据，最后得到的是搜索结果对象。NET 中的实现方法是：使用 WebClient 请求数据，然后使用 System.Xml 解析结果，最后使用一个对象封装器把搜索结果封装成对象。

第 5 章　Solr 分布式搜索引擎

SolrNet 使用 SOLID 原则构建而成。这里的 S 代表单一责任原则：当需要修改某个类的时候原因有且只有一个。换句话说就是让一个类只做一种类型责任，当这个类需要承担其他类型的责任的时候，就需要分解这个类；O 代表开放封闭原则：软件实体应该是可扩展，而不可修改的。也就是说，对扩展是开放的，而对修改是封闭的。这个原则是诸多面向对象编程原则中最抽象、最难理解的一个；L 表示里氏替换原则：当一个子类的实例应该能够替换任何其超类的实例时，它们之间才具有 is-A 关系；I 表示依赖倒置原则：高层模块不应该依赖于低层模块，二者都应该依赖于抽象；抽象不应该依赖于细节，细节应该依赖于抽象；D 表示接口分离原则：不能强迫用户去依赖那些他们不使用的接口，换句话说，使用多个专门的接口比使用单一的总接口要好。

可以从 https://github.com/mausch/SolrNet/downloads 下载源代码。

如下的属性定义索引一个大的浮点数或者双精度列：

```
[SolrField("price")]
    public decimal Price { get; set; }
```

这里的 decimal 类型在 solrNet 的 DecimalFieldParser.Parse() 方法中解析。除了 DecimalFieldParser，还有 FloatFieldParser、DateTimeFieldParser、DecimalFieldParser、IntFieldParser、LongFieldParser。

5.6.3　扩展 SolrNet

Solr 对于对象的序列化基本采用以下格式：

```
version+[tag+value]
```

第一个 byte 表示 Version，必须为"1"。下面就是对象的定义了。tag 为一个字节，前 3 位表示 tag 的名称，后 5 位表示数据长度。Solr 的 tag 定义如下：

```
NULL = 0,
BOOL_TRUE = 1,
BOOL_FALSE = 2,
BYTE = 3,
SHORT = 4,
DOUBLE = 5,
INT = 6,
LONG = 7,
FLOAT = 8,
DATE = 9,
MAP = 10,
SOLRDOC = 11,
SOLRDOCLST = 12,
BYTEARR = 13,
ITERATOR = 14,
END = 15,
TAG_AND_LEN = (byte)(1 << 5),
STR = (byte)(1 << 5),
```

```
    SINT = (byte)(2 << 5),
    SLONG = (byte)(3 << 5),
    ARR = (byte)(4 << 5), //
    ORDERED_MAP = (byte)(5 << 5),
    NAMED_LST = (byte)(6 << 5),
    EXTERN_STRING = (byte)(7 << 5);
```

对于 Solr 的序列化了解以后,就可以对 Solr 的 javabin 二进制格式数据进行解析了:

```
using System.IO;
namespace EasyNet.Solr
{
    public class FastInputStream
    {
        private BufferedStream stream;  public FastInputStream(Stream stream)
            : this(stream, 8192)
        {        }          public FastInputStream(Stream stream, int bufferSize)
        {
            this.stream = new BufferedStream(stream, bufferSize);
        }          public byte ReadByte()
        {
            return (byte)stream.ReadByte();
        }          public int Read()
        {
            return stream.ReadByte() & 0xff;
        }          public int ReadUnsignedByte()
        {
            return stream.ReadByte() & 0xff;
        }          public int Read(byte[] b, int off, int len)
        {
            return stream.Read(b, off, len);
        }
        public void Close()
        {
            stream.Close();
        }          public void ReadFully(byte[] b)
        {
            ReadFully(b, 0, b.Length);
        }          public void ReadFully(byte[] b, int off, int len)
        {
            while (len > 0)
            {
                int ret = Read(b, off, len);
                if (ret == -1)
                {
                    throw new EndOfStreamException();
                }
                off += ret;
                len -= ret;
            }
```

```
    }
    public bool ReadBoolean()
    {
        return ReadByte() == 1;
    }
    public short ReadShort()
    {
        return (short)((ReadUnsignedByte() << 8) | ReadUnsignedByte());
    }   public int ReadUnsignedShort()
    {
        return (ReadUnsignedByte() << 8) | ReadUnsignedByte();
    }   public char ReadChar()
    {
        return (char)((ReadUnsignedByte() << 8) | ReadUnsignedByte());
    }   public int ReadInt()
    {
        return ((ReadUnsignedByte() << 24)
            | (ReadUnsignedByte() << 16)
            | (ReadUnsignedByte() << 8)
            | ReadUnsignedByte());
    }   public long ReadLong()
    {
        return (((long)ReadUnsignedByte()) << 56)
            | (((long)ReadUnsignedByte()) << 48)
            | (((long)ReadUnsignedByte()) << 40)
            | (((long)ReadUnsignedByte()) << 32)
            | (((long)ReadUnsignedByte()) << 24)
            | (ReadUnsignedByte() << 16)
            | (ReadUnsignedByte() << 8)
            | (ReadUnsignedByte());
    }   public float ReadFloat()
    {
        FloatConverter floatConverter = new FloatConverter();
return FloatConverter.ToFloat(ReadInt(), ref floatConverter);
    }   public double ReadDouble()
    {
        DoubleConverter doubleConverter = new DoubleConverter();
return DoubleConverter.ToDouble(ReadInt(), ref doubleConverter);
    }
    }
}
```

5.7 Solr 的 PHP 客户端

PHP 索引数据有两种方法：用 PHP 模拟 curl 的方式做索引；直接发送 POST 的客户端。后一种方法更成熟一些，solr-php-client(https://github.com/PTCInc/solr-php-client)就是使用直接发送 POST 请求实现。

solr-php-client 中有 3 个最主要的 PHP 类：

(1) Apache_Solr_Service 代表一个 Solr 服务器。Apache_Solr_Service::search()的前 3 个参数分别对应 q、start 和 rows，这是查询中 3 个最常用的参数；

(2) Apache_Solr_Document 封装一个 Solr 文档。可以直接设置文档列中的值，例如 $document->title = 'Something';

(3) Apache_Solr_Response 封装一个 Solr 响应。

如下是一个简单的查询例子：

```php
//连接位于端口号 8983 的 Solr 服务
$solr = new Apache_Solr_Service('localhost', '8983', '/solr');
$query = "NBA";//查询词
$offset = 0;
$limit = 10;

//提供给查询方法的参数包括要返回文档的开始位置和要返回的最多文档数
$response = $solr->search( $query, $offset, $limit );
if ( $response->getHttpStatus() == 200 ) {
    if ( $response->response->numFound > 0 ) {
        echo "$query <br />";
        //遍历结果文档
        foreach ( $response->response->docs as $doc ) {
            echo "$doc->partno $doc->name <br />";
        }

        echo '<br />';
    }
}
else {
    echo $response->getHttpStatusMessage();
}
```

首先看一下创建 Solr 服务对象这行代码：

```php
$solr = new Apache_Solr_Service( 'localhost', '8983', '/solr' );
```

其中"/solr"这个参数是可以变的，可以对应不同的索引库。如果只要搜索一个栏目，只用一个索引库就够了，如果要搜索不同的栏目，而且每个栏目需要显示不一样的内容，才用多个索引库。

为了搜索一些零件，增加零件描述文档到 Solr 的例子：

```php
<?php
  require_once( 'Apache/Solr/Service.php' );
  // 根据服务器名、端口和 URL 地址连接服务器
  $solr = new Apache_Solr_Service( 'localhost', '8983', '/solr' );
  if ( ! $solr->ping() ) {
    echo 'Solr service not responding.';
    exit;
```

```php
}
//创建两个文档表示两个零件，实际上，文档更可能来源于数据库查询
$parts = array(
  'spark_plug' => array(      //第 1 个零件
        'partno' => 1,        //零件编号
        'name' => 'Spark plug',  //零件名称
     'model' => array( 'Boxster', '924' ), //零件型号
        'year' => array( 1999, 2000 ),  //年
       'price' => 25.00,      //价格
       'inStock' => true,     //是否在仓库中
  ),
  'windshield' => array(       //第 2 个零件
        'partno' => 2,
        'name' => 'Windshield',
        'model' => '911',
        'year' => array( 1999, 2000 ),
       'price' => 15.00,
       'inStock' => false,
  )
);

$documents = array();

foreach ( $parts as $item => $fields ) {
    $part = new Apache_Solr_Document();

    foreach ( $fields as $key => $value ) {
      if ( is_array( $value ) ) {
       foreach ( $value as $datum ) {
           $part->setMultiValue( $key, $datum );
       }
      } else {
       $part->$key = $value;
      }
    }

    $documents[] = $part;
}
//把文档加载到索引
try {
   $solr->addDocuments( $documents );
   $solr->commit();
  }
catch ( Exception $e ) {
   echo $e->getMessage();
}
?>
```

5.8 Solr 的其他客户端

一般通过各种动态页面生成服务器返回搜索结果给查询用户。常用的动态网页开发工具有 ASP.NET、JSP 或 PHP 等。Solr 提供对应的客户端来支持各种常用开发工具。Solr 返回的数据格式通常是 XML，返回 JSON，Python，Ruby 格式的数据也是支持的。Solr 客户端是对 HTTP 请求的封装，同时也包括对 XML 格式搜索结果的解析。已有的部分 Solr 客户端见表 5-3。

表 5-3 Solr 客户端列表

前台语言	Solr 客户端	网 址
JSP	SolrJ	http://svn.apache.org/repos/asf/lucene/solr/trunk/src/solrj/
PHP	solr-php-client	https://github.com/PTCInc/solr-php-client
ASP.NET	SolrNet	https://github.com/mausch/SolrNet
AJAX	ajax-solr	https://github.com/evolvingweb/AJAX-Solr/
Ruby on Rails	acts_as_solr	http://acts-as-solr.rubyforge.org/
Python	SolrClient	https://github.com/moonlitesolutions/SolrClient

5.9 为网站增加搜索功能

其他网页和搜索页面位于同一个 Web 容器内。如果其他网页使用 Spring 框架，为了风格统一，则搜索页面最好也使用 Spring 框架。

在 Spring 配置文件中创建一个 Bean：

```xml
<bean id="solrServer"
class="org.apache.solr.client.solrj.impl.HttpSolrServer">
<constructor-arg value="${solr.serverUrl}"/>
<property name="connectionTimeout" value="${solr.connectionTimeout}"/>
<property name="defaultMaxConnectionsPerHost"
value="${solr.defaultMaxConnectionsPerHost}"/>
<property name="maxTotalConnections"
value="${solr.maxTotalConnections}"/>
</bean>
```

在服务层使用这个叫作 "solrServer" 的 Spring Bean。

应用程序运行时依赖一些对象。将依赖对象的创建和管理交由 Spring 容器，这叫作控制反转。控制反转的英文缩写是 IOC。

使用 Spring IOC 机制，创建 org.apache.solr.client.solrj.SolrServer 对象，作为服务中的一个成员变量。使用如下的代码和 Solr 打交道：

查询：

```
SolrQuery query = new SolrQuery();
query.setQuery(search);
QueryResponse qr = **solrServer**.query(query);
return qr.getBeans(SearchItem.class);
```

只是在开发应用程序时使用 Roo，应用程序在运行时不需要 Roo。由于 Roo 在运行时不存在，所以它没有任何性能或内存开销。

5.10 SolrCloud

分布式搜索系统和其他的分布式系统一样，主要关注点不是单节点的异常，而是系统整体的稳定和健壮。

当网络由于发生异常情况，导致分布式系统中部分节点之间的网络延时不断增大，最终导致组成分布式系统的所有节点中，只有部分节点之间能够进行正常通信，而另一些节点则不能——我们将这个现象称为网络分区，就是俗称的"脑裂"。分布式系统需要解决不同节点间如何达成共识的问题。

5.10.1 Zab 协议

Zab(Zookeeper atomic broadcast protocol)是一个有序广播协议。Zab 有前缀顺序属性。它允许在同一个时刻主/备实现多个悬而未决的操作。在领导权改变的时候，Zab 做了一些特殊的处理，保证领导者提出的值只有唯一的顺序。

5.10.2 ZooKeeper

ZooKeeper 使用 Zab 分布式协议。ZooKeeper 最新的版本可以通过官网 http://zookeeper.apache.org/来获取。通过下面的网址选择最接近镜像：

http://www.apache.org/dyn/closer.cgi/zookeeper/

Zookeeper 的安装非常简单，下面将从单机模式和集群模式两个方面介绍 Zookeeper 的安装和配置。

单机安装非常简单，只要获取 Zookeeper 的压缩包并解压到某个目录(如 /home/zookeeper-3.4.9)下，Zookeeper 的启动脚本在 bin 目录下，Linux 下的启动脚本是 zkServer.sh。

只需获取 Zookeeper 的压缩包并解压到某个目录(如 /home/zookeeper-3.4.9)下，Zookeeper 的启动脚本在 bin 目录下，Linux 下的启动脚本是 zkServer.sh。在 Windows 下启动 zkServer.cmd。

Zookeeper 的配置文件在 conf 目录下，这个目录下有一个 zoo_sample.cfg 文件，要将 zoo_sample.cfg 改名为 zoo.cfg，因为 Zookeeper 在启动时会找这个文件作为默认配置文件。在 zoo.cfg 中指定 clientPort 端口号。

启动后要检查 Zookeeper 是否已经在服务，可以通过 netstat - ano 命令查看是否有配置的 clientPort 端口号在监听服务。也可以运行 ./zkServer.sh status 命令检查状态。

为了编写 Java 客户端代码，可以先手工下载一些必需的 jar 包。创建一个 pom.xml 文件：

```xml
<?xml version="1.0"?>
<project
     xmlns="http://maven.apache.org/POM/4.0.0"
     xmlns:xsi="http://www.w3.org/2001/XMLSchema-instance"
         xsi:schemaLocation="http://maven.apache.org/POM/4.0.0 http://maven.apache.org/xsd/maven-4.0.0.xsd">

    <modelVersion>4.0.0</modelVersion>
    <groupId>temp.download</groupId>
    <artifactId>temp-download</artifactId>
    <version>1.0-SNAPSHOT</version>

    <dependencies>
            <!-- 需要下载什么 jar 包，添加相应依赖，其余部分无须在意-->
            <dependency>
                <groupId>org.apache.zookeeper</groupId>
                <artifactId>zookeeper</artifactId>
                <version>3.4.9</version>
            </dependency>
    </dependencies>

</project>
```

编写批处理文件内容如下：

```
call mvn -f pom.xml dependency:copy-dependencies
@pause
```

运行完成后会新增 target 文件夹，其下有一个 dependency 文件夹，里面便是需要的 jar 包。

测试 Zookeeper 的 Java 客户端代码：

```java
public class Executor implements Watcher{
    private static ZooKeeper zooKeeper;

    @Override
    public void process(WatchedEvent watchedEvent) {
        System.out.println("接收内容："+watchedEvent.toString());
    }

    public static void main(String[] args) {
        try {
//创建一个 Zookeeper 实例，第一个参数为目标服务器地址和端口，第二个参数为 Session 超时
//时间，第三个参数为节点变化时的回调方法
            zooKeeper =
                new ZooKeeper("123.56.152.236:2181",5000, new Executor());
            System.out.println(zooKeeper.getState());
```

```
            Thread.sleep(5000);
        } catch (IOException e) {
            e.printStackTrace();
        } catch (InterruptedException e) {
            e.printStackTrace();
        }
    }
}
```

Zookeeper 的集群模式的安装和配置也不是很复杂,所要做的就是增加下面几个配置项:

```
initLimit=5
syncLimit=2
server.1=192.168.211.1:2888:3888
server.2=192.168.211.2:2888:3888
```

initLimit:这个配置项是用来配置 Zookeeper 接收客户端(这里所说的客户端不是用户连接 Zookeeper 服务器的客户端,而是 Zookeeper 服务器集群中连接到 Leader 的 Follower 服务器)初始化连接时最长能忍受的心跳时间间隔数。当已经超过 10 个心跳的时间(也就是 tickTime)长度后 Zookeeper 服务器还没有收到客户端的返回信息,那么表明这个客户端连接失败。tickTime 缺省值是 2000,总的时间长度就是 5×2000 毫秒=10 秒。

syncLimit:这个配置项标识 Leader 与 Follower 之间发送消息、请求和应答时间长度,最长不能超过多少个 tickTime 的时间长度,总的时间长度就是 2×2000 毫秒=4 秒。

server.A=B:C:D:其中 A 是一个数字,表示这个是第几号服务器;B 是这个服务器的 IP 地址;C 表示的是这个服务器与集群中的 Leader 服务器交换信息的端口;D 表示的是万一集群中的 Leader 服务器挂了,需要一个端口来重新进行选择,选出一个新的 Leader,而这个端口就是用来执行选择时服务器相互通信的端口。

除了修改 zoo.cfg 配置文件,集群模式下还要配置一个文件 myid,这个文件在 dataDir 目录下,这个文件里面就有一个数据,就是 A 的值,Zookeeper 启动时会读取这个文件,拿到里面的数据与 zoo.cfg 里面的配置信息比较,从而判断到底是哪个 server。

Zookeeper 作为一个分布式的服务框架,主要用来解决分布式集群中应用系统的一致性问题,它能提供基于类似于文件系统的目录节点树方式的数据存储,但是 Zookeeper 并不是用来专门存储数据的,它的作用主要是用来维护和监控存储的数据的状态变化。通过监控这些数据状态的变化,从而可以达到基于数据的集群管理。

5.10.3 使用 SolrCloud

在 SolrCloud 集群中,逻辑意义上的完整的索引叫作一个文档集合(Collection)。使用两个不同端口的 Jetty 启动两个不同的 Solr 服务,组成一个 Solr 集群。

```
>cd example
>java -Dbootstrap_confdir=./solr/collection1/conf
-Dcollection.configName=myconf -DzkRun -DnumShards=2 -jar start.jar
```

其中：
- -DzkRun：让嵌入式 Zookeeper 服务器运行。
- -Dcollection.configName=myconf：设置用于新的集合的配置名。
- -Dbootstrap_confdir=./solr/collection1/conf：因为我们还没有在 zookeeper 配置，这个参数会产生本地配置目录。把./solr/conf 加载成"myconf"。"myconf"这个名字来源于 collection.configName 参数。
- -DnumShards=2：计划把索引分开的逻辑分区的数量。

浏览 http://localhost:8983/solr/#/~cloud 察看集群的状态，也就是 Zookeeper 分布式文件系统的状态。

对于一个 SolrCloud 集群来说，用户可以往其中的任何一个 Solr 服务器增加文档。SolrJ 使用一个已经搭建好的 SolrCloud 集群代码如下：

```
//嵌入式Zookeeper服务器运行在Solr端口加1000，所以端口是9983
String zkHost = "127.0.0.1:9983";
CloudSolrClient server = new CloudSolrClient(zkHost);
SolrQuery parameters = new SolrQuery();
parameters.set("q", "*:*"); //查询所有的文档
parameters.set("qt", "/select");
parameters.set("collection", "Test");
QueryResponse response = server.query(parameters);
```

5.10.4 SQL 查询

首先在命令行测试 Solr 的 SQL 查询。SQL 特性目前仅支持 SolrCloud 集群方式，单机方式并不支持。需要在 SolrCloud 模式运行 Solr。我们可以使用 Solr 分布中的云配置：

```
>bin/solr -e cloud -noprompt
```

上面的命令将创建 gettingstarted 集合。SolrCloud 启动后索引一些数据：

```
>bin/post -c gettingstarted example/exampledocs/*.xml
```

默认情况下，Solr 中有一个叫作/sql 的专门的请求处理器。这个处理器能让我们在 Solr 中使用并行 SQL 功能。

在标准的 SELECT 语句如 "SELECT <expressions> FROM <table>"中，表名对应 Solr 集合的名字。表名不区分大小写。SQL 查询的列名称直接映射到待查询集合在 Solr 索引中的字段。这些标识符是区分大小写的。

让我们从 gettingstarted 集合中检索所有 inStock 字段等于 true 的文档。使用 Solr 中默认的查询解析器，可以运行以下命令：

```
>curl 'localhost:8983/solr/gettingstarted/select?q=inStock:true&fl=id,title&indent=true'
```

这个查询返回 17 个文档。现在让我们使用 SQL 做同样的查询：

```
>curl --data-urlencode 'stmt=select id,name from gettingstarted where
inStock = 'true'' http://localhost:8983/solr/gettingstarted/sql
```

这个功能使用 Facebook 开源的大数据 SQL 检索框架 Presto 的 SQL Parser 实现。它把 SQL 语句编译成流式表达式。

使用 SolrJ 完成同样的查询：

```
String sql = "select id,name from gettingstarted where inStock = 'true'";
Properties info = new Properties();
info.setProperty("lex", "JAVA");
Connection connection = DriverManager.getConnection(
        "jdbc:calcite:model=src/java/test/model.json", info); //建立连接
Statement statement = connection.createStatement();
ResultSet resultSet = statement.executeQuery(sql);
```

当然，还可以限制返回的文档数量，使用 limit 关键字：

```
>curl --data-urlencode 'stmt=select id,name from gettingstarted where
inStock = 'true' limit 2' http://localhost:8983/solr/gettingstarted/sql
```

也可以检索匹配文档的数量：

```
>curl --data-urlencode 'stmt=select count(*) from gettingstarted where
inStock = 'true'' http://localhost:8983/solr/gettingstarted/sql
```

或者根据文档得分或字段值排序：

```
>curl --data-urlencode 'stmt=select id,name from gettingstarted where
inStock = 'true' order by id desc'
```

5.11 Solr 原理

介绍如何开发支持 Solr 的中文分词和 Solr 中的缓存技术以及相关参数调整。

5.11.1 支持 Solr 的中文分词

因为 Solr 会重用 Tokenizer 的实例，所以首先保证 NgramTokenizer 正确地重写 reset() 方法。例如：

```
public final class NgramTokenizer extends Tokenizer {
    @Override
    public void reset() throws IOException {
        super.reset();
        result.clear();
        this.done = false;
        this.upto = 0;
    }
}
```

写个分词工厂类，这个类必须扩展 TokenizerFactory。TokenizerFactory 这个类位于 org.apache.lucene.analysis.util 包，此外 ResourceLoader 也位于 org.apache.lucene.analysis.util。词典当作资源加载。

```java
public class CnTokenizerFactory extends TokenizerFactory {
    static final Logger log = 
LoggerFactory.getLogger(CnTokenizerFactory.class);

    BigramDictioanry dict; //分词词典

    public Tokenizer create(Reader input) {
        return new NgramTokenizer(input, dict);
    }

    public void inform(ResourceLoader loader) {
        String dicPath = args.get("dicDir");
        log.info("词典路径=" + dicPath);
        dict = BigramDictioanry.getInstance(dicPath);
    }
}
```

这里的 CnTokenizerFactory 是一个分词类，它扩展了 org.apache.lucene.analysis.util.TokenizerFactory。当然也可以定义自己的 Tokenizer 类。所有这样的类都必须是 TokenizerFactory 的子类。需要注意的是：Lucene 4.3 版本的 TokenizerFactory 写法和以前不一样，其中的 create() 方法增加了一个 AttributeFactory 参数。CnTokenizerFactory 改成：

```java
@Override
public CnTokenizer create(AttributeFactory factory, Reader input) {
    return new CnTokenizer(factory, input, dict);
}
```

Solr 通过 Tokenizer 工厂类创建一个 Tokenizer，然后通过 reset 重用，所以 CnTokenizer 需要已经实现 reset() 方法。

下面定义一个基本的文字搜索列类型。在配置文件中指定词典路径。

```xml
<fieldType name="textSimple" class="solr.TextField" positionIncrementGap="100" >
    <analyzer>
        <tokenizer class="com.lietu.seg.CnTokenizerFactory"
                dicDir="C:/apache/apache-solr-3.1.0/example/solr/dic/"/>
    </analyzer>
</fieldType>
```

在 Solr 的控制界面测试索引和查询时的分词效果。

5.11.2 缓存技术

减少搜索过程中的磁盘 IO 对提升搜索响应速度有很大的帮助。Solr 中主要有以下 3 种缓存：

- Filter cache(过滤器缓存)：用于保存过滤器(fq 参数)和刻面搜索的结果；
- Document cache(文档缓存)：用于保存 Lucene 文档存储的字段；
- Query result(查询结果缓存)：用于保存查询的结果。

缓存使用的内存总是有限的。需要使用一种算法来检测和替换不值得的东西。有一些算法用于缓存项替换。包括：

- Least Recently Used (LRU)：最近最少使用；
- Least Frequently Used (LFU)：最不经常使用；
- First In First Out (FIFO)：先进先出。

最受欢迎的一个算法是最近最少使用(LRU)算法。研究表明，相比旧物品来说，更可能使用新的项目。LRU 就是基于这个观察。该算法保持跟踪项目的最后访问时间。它清除有最古老的访问时间戳的项目。

如果想实现 LRU 缓存，LinkedHashMap 确实很有用。即使是 Sun 的 Java 框架也使用这个类来实现 com.sun.tdk.signaturetest.util.LRUCache 和 sun.security.ssl.X509KeyManagerImpl.SizedMap。

对于这个实现，应该重写 removeEldestEntry()方法。在 put()和 putAll()之后调用这个方法。基于其返回值 Map 删除旧的条目。如果这个方法返回 true，那么旧条目删除。否则它会留在 Map 里面。这个方法的默认实现返回 false。在这种情况下，旧条目保留在 Map，不会被删除，它只是和普通的 Map 集合类一样。在这个方法的大多数实现中，如果 Map 中的条目的数量大于初始容量，就返回 true。

```
public class LRUCacheImpl extends LinkedHashMap<Integer, String> {
  private static final long serialVersionUID = 1L;
  private int capacity;

  public LRUCacheImpl(int capacity, float loadFactor){
    super(capacity, loadFactor, true);
    this.capacity = capacity;
  }

  /**
   * removeEldestEntry() 用户应该重写这个方法
   */
  @Override
  protected boolean removeEldestEntry(Map.Entry<Integer, String> eldest){
    return size() > this.capacity;
  }

  public static void main(String arg[]){
```

```
        LRUCacheImpl lruCache = new LRUCacheImpl(4, 0.75f);

        lruCache.put(1, "Object1");
        lruCache.put(2, "Object2");
        lruCache.put(3, "Object3");
        lruCache.get(1);
        lruCache.put(4, "Object4");
        System.out.println(lruCache);
        lruCache.put(5, "Object5");
        lruCache.get(3);
        lruCache.put(6, "Object6");
        System.out.println(lruCache);
        lruCache.get(4);
        lruCache.put(7, "Object7");
        lruCache.put(8, "Object8");
        System.out.println(lruCache);
    }
}
```

println()方法按过时性打印对象。输出如下：

```
{2=Object2, 3=Object3, 1=Object1, 4=Object4}
{4=Object4, 5=Object5, 3=Object3, 6=Object6}
{6=Object6, 4=Object4, 7=Object7, 8=Object8}
```

正如在上面的代码中所看到的，插入 Object1、Object2 和 Object3，然后在插入 Object4 之前访问 Object1。因此在输出的第一行中，在 Object4 之前打印 Object1。当插入 Object5 以后，从列表中去掉了 Object2，因为这个对象在列表中是最古老的。当访问 Object3 时，会把它提升到高于 Object5。当插入 Object6 以后，删除 Object1。

Solr 实现了两种缓存机制，分别是 LRUCache 和 FastLRUCache。LRUCache 采用基于线程安全的 LinkedHashMap 实现。而 FastLRUCache 则基于 ConcurrentHashMap 实现。

单线程的情况下 FastLRUCache 具有更快的 gets 操作和比较慢的 puts 操作，因此使用它的查询命中率会比 LRUCache 高出 75%，当然在多线程的情况下差距可能更大。

5.12 本章小结

Solr 是网站的内部系统演化出来的一个基于 Lucene 的开源项目。2004 年秋天，CNET 启动 Solr 项目的前身 Solar。2005 夏天，CNET 产品目录搜索开始使用 Solar。2006 年 1 月 CNET 把这个搜索项目代码捐赠给 Apache 并命名为 Solr。2007 年 1 月 Solr 成为 Lucene 的子项目并发布 1.2 版本。2008 年 9 月发布 1.3.0 版本。2009 年 11 月发布 1.4 版本。Solr 当前主要由 Yonik 维护代码，Yonik 是斯坦福大学计算机专业硕士毕业。关于 Solr 的文档介绍参考 http://wiki.apache.org/solr/。

Solr 的索引主-从分发结构从 2006 年开始就已经出现，而 SolrCloud 则在 2012 年才在 4.0 版本引入。

第 5 章　Solr 分布式搜索引擎

本章介绍了企业级的搜索服务器 Solr。它的基本用法包括如下几个步骤：
- 修改索引模式定义文件；
- 把数据灌入 Solr；
- 实现 Solr 客户端界面。

对于简单的网站搜索，可以在索引阶段使用 Solr，在搜索界面直接使用 Lucene。也就是说，在搜索阶段不使用 Solr 服务。

因为 SolrJ 除了支持 XML 格式，还可以采用 Java 二进制(wt=javabin)方式返回搜索结果，所以性能相对其他客户端更好。

Solr 的 PHP 客户端除了 solr-php-client，还可以使用 Solarium (http://www.solarium-project.org/)。

和 Lucene 相比，Solr 有更多的缓存支持，并且支持主从复制和分片两种分布式部署方式。2009 年年底发布的 Solr 1.4 集成了在搜索结果中聚类的功能。聚类功能本身使用另外一个开源项目 Carrot2(http://project.carrot2.org/)完成。

雅虎开发了高性能高可靠的分布式协调服务 ZooKeeper。ZooKeeper 最初是作为 Hadoop 的一个子项目存在。2011 年 1 月，ZooKeeper 脱离 Hadoop，成为 Apache 顶级项目。

Presto 是一个用于大数据的分布式 SQL 查询引擎。

除了使用 Solr，还可以使用 Elasticsearch 搭建搜索服务器集群。

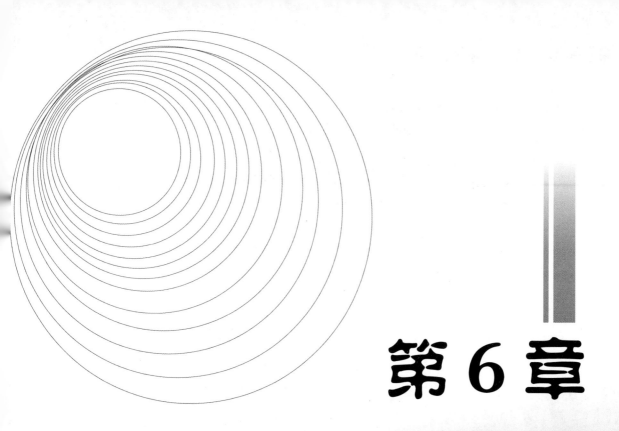

第6章

ElasticSearch 分布式搜索引擎

Lucene 穿了一件 JSON 的外衣，就是 ElasticSearch。ElasticSearch 的强大之处在于对分布索引的管理。使用 ElasticSearch 的搜索系统整体架构如图 6-1 所示。

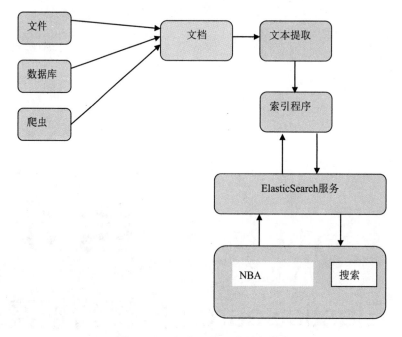

图 6-1　ElasticSearch 的外部结构

ElasticSearch 代表了大数据和搜索集群的结合体。因为创建索引耗时，所以文档预先写入到一个日志目录。

ElasticSearch 比 Solr 支持两个额外的东西——嵌套的文档结构和一个索引内多种文档类型。ElasticSearch 与 Solr 的对比见表 6-1。

表 6-1　ElasticSearch 与 solr 的对比

项　目	Solr	ElasticSearch
配置文件	schema.xml	elasticsearch.yml
整个删除索引库	直接删除目录	必须调用删除 API
缓存	*:*这样的查询语句会影响性能	可以作为文档数据库和 k-v 数据库来用
Web 容器	使用 Jetty	使用 Netty

Netty 是一个高性能、事件驱动的异步的非堵塞的 IO 框架。ElasticSearch 默认是使用 Netty 作为 HTTP 的容器，由于 Netty 并没有权限模块，所以默认 ElasticSearch 没有任何的权限控制，直接通过 HTTP 就可以进行任何操作，除非把 HTTP 禁用。如果使用 elasticsearch-jetty 插件，就可以使用 Jetty 自带的权限管理进行一些权限的控制，同时也可以支持通过 HTTPS 协议来访问 ElasticSearch，还有就是支持 gzip 压缩响应信息。

第 6 章 ElasticSearch 分布式搜索引擎

6.1 安　　装

首先介绍 ElasticSearch 在 Windows 下的安装，然后介绍在 Linux 下的安装。

安装包下载网址：http://www.elasticsearch.org/download/。当前最新版本为 5.1.2。得到文件：elasticsearch-5.1.2.zip。

直接解压至某目录，例如：D:\elasticsearch-5.1.2。下载完并解压后有以下几个文件夹：bin——运行的脚本；config——设置文件；lib——用于存放依赖的包。

到目录 D:\elasticsearch-5.1.2\bin 下，运行 elasticsearch.bat。

如果显示 Java 虚拟机内存不够，则可以在 D:\elasticsearch-5.1.2\config\jvm.options 配置文件中调整内存大小。

在浏览器中打开网址：http://localhost:9200/。

启动成功后，会在解压目录下增加 2 个文件夹：data，用于存储索引数据；logs，用于日志记录。

为了看到索引内容，需要安装 head 插件。Elasticsearch 5.x 以前的版本安装插件的过程：

(1) 首先创建目录 D:\elasticsearch-2.*\plugins；

(2) 执行命令 D:\elasticsearch-2.*\bin>.\plugin.bat -install mobz/elasticsearch-head。

在浏览器中打开网址：http://localhost:9200/_plugin/head/，可以看到一个管理界面。可以在浏览器中查看到索引元数据信息。如果没有安装插件，就会返回 404 错误。

Elasticsearch 5.x 安装 head 插件需要随同 NodeJS 一起安装包管理工具 npm。

在 Linux 下安装则下载并解压缩 elasticsearch-5.1.2.tar.gz。启动：

sh elasticsearch start

执行 elasticsearch，默认会以控制台的方式执行，如果想脱离控制台运行，则加上参数 -d 即可(./elasticsearch -d)。

如果访问不了，先检查相关端口号是否被防火墙屏蔽了。方法是用 wget http://localhost:9200/。用 iptables 修改防火墙设置。

在 Linux 下安装 head 插件的过程：

首先在 elasticsearch-Home 路径下创建 plugins 目录：

```
# mkdir plugins
```

然后执行命令：

```
# ./plugin -install mobz/elasticsearch-head
-> Installing mobz/elasticsearch-head...
Trying https://github.com/mobz/elasticsearch-head/archive/master.zip...
```

发送 POST 请求停止节点：

```
#停止本地节点
$ curl -XPOST 'http://localhost:9200/_cluster/nodes/_local/_shutdown'
```

```
#停止集群中所有的节点
$ curl -XPOST 'http://localhost:9200/_shutdown'
```

6.2 搜索集群

分布式索引需要考虑的问题包括：
- 确定一个集群中应该包括哪些机器；
- 数据如何分布；
- 快速地分布式打分。

数据分布有两种方法：按文档分片或者按词分片。不同的文档集合放在不同的机器上。不同的词列表集合放在不同的机器上。ElasticSearch 仍然是把不同的文档集合放在不同的机器上。

最简单的方法是手工管理集群。修改 elasticsearch.yml 文件：

对于主机 esa.lietu.com：

```
cluster.name: OurCluster
node.name: "esa"
discovery.zen.ping.multicast.enabled: false
discovery.zen.ping.unicast.hosts: ["esb.lietu.com"]
```

对于主机 esb.lietu.com：

```
cluster.name: OurCluster
node.name: "esb"
discovery.zen.ping.multicast.enabled: false
discovery.zen.ping.unicast.hosts: ["esa.lietu.com"]
```

一般使用多播自动建立集群。cluster name 一样，而 nodename 不一样，就自动加入集群。ElasticSearch 集群使用什么管理？默认使用 Zen Discovery。

6.2.1 Zen 发现机制

Zen 发现机制是 ElasticSearch 默认的内建模块。它提供了多播和单播两种发现方式，能够很容易地扩展至云环境。

Zen 发现机制是和其他模块集成的，例如所有节点间通信必须用 Transport 模块来完成。Transport 这层是自己可以扩展的，thrift 也是一个插件。

ElasticSearch 运行时会启动两个探测进程。一个进程用于从主节点向集群中其他节点发送 ping 请求来检测节点是否正常可用。另一个进程的工作反过来，由其他的节点向主节点发送 ping 请求来验证主节点是否正常且忠于职守。

一个集群有唯一的一个名字，包含一个或者多个节点。集群会在所有的节点中自动选择一个作为主节点，如果主节点宕机了，则会自动选择另外一个节点作为主节点。一个经典的主节点选举算法是同行评审出版算法(peer-reviewed published algorithm)。

第 6 章　ElasticSearch 分布式搜索引擎

ElasticSearch 采用了一个简单的方法选出主节点：它根据编号来选择节点，较小的编号更有可能成为主节点。DiscoveryNode 类中记录了节点编号。选举算法的实现代码在 ElectMasterService.electMaster()方法中。

为了避免一个集群中存在不同的主节点，也就是避免脑裂，需要合理地设置 elasticsearch.yml 配置文件。

假设可以成为集群一部分的 ES 节点的数量(ES 进程而不是物理机器的数量)是 N，那么在一个有 $N>2$ 个节点的集群上，可以设置 discovery.zen.minimum_master_nodes 的值不小于 $(N/2)+1$。

理想的拓扑结构是有 3 个专用的主节点，即 master: true，data:false，discovery.zen。minimum_master_nodes 设置为 2。这样无论有多少数据节点应该是集群的一部分，都不需要改变设置。

每个文档都保存在单独的主分片里。当对一个文档做索引的时候，首先对主分片做索引，然后在所有主分片的副本里做索引。默认一个索引有 5 个主分片，可以调整主分片的数量以控制一个索引中容纳文档的数量。索引创建之后，不可以更改主分片数。即使只在一台机器上安装 ES，也可能会有 5 个独立的索引库。

每个主分片可以有 0 个或者多个副本。副本是主分片的备份，有两个作用：

(1) 提高容错能力：如果主分片宕机，副本分片可以被提升至主分片。
(2) 提高性能：搜索访问可以分布在主分片和副本分片之间。

默认每个主分片有一个副本分片，但副本分片数量可以在已经存在的索引上动态调整。在同一个节点上，副本分片不会被当作主分片启动。

举例说明索引分片的用处：第一台机器中存放索引分片 a、b、c，第二台机器中存放索引分片 a、b、d，第三台机器中存放索引分片 b、c、d。提升索引整体容量的同时，提升性能和容错能力。

新增一个节点，ElasticSearch 会自动把索引数据同步到这个新增的节点上。控制界面中显示的紫色的块表示正在迁移这部分数据。

主控节点管理 shard 的分配。当新机器进来或者有旧机器失效的时候，就会重新分配 shard。

依赖注入(Dependency Injection，DI)很好，因为一个节点有很多个索引，每个索引有很多个分片，每个分片是不同的 Guice 模块。ElasticSearch 使用 Google 开源的依赖注入框架 Guice(https://github.com/google/guice)，没有使用 Spring 实现依赖注入的原因是：Spring 需要配置文件，用起来太笨重。ElasticSearch 直接把 Guice 的源码放到自己的 org.elasticsearch.common.inject 包内。

6.2.2　JGroups

ElasticSearch 早期的版本使用 JGroups 实现多播。JGroups 就是一个用于方便集群开发的组件。它依赖组播。使用的地址是组播段的 IP 地址，协议是 UDP 格式的协议。这个需要路由交换支持。Windows 不用关心这些，因为 Windows 自身有一个编写得比较好的 NetworkManager 服务，能够根据应用程序的请求灵活配置路由规则。若同样的网络环境，

Linux 收组播流有路由配置的问题。

首先创建一个 JChannel 对象，然后通过调用 connect()方法加入一个集群：

```
JChannel channel=new JChannel();
channel.connect("SearchCluster");
```

如果 SearchCluster 这个集群已经存在，则加入该集群。不需要先定义集群，再加入这个集群。如果这个集群还不存在，则自动创建。

连接到集群中的所有实例(instance)被称为一个视图(org.jgroups.View)。通过 View.getMembers()可以得到所有实例的地址。org.jgroups.Address 接口及其实现类封装了地址信息，它通常包含 IP 地址和端口号。

使用 JGroups 实现领导者选举的代码如下：

```
import org.jgroups.Address;
import org.jgroups.JChannel;
import org.jgroups.View;

public class JGroupsLeaderElectionExample {
  private static final int MAX_ROUNDS = 1_000;
  private static final int SLEEP_TIME_IN_MILLIS = 1000;

  public static void main(String[] args) throws Exception {
    JChannel channel = new JChannel();
    channel.connect("The Test Cluster");
    for (int round = 0; round < MAX_ROUNDS; round++) {
      checkLeaderStatus(channel);
      sleep();
    }

    channel.close();
  }

  private static void sleep() {
    try {
      Thread.sleep(SLEEP_TIME_IN_MILLIS);
    } catch (InterruptedException e) {
      // Ignored
    }
  }

  private static void checkLeaderStatus(JChannel channel) {
    View view = channel.getView();  // 通过 View 得到集群中的成员名单
    Address address = view.getMembers()
                 .get(0);
    if (address.equals(channel.getAddress())) {
      System.out.println("I'm (" + channel.getAddress() + ") the leader");
    } else {
```

```
            System.out.println("I'm (" + channel.getAddress() + ") not the leader");
        }
    }
}
```

6.3 创建索引

可以在 Solr 的配置文件中预先定义好索引库的结构，不能在 elasticsearch.yml 中预先定义。表的结构和相关设置的信息在 mapping 中设置。

索引模板也可以放置在模板目录下的配置位置(path.conf)。注意：所有有主节点资格的节点都要放。例如：一个叫作 template_1.json 的文件可以放置在 config/templates 目录下，如果它能够匹配上一个索引，就会把它添加进去。放在 config/[index_name]/[some_name].json 的例子如下：

```
{
    "[type]" : {
        "properties" : {
            "title" : {
                "type" : "string",
                "boost" : 2.0
            }
            "tags": {
                "type" : "string"
            }
        }
    }
}
```

在命令行设置 Mapping 的例子：

```
curl -XPOST localhost:9200/wf_mds_org(索引名称) -d '{
    "settings": {
        "number_of_shards": 1,
        "number_of_replicas": 0,
        "index.refresh_interval": "-1",
        "index.translog.flush_threshold_ops": "100000"
    },
    "mappings": {
        "org": {     //(类型)
            "_all": {
                "analyzer": "ike"
            },
            "_source": {
                "compress": true
            },
            "properties": {
```

```
            "_ID": {
    "type": "string",
    "include_in_all": true,
    "analyzer": "keyword"
},
                "NAME": {
    "type": "multi_field",
    "fields": {
        "NAME": {
            "type": "string",
            "analyzer": "keyword"
        },
        "IKO": {
            "type": "string",
            "analyzer": "ike"
        }
    }
},
                "SHORTNAME": {
    "type": "string",
    "index_analyzer": "pct_spliter",
    "search_analyzer": "keyword",
    "store": "no"
},
                "OLDNAME": {
    "type": "multi_field",
    "fields": {
        "OLDNAME": {
            "type": "string",
            "analyzer": "keyword"
        },
        "IKO": {
            "type": "string",
            "analyzer": "ike"
        }
    }
},
                "TNAME": {
    "type": "string",
    "analyzer":"custom_snowball_analyzer",
    "store": "no"
},
                "TSNAME": {
    "type": "string",
    "index": "no",
    "store": "no"
},
```

```
                    "TONAME": {
            "type": "string",
            "index": "no",
            "store": "no"
        }
      }
    }
  }
}'
```

上面给出了一个完整 Mapping，我们可将 Mapping 信息大致分为 settings 和 mappings 两个部分，settings 主要是作用于 index 的一些相关配置信息，如分片数、副本数等(分片和副本在 ElasticSearch 简介中讲过，更加详细的东西会在索引优化篇中讲)、tranlog 同步条件、refresh 条条等；Mappings 部分主要是索引结构的一些说明。mappings 主体上大致又分成 _all、_source、properites 3 个部分。

(1) _all：主要指的是 All Field 字段，可以将一个或多个包含进来，检索时可在无须指定字段的情况下检索多个字段。前提是要开启 All Field 字段：

```
"_all" : {"enabled" : true}
```

(2) _source：主要指的是 Source Field 字段。Source 可以理解为 ES 除了将数据保存在索引文件中，另外还有一份源数据。_source 字段在进行检索时相当重要，如果在{"enabled": false}情况下默认检索只会返回 ID，需通过 Fields 字段去倒排索引中取数据，当然效率不是很高。如果设置 enabale:true，索引的膨胀率比较大时，可以通过下面一些辅助设置进行优化：

```
Compress:是否进行压缩，建议一般情况下将其设为 true
"includes" : ["author", "name"],
"excludes" : ["sex"]
```

上面的 includes 和 excludes 主要是针对默认情况，_source 一般是保存全部 Bulk 过去的数据，我们可以通过 include,excludes 在字段级别上做出一些限制。

(3) properites 部分是最重要的部分，主要是针对索引结构和字段级别上面的一些设置。

```
"NAME": { //字段项名称对应 Lucene 里面 FiledName
    "type": "string",//type 为字段项类型
    "analyzer": "keyword"//字段项分词的设置对应 Lucene 里面的 Analyzer
},
```

在 ES 中字段项的 type 是一个很重要的概念，在 ES 中，在 Lucene 的基础上提供了比较多的类型，而这些类型对应一些相关的检索特性，如 Date 型，可以使用[2001 TO 2012]的方式进行范围检索等，ES 的类型有如下一些：

简单类型：

```
String:字符型最常用的
Integer:整型
Long:长整型
```

```
Float:浮点型
Double:双字节型
Boolean:布尔型
```

复杂类型：

```
Array：数组型
"lists":{{"name":"…"},{"name":"…"}}
Object:对象类型
"author":{"type":"object","perperites":{"name":{"type":"string"}}}
Multi_field：多分词字段，针对一个字段提供多种分词方式
Nested：嵌入类型用得还是比较多的
```

Analyzer，在 Lucene 中是一个分词器的概念，我们知道 ES 是建立在 Lucene 之上的，所以这里的 Analzyer 同样也适用，Mapping 中的 Analyzer 是指定字段采用什么分词器，具体的程序和配置分词在插件和配置部分都有过一些说明。

```
Analyzer 在 Es 中分为 index_analyzer 和 search_analyzer
Index_analzyer：指的是索引过程中采用的分词器
Search_analyzer:指的是检索过程中采用的分词器
```

我们知道 index 和 search 是两个过程，但是尽量保证这两个过程和分词方式一致，这样可以保证查全和查准，否则再厉害的分词，index 和 search 采用的不相同也是徒劳。

与 Analyzer 相关的就是另外一项 index。例如：

```
"HC":{ "type":"string", "index":"no", "store":"no"}
```

Index 表示该字段是否索引，如果 index 为 "no" 则 Analyzer 设置为任何值也没用。
最后是 "store" 项，它表示该项是否存储到倒排索引中去，并不是_source。
可以查询所有的文档类型或只选择一个单一的文档类型来查询。
自定义的分词器，修改配置文件，位于 config 目录下的 elasticsearch.yml。
如何指定不同的列，使用不同的分词器？以前 Solr 在 schema.xml 中指定很方便。
调用 http://localhost:9200/index_name/_mapping，这样的接口提供了列所使用的分析器。
例如：

```
http://localhost:9200/news/_mapping
```

返回结果：

```
{"news":{"mappings":{"type1":{"properties":{"body":{"type":"string"},"title":{"type":"string"}}}}}}
```

6.4 Java 客户端接口

虽然可以用 curl 这样的命令行工具与 ElasticSearch 搜索服务器打交道，但实际开发中一般使用 ElasticSearch 的 Java API。

首先保证客户端的 JDK 和服务器的版本一致。然后在 Eclipse 中创建一个项目，用 Maven

第 6 章 ElasticSearch 分布式搜索引擎

管理项目，或者把 D:\elasticsearch-5.1.2\lib 目录下的 jar 包复制到 lib 目录下。ElasticSearch 的 Java 客户端 API 最少依赖两个包：elasticsearch-5.1.2.jar 和 lucene-core-6.3.0.jar。

通过 TransportClient 这个接口和 ES 集群进行通信时，需要指定 ES 集群中其中 1 台或多台机器的 IP 地址和端口号。连接本地集群的例子代码：

```java
String HOST_NAME = "localhost";
int PORT = 9300;
Client client= new TransportClient().addTransportAddress(
        new InetSocketTransportAddress(HOST_NAME, PORT));
```

指定多个机器连接的例子：

```java
Client client = new TransportClient()
    .addTransportAddress(new InetSocketTransportAddress("host1", 9300))
    .addTransportAddress(new InetSocketTransportAddress("host2", 9300));
//使用连接…
client.close();
```

集群名默认是"elasticsearch"。如果需要更改集群名，需要如下设置：

```java
Settings settings = ImmutableSettings.settingsBuilder()
            .put("cluster.name", "myClusterName").build();
Client client = new TransportClient(settings);
```

可以设置 client.transport.sniff 为 true 来使客户端去嗅探整个集群的状态，把集群中其他机器的 IP 地址加到客户端中，这样做的好处是不用手动设置集群里所有集群的 IP 到连接客户端，它会自动帮你添加，并且自动发现新加入集群的机器。代码实例如下：

```java
Settings settings = ImmutableSettings.settingsBuilder()
            .put("client.transport.sniff", true).build();
TransportClient client = new TransportClient(settings);
```

创建索引：

```java
Settings settings =
   ImmutableSettings.settingsBuilder().put("cluster.name",
    "myClusterName").put("client.transport.sniff", true).build();
Client client =
 new TransportClient(settings).addTransportAddress(
   new InetSocketTransportAddress("localhost", 9300));
CreateIndexRequestBuilder createIndexRequestBuilder =
    client.admin().indices().prepareCreate("test1");  //创建一个叫作test1的索引
CreateIndexResponse response =
createIndexRequestBuilder.execute().actionGet();
System.out.println(response.isAcknowledged());
```

使用 IndicesAdminClient 删除索引。

```java
IndicesAdminClient admin = client.admin().indices();
return admin.prepareDelete(indexName).execute().actionGet().isAcknowledged();
```

6.4.1 创建索引

使用 IndicesAdminClient.prepareCreate()方法设置好创建索引需要的参数。

```
Client client = ESUtils.getClient();
Settings settings = ImmutableSettings.settingsBuilder()
        // 1个主分片
        .put("number_of_shards", 1)
        // 测试环境，减少副本提高速度
        .put("number_of_replicas", 0).build();
// 首先创建索引库
CreateIndexResponse indexresponse = client.admin().indices()
        // 这个索引库的名称还必须不包含大写字母
        .prepareCreate("testindex").setSettings(settings).execute()
        .actionGet();
System.out.println(indexresponse.isAcknowledged()); //看是否成功创建索引
```

也可以用 JSON 字符串设置参数。使用 XContentFactory 提供的方法构建 JSON 字符串。

```
String jsonStr = XContentFactory.jsonBuilder().startObject()
.field("number_of_shards", 1).endObject().string();
System.out.println(jsonStr);   //输出{"number_of_shards":1}
```

设置参数完整的代码如下：

```
Builder setting = ImmutableSettings.settingsBuilder().loadFromSource(
    XContentFactory.jsonBuilder().startObject()
            .field("number_of_shards", 1).endObject().string());

IndicesAdminClient ac = client.admin().indices();
CreateIndexRequestBuilder builder =
ac.prepareCreate(index).setSettings(setting);
```

6.4.2 插入数据

可以单条或者批量插入数据。还可以用 JDBC River 插件配置数据源来插入数据。插入单条记录的方法：

```
public void insert(String index, String type, Map<String, String> data) {
    Gson gson = new Gson();
    String json = gson.toJson(data);   //使用 GSON 把 Map 转换成 JSON 字符串
    Client client = getClient();
    IndexRequestBuilder indexBuilder = client.prepareIndex(index, type)
            .setSource(json);
    IndexResponse response = indexBuilder.execute().actionGet();
}
```

索引一个 JSON 字符串表示的文档。

```
String json = "{" +
    "\"user\":\"kimchy\"," +
    "\"postDate\":\"2013-01-30\"," +
    "\"message\":\"trying out Elasticsearch\"," +
  "}";

IndexResponse response = client.prepareIndex("twitter", "tweet")
    .setSource(json)
    .execute()
    .actionGet();
```

使用 XContentFactory 得到 JSON 字符串：

```
XContentBuilder b = jsonBuilder().startObject();
b.field("title", "标题");
b.field("body", "内容");
b.endObject();

String id="http://xxx";    //唯一列的值

IndexRequestBuilder irb = client.prepareIndex(getIndexName(), getIndexType(), id).
      setSource(b);
irb.execute().actionGet();
```

批量插入数据：

```
public void bulkInsert(String index, String type, Map sourceMap) {
    Client client = getClient();
    BulkRequestBuilder bulkRequestBuilder = client.prepareBulk();
    bulkRequestBuilder.add(client.prepareIndex(index, type).setSource(
          sourceMap));
    BulkResponse bulkResponse = bulkRequestBuilder.execute().actionGet();
}
```

批量插入数据可能出错：

```
if (bulkResponse.hasFailures()) {
 this.logger.error(bulkResponse.buildFailureMessage());
}
```

ElasticSearchClient 类简单封装了增加删除文档的方法。

```
ElasticSearchClient esc = new ElasticSearchClient();
String index="test1";   //索引名
String type="type1";    //类型，类似表名

Map<String , String> sourceMap = new HashMap<String , String>();
sourceMap.put("title", "value1");   //列名是 title，值是 value1
sourceMap.put("body", "value2");    //列名是 body，值是 value2
```

```
esc.insertRecord(index, type, sourceMap);
```

插入多值列:

```
String[] cn = { "意义1", "意义2"};
XContentBuilder qBuilder = jsonBuilder().startObject();
qBuilder.field("word", "hidden");
qBuilder.field("pos", "adj");
qBuilder.startArray("cn");
for (String field : cn) {
    qBuilder.value(field);
}
qBuilder.endArray();
qBuilder.endObject();

IndexRequestBuilder irb = client.prepareIndex(getIndexName(),
        getIndexType()).setSource(qBuilder);
```

通过重试解决 no node available 错误:

```
while (true) {
    try {
        bulk.execute().actionGet(getRetryTimeout());
        break;
    }
    catch (NoNodeAvailableException cont) {
        Thread.sleep(5000);
        continue;
    }
}
```

如果要存储大的二进制文件,可以使用 attachments 插件。

6.4.3 索引库结构

在 Mapping 中声明索引库的结构。设置索引库的结构的代码如下:

```
IndicesAdminClient ac = client.admin().indices();
CreateIndexRequestBuilder builder =
ac.prepareCreate(index).setSettings(setting);

String type = "type1";
builder = builder.addMapping(type, getMapping());
CreateIndexResponse indexresponse = builder.execute().actionGet();
```

XContentFactory 是 ES 内部的一个工具类。使用这个类得到 JSON 格式的字符串。

6.5 查　　询

ElasticSearch 提供基于 JSON 的 Query DSL 查询表达式，DSL 即领域专用语言。可以把 Query DSL 当作是一系列的抽象的查询表达式树。某个查询能够包含其他的查询(如布尔查询)，有些查询能够包含过滤器(如 constant_score)，还有的可以同时包含查询和过滤器(如 filtered)。

Query DSL 是一个通用的查询框架。可以通过 Java API 向搜索服务器发送 JSON 格式的 Query DSL。

基本词查询：

```
String keyWords = "value1";   //查询词
TermQueryBuilder qb=new TermQueryBuilder("title", keyWords);
//或者用 QueryBuilder qb = QueryBuilders.termQuery("title", keyWords);

String index = "test1"; // 索引名

Client client = getClient();
SearchResponse searchResponse =
        client.prepareSearch(index).setQuery(qb).execute().actionGet();
```

返回结果：

```
SearchHits hits = response.getHits();
for (SearchHit hit : hits) {
    System.out.println("id "+hit.getId()); // 文档id
    Map<String, Object> result = hit.getSource(); // 键是列名，值是文档中该列的值
    for (final Entry<String, Object> entry : result.entrySet()) {
        System.out.println(entry.getKey() + " : " + entry.getValue());
    }
}
```

如果不指定列，则可以通过 SearchHit.getSource()方法返回所有列的值，也可以只返回指定返回查询的列，例如只返回 word 列的内容：

```
SearchResponse searchResponse =
client.prepareSearch(DictCrawler.getIndexName())
        .setQuery(query).addFields("word").execute().actionGet();
SearchHits hits = searchResponse.getHits();
for (SearchHit hit : hits) {
    Map<String, SearchHitField> map = hit.getFields();

    SearchHitField field = map.get("word");
    System.out.println(field.value());
}
```

可以指定索引库中的类型：

```
String index = "test1"; // 索引名
SearchRequestBuilder builder = client.prepareSearch(index);
String type = "type1"; //类型名
builder.setTypes(type);
```

短语查询：

```
String keyWords = "自2006年8月2日北京市";
String field = "body";
MatchQueryBuilder qb = QueryBuilders.matchPhraseQuery(field, keyWords);
```

使用 BoolQueryBuilder 把多个查询条件串联起来。例如同时查询标题和内容列：

```
String keyWords = "工作顺利完成";
String field = "body";
MatchQueryBuilder pqBody = QueryBuilders.matchPhraseQuery(field, keyWords);

field = "title";
MatchQueryBuilder pqTitle = QueryBuilders.matchPhraseQuery(field, keyWords);

QueryBuilder qb =
QueryBuilders.boolQuery().should(pqBody).should(pqTitle);
```

同时查询多列：

```
MultiMatchQueryBuilder qb = QueryBuilders.multiMatchQuery(keyWords,
"title","body");
```

可以把连续匹配的文档排前面，也可以返回不连续匹配的文档，组合模糊匹配和短语查询。

```
//让连续出现查询串的文档相关度高
String field = "body";
MatchQueryBuilder pqBody = QueryBuilders.matchPhraseQuery(field, keyWords);

field = "title";
MatchQueryBuilder pqTitle = QueryBuilders.matchPhraseQuery(field,
keyWords);
//可以匹配上不连续出现查询串的文档
QueryStringQueryBuilder fuzzyQb = new QueryStringQueryBuilder(keyWords);

//用 OR 关系连接多个查询条件
QueryBuilder qb =
QueryBuilders.boolQuery().should(pqBody).should(pqTitle).should(fuzzyQb);
```

测试得到的 Query DSL。

```
String keyWords = "自2006年8月2日北京市";
String field = "body";
```

```
MatchQueryBuilder pqBody = QueryBuilders.matchPhraseQuery(field, keyWords);

field = "title";
MatchQueryBuilder pqTitle = QueryBuilders.matchPhraseQuery(field, keyWords);

QueryStringQueryBuilder fuzzyQb = new QueryStringQueryBuilder(keyWords);

QueryBuilder qb =
QueryBuilders.boolQuery().should(pqBody).should(pqTitle).should(fuzzyQb)
;
System.out.println(qb);
```

输出结果：

```
{
  "bool" : {
    "should" : [ {
      "match" : {
        "body" : {
          "query" : "自 2006 年 8 月 2 日北京市",
          "type" : "phrase"
        }
      }
    }, {
      "match" : {
        "title" : {
          "query" : "自 2006 年 8 月 2 日北京市",
          "type" : "phrase"
        }
      }
    }, {
      "query_string" : {
        "query" : "自 2006 年 8 月 2 日北京市"
      }
    } ]
  }
}
```

结果中只返回指定的几列。

在 http://localhost:9200/_plugin/head/运行：

```
{
  "query": {
    "match_all": {}
  },
  "fields": ["_id"]
}
```

通过 Java 客户端执行：

```
        String json = "{\n" +
            "    \"query\" : {\n" +
            "        \"match_all\" : {}\n" +
            "    },\n"+
            "    \"fields\" : [\"_id\"]\n" +
        "}";
    System.out.println(json);
    Client client = getClient();
    SearchResponse searchResponse =
    client.prepareSearch(ESConfig.indexName)
            .setSource(json).execute().get();
    SearchHits hits = searchResponse.getHits();



String text = "地质"; // 查询词
String[] fields = { "repname", "repsubdate" };

// 生成 JSON 内容
XContentBuilder qBuilder = XContentFactory.jsonBuilder().startObject();

qBuilder.startObject("query");
qBuilder.startObject("term");
qBuilder.field("repname", text);
qBuilder.endObject();
qBuilder.endObject();
// 指定返回列
qBuilder.startArray("fields");
for (String field : fields) {
    qBuilder.value(field);
}
qBuilder.endArray();
qBuilder.endObject();

//生成的查询 DSL {"query":{"term":{"repname":"地质
"}},"fields":["repname","repsubdate"]}
System.out.println(qBuilder.string());

Client client = getClient();
SearchResponse searchResponse = client
        .prepareSearch(ESConfig.indexName).setSource(qBuilder)
        .execute().actionGet();
SearchHits hits = searchResponse.getHits();

long totalHits = hits.getTotalHits(); // 得到结果总数
System.out.println("totalHits:" + totalHits);
```

```java
for (SearchHit hit : hits) {
    SearchHitField j = hit.field("repname");
    System.out.print(j.value());
    j = hit.field("repsubdate");
    System.out.println("  " + j.value());
    System.out.println("score: " + hit.getScore());
}
```

在客户端拼 JSON 查询串：

```java
String queryString="我爱北京天安门";
System.out.println(QueryBuilders.queryString(queryString).analyzer("cn"));
```

输出：

```
{
  "query_string" : {
    "query" : "我爱北京天安门",
    "analyzer" : "cn"
  }
}
```

如果只有一个过滤器，可以只使用 match_all 查询搭配这个过滤器。

```java
MatchAllQueryBuilder matchQqueryBuilder = QueryBuilders.matchAllQuery();

//Add filter on the query based on filtered query
AndFilterBuilder andFilterBuilder = FilterBuilders.andFilter();
andFilterBuilder.add(FilterBuilders.termFilter("version", "3"));
requestBuilder.setQuery(QueryBuilders.filteredQuery(matchQqueryBuilder,
            andFilterBuilder));
```

6.6 高亮显示

首先指定哪些列需要高亮，然后在命中结果中得到高亮段。SearchRequestBuilder 中的 addHighlightedField() 方法可以定制在哪个域值的检索结果的关键字上增加高亮。

```java
String keyWords = "hello"; // 查询词
// 设置查询关键词
QueryStringQueryBuilder qb = new QueryStringQueryBuilder(keyWords);
searchRequestBuilder.setQuery(qb);

// 设置高亮显示
searchRequestBuilder.addHighlightedField("title");
searchRequestBuilder.setHighlighterPreTags("<span style=\"color:red\">");
searchRequestBuilder.setHighlighterPostTags("</span>");
// 执行搜索,返回搜索响应信息
SearchResponse response = searchRequestBuilder.execute().actionGet();
```

```java
// 获取搜索的文档结果
SearchHits searchHits = response.getHits();
for (SearchHit hit : searchHits) {
    // 将文档中的每一个对象转换 JSON 串值
    String json = hit.getSourceAsString(); //搜索结果用 Gson 解析,解析要自己写

    // 获取对应的高亮域
    Map<String, HighlightField> result = hit.highlightFields();
    // 从设定的高亮域中取得指定域
    HighlightField titleField = result.get("title");
    // 取得定义的高亮标签
    Text[] titleTexts = titleField.fragments();
    // 为 title 串值增加自定义的高亮标签
    StringBuilder title = new StringBuilder();
    for (Text text : titleTexts) {
        title.append(text);
    }
    System.out.println("title highlighter " + title);
}
```

显示结果的时候,首先可以检查 highlightFields(),如果一列没有出现在那里,用 fields() 代替。

```java
public static String getTitle(HighlightField titleField,Map<String, Object> source){
    if(titleField==null){  //如果没有高亮值,就显示原来的值
        return (String) source.get("title");
    }

    StringBuilder title = new StringBuilder();

    // 取得定义的高亮标签
    Text[] titleTexts = titleField.fragments();
    // 为 title 串值增加自定义的高亮标签
    for (Text text : titleTexts) {
        title.append(text);
    }

    return title.toString();
}
```

6.7 分　　页

设置两个参数:从第几个结果开始返回文档,以及最多返回多少个文档。返回一个结果总数,用于知道有多少页可以翻。

第 6 章　ElasticSearch 分布式搜索引擎

在 SearchRequestBuilder 中设置分页参数：

```
searchRequestBuilder.setFrom(0).setSize(60);
```

用于分页显示的代码如下：

```
int rows=10; //一页显示多少条搜索结果
int offset=0; //开始行

// ...
String keyWords = "hello"; // 查询词
// 设置查询关键词
QueryStringQueryBuilder qb = new QueryStringQueryBuilder(keyWords);
searchRequestBuilder.setQuery(qb);

// 分页应用
searchRequestBuilder.setFrom(offset).setSize(rows);

// 执行搜索，返回搜索响应信息
SearchResponse response = searchRequestBuilder.execute().actionGet();

// 获取搜索的文档结果
SearchHits searchHits = response.getHits();
long totalHits = searchHits.getTotalHits();   //得到结果总数
for (SearchHit hit : searchHits) {
    System.out.println("hit " + hit);
}
```

6.8　中 文 搜 索

实现一个继承 AbstractIndexAnalyzerProvider 的类，提供自己的分析器实现。然后在配置文件的 type 值中声明类的全名。

6.8.1　中文 AnalyzerProvider

编写支持中文的 AnalyzerProvider 类：CnAnalyzerProvider：

```
public class CnAnalyzerProvider
        extends AbstractIndexAnalyzerProvider<NgramAnalyzer>{
    private final NgramAnalyzer analyzer;

    @Inject
    public CnAnalyzerProvider(Index index, @IndexSettings Settings indexSettings,
        Environment env, @Assisted String name, @Assisted Settings settings){
        super(index, indexSettings, name, settings);
        //得到插件目录，词典文件放在插件目录的子目录下
```

```
        File pluginDir = env.pluginsFile();
        String dicPath=new File(pluginDir,"seg/dic").getPath();
        analyzer = new NgramAnalyzer(dicPath+"/");
    }

    @Override
    public NgramAnalyzer get() {
        return analyzer;
    }
}
```

注意这里的注解"@Inject"是必需的。把生成出来的 seg.jar 放入插件路径下：D:\elasticsearch-5.1.2 \plugins\seg。

在 elasticsearch.yml 配置文件中增加中文分析器：

```
index:
  analysis:
    analyzer:
      cn:
          alias: [cn_analyzer]
          type: com.lietu.ds.CnAnalyzerProvider

index.analysis.analyzer.default.type : "com.lietu.ds.CnAnalyzerProvider"
```

在 head 插件中就可以测试这个分析器。索引下面有个"test analyze"选项。设置全局默认分词会看到效果，如果没有，还是会按照 ES 默认的一元分词。

首先要创建一个索引库，然后才能测试分词器。

查看分词效果的命令：

```
_analyze?text='我爱北京天安门'&analyzer=standard
_analyze?text='我爱北京天安门'&analyzer=cn
```

例如，news 索引使用如下的测试地址：

```
http://localhost:9200/news/_analyze?analyzer=cn&text='我爱北京天安门'
http://localhost:9200/news/_analyze?analyzer=standard&text='我爱北京天安门'
```

inquisitor(https://github.com/polyfractal/elasticsearch-inquisitor)是一个测试分词的插件。

index_analyzer 是代表这个字段建立索引时使用的分词方式，search_analyzer 代表对这个字段搜索时使用的分词。

Mappings 配置的例子：

```
{
  "recruitinfo":{
    "properties":{
      "id":{
        "type":"string",
        "index":"not_analyzed"
      },
```

```
        "title":{
           "type":"string",
           "term_vector": "with_positions_offsets",
           "index_analyzer": "cn",
           "search_analyzer": "cn",
           "store":"yes"
        }
     }
  }
}
```

使用这个 Mappings：

```
client.admin().indices().prepareCreate(indexName)
.setSettings("... your JSON settings..")
 .addMapping(type, "... your mapping...")
```

6.8.2 字词混合索引

查询某些短语时，按字分词列和按词分词列的返回结果数量不一样。测试代码如下：

```
public static void searchField(String field,String keyWords ){
    MatchQueryBuilder qb = QueryBuilders.matchPhraseQuery(field, keyWords);

    Client client = getClient();
    SearchResponse searchResponse = client
            .prepareSearch(ESConfig.indexName).setQuery(qb).execute()
            .actionGet();
    SearchHits hits = searchResponse.getHits();

    long totalHits = hits.getTotalHits();  // 得到结果总数
    System.out.println("totalHits:" + totalHits);
}
```

测试：

```
String keyWords = "一九九六年三月～二 OO 一年";
String field = "contentsS";
searchField(field ,keyWords);   //按字查询

field = "contents";
searchField(field ,keyWords);   //按词查询
```

输入相同的查询词返回结果数量不一样。

为了保证搜索的查全和查准，对全文查询列用单字索引和词索引两列，索引同样的内容。对于标题，按字索引的列在 Mapping 中定义如下：

```
        "stitle":{
           "type":"string",
```

```
        "term_vector": "with_positions_offsets",
        "index_analyzer": "standard",
        "search_analyzer": "standard",
        "store":"yes"
    },
```

按词索引的列在 Mapping 中定义如下：

```
    "title":{
        "type":"string",
        "term_vector": "with_positions_offsets",
        "index_analyzer": "cn",
        "search_analyzer": "cn",
        "store":"yes"
    }
```

这里的"cn"是在配置文件中指定的，"standard"是 ES 自带的。

打开 Index Metadata，可以看到索引库的结构。可以用 Java 代码修改索引库的结构，增加字索引列。

可以定义 Mapping，一列是按字索引，另一列是按词索引。D:\elasticsearch-1.0.0\config\mappings\news 目录下的 type1.json 内容如下：

```
{
  "type1":{
    "properties":{
      "title":{
        "type":"string",
        "term_vector": "with_positions_offsets",
        "index_analyzer": "cn",
        "search_analyzer": "cn",
        "store":"yes"
      },
      "body":{
        "type":"string",
        "term_vector": "with_positions_offsets",
        "index_analyzer": "cn",
        "search_analyzer": "cn",
        "store":"yes"
      },
      "stitle":{
        "type":"string",
        "term_vector": "with_positions_offsets",
        "index_analyzer": "standard",
        "search_analyzer": "standard",
        "store":"yes"
      },
      "sbody":{
        "type":"string",
```

```
                "term_vector": "with_positions_offsets",
                "index_analyzer": "cn",
                "search_analyzer": "cn",
                "store":"yes"
            }
        }
    }
}
```

使用 API 创建 JSON 内容。

```
XContentBuilder mapping = XContentFactory
    .jsonBuilder()
    .startObject()
    .startObject(indexType)
    //
    .startObject("properties")
    //
    .startObject("title")
    .field("type", "string")
    // start title
    .field("store", "yes")
    .field("analyzer", "cn")   //词
    //
    .endObject()
    // end title
    .startObject("postDate")
    //
    .field("type", "date").field("store", "yes")
    .field("index", "analyzed")
    //
    .endObject()
    // end post date
    .startObject("body")
    //
    .field("type", "string").field("store", "yes")
    //
    .field("index_analyzer", "standard")    //字
    .field("search_analyzer", "standard")
    .endObject() // end field
    .endObject() // end properties
    .endObject() // end index type
    .endObject();
```

为了保证连续匹配的文档得分高，把短语查询和普通的模糊查询组合成布尔查询。写法如下：

```
//短语查询
MatchQueryBuilder pqTitle = QueryBuilders.matchPhraseQuery(title, qString);
```

```
MatchQueryBuilder pqTitle2 = QueryBuilders.matchPhraseQuery(title2,
qString);
MatchQueryBuilder pqBody = QueryBuilders.matchPhraseQuery(body, qString);
MatchQueryBuilder pqBody2 = QueryBuilders.matchPhraseQuery(body2, qString);

//普通模糊查询
QueryStringQueryBuilder fuzzyQb = new QueryStringQueryBuilder(qString)
        .field(title2).field(body2).field(title).field(body);

//把上面的查询组合到布尔查询得到最终的查询
QueryBuilder qb =
  queryBuilders.boolQuery().should(pqTitle2).should(pqTitle).
            should(pqBody2).should(pqBody).should(fuzzyQb);
```

6.9 分组统计

搜索结果按某个不分词的列统计:

```
SearchRequestBuilder requestBuilder = client.prepareSearch(index)
        .setQuery(qb);

// 增加 facet 到搜索请求
TermsFacetBuilder termsFacetBuilder = new TermsFacetBuilder("version");
//分类列
termsFacetBuilder.field("version");
termsFacetBuilder.size(100);    //只显示前 100 个统计结果
requestBuilder.addFacet(termsFacetBuilder);
```

使用过滤器实现下钻:

```
//得到 QueryBuilder
String keyWords = "的";
String field = "body";
MatchQueryBuilder pqBody = QueryBuilders.matchPhraseQuery(field, keyWords);

field = "title";
MatchQueryBuilder pqTitle = QueryBuilders.matchPhraseQuery(field, keyWords);

QueryBuilder qb = QueryBuilders.boolQuery().should(pqBody)
        .should(pqTitle);

String index = "news";  // 索引名

Client client = QueryTest.getClient();

SearchRequestBuilder requestBuilder = client.prepareSearch(index);
```

```
//在查询上增加过滤器
AndFilterBuilder andFilterBuilder = FilterBuilders.andFilter();
andFilterBuilder.add(FilterBuilders.termFilter("version", "3"));
requestBuilder.setQuery(QueryBuilders.filteredQuery(qb,
andFilterBuilder));

//Execute search
SearchResponse searchResponse = requestBuilder.execute().actionGet();

//parse search resuls and response.
SearchHits hits = searchResponse.getHits();

long totalHits = hits.getTotalHits(); // 得到结果总数
System.out.println("totalHits:" + totalHits);
```

6.10 与爬虫集成

把爬虫抓的数据直接放到 ElasticSearch。

```
public static void insert(News news){
    ElasticSearchClient esc = new ElasticSearchClient();
    String index = "test1"; // 索引名
    String type = "type1"; // 类型,类似表名

    Map<String, String> sourceMap = new HashMap<String, String>();
    sourceMap.put("title", news.title);
    sourceMap.put("body", news.body);
    esc.insert(index, type, sourceMap);
}
```

6.11 Percolate

Percolate 是一个根据文档倒过来查询的功能。可以把 Percolate 看成是索引后搜索的反向操作。

不是发送文档，索引它们，然后运行查询。Percolate 是发送一些查询，注册这些查询，然后发送文档并找出哪些查询匹配这个文档。

举一个例子：一个用户可以注册一个兴趣(查询)到所有的微博上。这个兴趣就是包含单词"雾霾"的文档。对于每一个微博，可以 Percolate 这个微博到所有注册的用户查询，并找出哪些查询匹配上了这个微博。

```
//要注册到过滤器的查询
QueryBuilder qb = termQuery("content", "amazing");
```

```
//索引这个查询 = 注册这个查询到过滤器
client.prepareIndex("_percolator", "myIndexName",
"myDesignatedQueryName")
   .setSource(qb)
   .setRefresh(true) //当需要查询立即可用时,需要这个
   .execute().actionGet();
```

上面的代码以 myDesignatedQueryName 的名字索引一个词查询。

为了在注册的查询上检查一个文档,使用如下的代码:

```
//构建一个文档来检查过滤器
XContentBuilder docBuilder = XContentFactory.jsonBuilder().startObject();
docBuilder.field("doc").startObject(); //需要用它来指定文档
docBuilder.field("content", "This is amazing!");
docBuilder.endObject(); //doc 列结束
docBuilder.endObject(); //JSON 根对象结束
//过滤
PercolateResponse response =
        client.preparePercolate("myIndexName",
"myDocumentType").setSource(docBuilder).execute().actionGet();
//遍历结果
for(String result : response) {
    //处理结果,结果是过滤器中查询的名字
}
```

6.12 权限

如果能让 ES 提供只读,而不可以写入端口号,就能够实现权限控制。一个方法是把 ES 部署在 Jetty(https://github.com/sonian/elasticsearch-jetty);另外一个方法是把 Nginx 放在 ES 前端。只允许 GET 请求的一个例子配置如下:

```
worker_processes  1;
pid               nginx.pid;

events {
   worker_connections  1024;
}

http {

  server {

    listen       8080;
    server_name  search.example.com;

    error_log    elasticsearch-errors.log;
```

```
    access_log elasticsearch.log;

    location / {
      if ($request_method !~ "GET") {
        return 403;
        break;
      }

      proxy_pass http://localhost:9200;
      proxy_redirect off;

      proxy_set_header   X-Real-IP $remote_addr;
      proxy_set_header   X-Forwarded-For $proxy_add_x_forwarded_for;
      proxy_set_header   Host $http_host;
    }

  }
}
```

使用这个配置:

```
$ nginx -c path/to/this/file
```

然后测试它:

```
curl -i -X GET http://localhost:8080/_search -d '{"query":{"match_all":{}}}'
HTTP/1.1 200 OK

curl -i -X POST http://localhost:8080/test/test/1 -d '{"foo":"bar"}'
HTTP/1.1 403 Forbidden

curl -i -X DELETE http://localhost:8080/test/
HTTP/1.1 403 Forbidden
```

安装 X-Pack 插件后，所有对 ES 的访问都增加了安全机制，即需要用户名和密码，默认分别为 elastic 和 changeme。使用 sense 插件访问的时候可以输入，如果是使用 curl 等方式访问，则需要在 HTTP 的 header 中增加 Authentication 参数。

安全的客户端可以参考 https://github.com/elastic/found-shield-example。

6.13 SQL 支持

ElasticSearch 公司的主要方向目前还在搜索上，并不自带对 SQL 的支持。

可以通过 Presto 或 Drill 的插件来实现 SQL 查询支持。Presto Elasticsearch Connector 的基于插件的方案难以充分发挥 ElasticSearch 的性能优势。因此，基于 Calcite 让 Elasticsearch 支持 SQL 的项目开发活跃。

Apache Calcite 是一个用来建立数据库和数据管理系统的开源框架。它包括一个 SQL 解析器，一个在关系代数上构建表达式的 API，以及一个查询计划引擎。作为一个框架，Calcite 不存储自己的数据或元数据，而是通过插件的方式允许访问外部数据和元数据。

对于一个查询操作，会经历如下的流程：Calcite 会解析 SQL 并将其转换成逻辑执行计划，在此期间会根据当前连接中 Schema 定义的信息初始化每一个 Schema，然后根据查询中指定的 Schema 调用对应的 getTableMap 函数获取元数据，根据这个信息判断查询中出现的表名、字段名是否正确以及检查 SQL 语法是否符合规范。然后再使用 Calcite 内部默认的实现生成物理执行计划，这个查询计划是树状结构的，最底层的节点是 ScanTable 操作(类似于 SQL 执行过程中首先执行 FROM 子句)，对每一个表获取该表的数据，这时候使用的算子为默认的 EnumerableTableAccessRel，然后再去调用具体 ScannableTable 的 scan()方法获取表的数据。之后再根据原始表的数据进行上层的 JOIN、FILTER、GROUP BY、SORT、LIMIT 甚至子查询等操作。

首先用命令行工具 sqlline 查询 csv 文件。然后使用代码测试。

example.json 文件内容如下：

```
{
  version: '1.0',
  defaultSchema: 'STREAM',
  schemas: [
    {
      name: 'SS',
      tables: [
        {
          name: 'ORDERS',
          type: 'custom',
          factory: 'org.apache.calcite.adapter.csv.CsvStreamTableFactory',
          stream: {
            stream: true
          },
          operand: {
            file: 'sales/SORDERS.csv',
            flavor: "scannable"
          }
        }
      ]
    }
  ]
}
```

schemas 定义了一些 schema，也就是数据库。每一个 schema 指定了 name、type(可以分为 Map Schema、Custom Schema 和 JDBC Schema)。

这里的 Custom Schema 意味着只需要指定 factory 和可选的 operand 参数(map 结构)，schema 都是通过指定的 factory 类创建出来的(它需要实现 org.apache.calcite.schema.SchemaFactory 接口)，具体这个 schema 下面有哪些表可以通过

schema 的 name 和 operand 变量决定生成。

SORDERS.csv 文件内容如下：

```
PRODUCTID:int,ORDERID:int,UNITS:int
3,4,5
2,5,12
2,1,6
```

查询测试代码如下：

```java
Class.forName("org.apache.calcite.jdbc.Driver");
Properties info = new Properties();
info.setProperty("lex", "JAVA");
Connection connection =
    DriverManager.getConnection("jdbc:calcite:model="
      +
"d:/Downloads/calcite-master/example/csv/target/test-classes/example.json",info);
CalciteConnection calciteConnection =
    connection.unwrap(CalciteConnection.class);
Statement statement = connection.createStatement();
ResultSet resultSet =
        statement.executeQuery("select STREAM * from SS.ORDERS where SS.ORDERS.UNITS > 5");
final StringBuilder buf = new StringBuilder();
while (resultSet.next()) {
  int n = resultSet.getMetaData().getColumnCount();
  for (int i = 1; i <= n; i++) {
    buf.append(resultSet.getMetaData().getColumnLabel(i))
       .append("=")
       .append(resultSet.getObject(i));
  }
  System.out.println(buf.toString());
  buf.setLength(0);
}
resultSet.close();
statement.close();
connection.close();
```

这里的 STREAM 关键字是流式 SQL 的主要扩展。它告诉系统，你感兴趣的是输入的订单，而不是现有的。

https://github.com/apache/calcite 中包含了 ElasticSearch 插件。

```
{
 "version": "1.0",
 "defaultSchema": "elasticsearch",
 "schemas": [
   {
     "type": "custom",
```

```
    "name": "elasticsearch_raw",
    "factory":
"org.apache.calcite.adapter.elasticsearch.ElasticsearchSchemaFactory",
    "operand": {
      "coordinates": "{'127.0.0.1': 9300}",
      "userConfig": "{'bulk.flush.max.actions': 10,
'bulk.flush.max.size.mb': 1}",
      "index": "usa"
    }
  },
  {
    "name": "elasticsearch",
    "tables": [
     {
       "name": "ZIPS",
       "type": "view",
       "sql": [
         "select cast(_MAP['city'] AS varchar(20)) AS \"city\",\n",
         " cast(_MAP['loc'][0] AS float) AS \"longitude\",\n",
         " cast(_MAP['loc'][1] AS float) AS \"latitude\",\n",
         " cast(_MAP['pop'] AS integer) AS \"pop\",\n",
         " cast(_MAP['state'] AS varchar(2)) AS \"state\",\n",
         " cast(_MAP['id'] AS varchar(5)) AS \"id\"\n",
         "from \"elasticsearch_raw\".\"zips\""
       ]
     }
   ]
  }
 ]
}
```

查询代码如下：

```
Class.forName("org.apache.calcite.jdbc.Driver");
Properties info = new Properties();
info.setProperty("lex", "JAVA");
Connection connection =
    DriverManager.getConnection("jdbc:calcite:model="
     +
"d:/Downloads/calcite-master/example/elasticsearch-zips-model.json",info
);
CalciteConnection calciteConnection =
    connection.unwrap(CalciteConnection.class);
Statement statement = connection.createStatement();
ResultSet resultSet =
      statement.executeQuery("select * from zips where \"pop\" in (20012,
15590)");
```

6.14 本章小结

可以通过 Mapping 定义索引结构，然后填充数据到索引，最后再做搜索部分。

Shay Banon 在开发 ElasticSearch 的前身 Compass 时，意识到需要开发一个分布式的搜索解决方案。ElasticSearch 的第一个版本在 2009 年发布。ElasticSearch 公司成立于 2012 年。这家公司专门做开源分布式搜索引擎。ElasticSearch 公司是一个分布式的组织。有两个主要的地点：位于美国加利福尼亚的洛杉矶和位于荷兰的阿姆斯特丹。还有小的办公室：美国亚利桑那州的凤凰城，巴黎，布拉格，奥斯汀，波士顿，巴塞罗那，柏林和罗马尼亚。

建立一个国际化的商业组织是复杂和昂贵的。ElasticSearch 仅仅创立 6 个月后，就获得了 1000 万美元的投资。2012 年，Elasticsearch 公司创始人 Steven Schuurman 和 Shay Banon 宣布，他们从风投公司 Benchmark Capital、Rod Johnson 和 Data Collective 筹得第一轮融资。在此之前，Schuurman 与 Johnson 共同创建了 SpringSource 公司，并且这家公司在 2009 年以 4 亿多美元卖给 VMware。SpringSource 的几个创始人在创建这个公司之前从来没见过面。

ElasticSearch 的引擎是一个运行时环境，使人们能够实时地分析和处理大量数据。它"不要求你必须是一位数据科学家才能把它用好"，这也是该产品一个主要的亮点。ElasticSearch 存在的理由是把大数据的复杂性化解成像苹果产品那般简单。

Mailgun 收发大量电子邮件，跟踪和存储每封邮件发生的每个事件。每个月会新增数十亿事件，需要展示给我们的客户，方便他们很容易地分析数据，也就是全文搜索。利用 ElasticSearch 和 Logstash 技术完成这个需求。

ElasticSearch 的服务器端代码开始成熟。但客户端代码仍然在发展和调整之中。Java API 与 ES 服务器在端口 9300 上打交道，而 RESTful 的 HTTP 客户端 Jest 使用端口 9200。Jest 提供自己的 Java API，还可以使用 ES Java API 来构建查询，然后提交给 RESTful 端点。

除了使用 Java API，还可以使用 Net 开发 ElasticSearch 搜索客户端。

2014 年出现了 Apache Calcite 项目。之后出现了数个支持 ElasticSearch 的 SQL 查询的插件版本。在将来，自然语言查询文本大数据的技术会逐渐走向成熟。